Modeling and Optimization of Energy Systems

Modeling and Optimization of Energy Systems

Edited by Alexia Fisher

CLANRYE
INTERNATIONAL
www.clanryeinternational.com

Clanrye International,
750 Third Avenue, 9th Floor,
New York, NY 10017, USA

ISBN: 978-1-64726-671-4

Cataloging-in-Publication Data

Modeling and optimization of energy systems / edited by Alexia Fisher.
p. cm.
Includes bibliographical references and index.
ISBN 978-1-64726-671-4
1. Power resources. 2. Mathematical optimization. 3. Energy facilities. I. Fisher, Alexia.
TJ163.2 .M63 2023
333.79--dc23

For information on all Clanrye International publications
visit our website at www.clanryeinternational.com

Clanrye International

Contents

Preface

Energy optimization refers to the controlled and mindful use of energy in the built environment as well as avoiding use of energy in order to maximize the benefits for people and the environment. Modeling involves interpreting a set of physical phenomena and developing a reasonably complete, closed and comprehensive phenomenological and mathematical formulation that describes a problem. Modeling usually results in a system of equations that can be solved to derive the values of variables that describe a phenomenon. Optimization of energy systems includes thermodynamic modeling, and analyzing and optimizing different types of energy systems in a variety of applications. It offers a fresh perspective on the systems and procedures of defining appropriate objective functions for choosing the best design parameters to achieve increased efficiency, sustainability, and cost effectiveness. All energy-intensive human activities must be revisited and modified to make them energy-efficient. This is essential to reduce the ecological footprint of humans on the Earth. This book unravels the recent advancements in modeling and optimization of energy systems. It will also provide interesting topics for research, which interested readers can take up. This book, with its detailed analyses and data, will prove immensely beneficial to professionals and students involved in this area of study at various levels.

After months of intensive research and writing, this book is the end result of all who devoted their time and efforts in the initiation and progress of this book. It will surely be a source of reference in enhancing the required knowledge of the new developments in the area. During the course of developing this book, certain measures such as accuracy, authenticity and research focused analytical studies were given preference in order to produce a comprehensive book in the area of study.

This book would not have been possible without the efforts of the authors and the publisher. I extend my sincere thanks to them. Secondly, I express my gratitude to my family and well-wishers. And most importantly, I thank my students for constantly expressing their willingness and curiosity in enhancing their knowledge in the field, which encourages me to take up further research projects for the advancement of the area.

Editor

1

Multi-Objective Optimization of Solar Thermal Systems Applied to Portuguese Dwellings

Ana Cristina Ferreira [1,*], Angela Silva [2], José Carlos Teixeira [1] and Senhorinha Teixeira [3]

1 MEtRICs Research Centre, University of Minho, 4804-533 Guimarães, Portugal; jt@dem.uminho.pt
2 Escola Superior de Ciências Empresariais, Instituto Politécnico de Viana do Castelo, 4930-600 Valença, Portugal; angela.a@esce.ipvc.pt
3 ALGORITMI Research Centre, University of Minho, 4804-533 Guimarães, Portugal; st@dps.uminho.pt
* Correspondence: acferreira@dps.uminho.pt

Abstract: Solar thermal systems have been widely used to increase energy efficiency in the building sector, since the use of renewable energy sources became one of the top priorities to meet environmental targets. The main objective of this study is the thermo-economic optimization of solar thermal systems for residential building applications, considering a multi-objective approach. The simulations were performed through a MatLab code by implementing an elitist variant of Non-dominated Sorting Genetic Algorithm-II (NASGA-II). The solar collection area and the linear loss coefficient as well as the tank storage volume were defined as decision variables. A two-dimensional Pareto front was obtained, considering as objective functions the minimization of the annualized investment cost and the maximization of the solar collection efficiency. Based on the best trade-off between both objectives and considering that the solar thermal systems can operate for a period of at least 15 years, the Pareto analysis led to the conclusion that a system with an annualized investment cost between 270 and 280 €/year allows reaching a collection efficiency of 60%. After the analysis of the optimal solution points, a configuration was selected to estimate the system total purchasing cost: a panel with a solar area of 4.17 m^2 and with a linear coefficient loss of 3.684 $W/m^2.K$; a storage volume of 0.275 m^3; and a pump flow rate of 0.1364 m^3/h. For this configuration, we estimated a total purchasing cost of 2545.0 €, whereas the solar collector and the storage tank are the most expensive components, representing a share of 42% and 43%, respectively. These results represent a specific cost of 610.3 €/m^2 per solar collection area.

Keywords: solar thermal systems; optimization; residential dwellings

1. Introduction

Solar energy is the most abundant and cleaner renewable energy resource and it can be supplied with minimal environmental impact. Sun uniformly sends solar radiation through space, but only about 1367 W/m^2 reaches the Earth atmosphere [1]. The amount of solar energy received on a 1 m^2 surface during a time interval can be calculated by integrating the irradiance over that time interval. This energy, also called solar irradiation or insolation, has two main components: the direct and the diffuse solar radiation [1–3]. The geographical coordinates, the topography, and the climate conditions of a certain location are the three main parameters that affect the effectiveness of solar thermal systems in converting direct and diffuse radiation into useful energy [4].

The demand of solar thermal systems is estimated to grow by 20% per year, depending on the collector working area, availability of the solar radiation throughout the year, and the type of solar thermal systems [5].

1.1. Technology Background

A solar thermal system converts the solar energy into thermal energy, which can be used for domestic hot water (DHW) production, space heating and cooling, or to heat swimming pools' water. These systems can be classified according to the type of thermal fluid used to transport the heat from the collector (air, pure water, refrigerants, and water–glycol mixtures). Usually, a mixture of water and glycol is mainly used for DHW and to partly cover the heating energy demand of a building. Systems should be able to provide the required heat demand at low and medium- temperature ranges, from 40 to 80 °C, which corresponds to the largest share of heat consumption in DHW and space heating [6,7]. Therefore, the solar collectors are the key elements of the solar thermal systems because the collector technology should be able to meet the application needs at the lowest cost [6]. The thermal collector converts the solar radiation in thermal energy, which is further transferred to a thermal fluid circulating through the panel. They are categorized into two categories, the non-concentrating (flat plate or evacuated tubes) and concentrating collectors, in the solar radiation concentration devices. Concerning the non-concentrating thermal collector, the flat ones (pipes are required to connect the two components) include several components: a transparent glazed cover, pipes, absorber, and a thermal insulation layer. In turn, the evacuated tubes' solar thermal collector consists of absorbers, glazed tubes, heat pipes, and insulated casing [1,2,8]. For both systems, the absorber plays an important role in heat transfer process, requiring a good thermal conductivity. In addition, the insulation between the absorber and the collector casing requires optimization to reduce the thermal losses. Such aspects require the use of more expensive materials, increasing the investment costs of solar thermal collectors [9,10].

In regions where the outdoor ambient temperature decreases below the water freezing point, a mixture of water and ethylene or propylene glycol (antifreeze) has to be used [1]. However, due to the organic component in the antifreeze mixture, these thermal fluids need to be replaced within 3 to 5 years, as a result of its degradation because of the higher temperatures in stagnation periods [11]. Air can also be used as a thermal fluid in solar thermal systems, but its use demands systems with large ducts to guarantee the higher the heat transfer efficiency. For systems operating with air, issues such as leakages, system location, or refilling are not a problem [12]. Some refrigerants, molten salts, and oil-based fluids are used by hybrid or concentrated solar thermal systems, which operate at high temperatures (varying between 350 and 900 °C) [13].

Solar thermal systems can be classified into natural and forced circulation systems. Natural circulation systems consist of a storage tank and a solar thermal collector, where the thermal fluid circulates by density gradients. The solar thermal collector can be either a flat plate or a different configuration having evacuated tubes that are directly inserted in the storage tank [9]. Forced circulation systems include the solar thermal collector, the storage tank, and a hydraulic pump used to force the thermal fluid circulation between these components (Figure 1). Usually, the pump is switched on/off by a solar controller that compares the measured water temperature in the collector with the temperature measured in the storage tank [14]. For mild climate locations, these systems can operate with a closed-loop, but for locations where the outdoor temperature can reach temperatures below 0 °C, the solar thermal systems operates as an open-loop circuit. In those circumstances, a solution is added to the thermal fluid to avoid water freezing while flowing in the external components. The flow is controlled by a set of valves that depend on the temperature measured at the collector. In practical terms, when the temperature reaches 0 °C, the water from the storage tank is prevented from passing to the solar collector by the valves, and the water from the solar collector is drained (Figure 1a). When the solar collector temperature increases and becomes higher than the storage tank temperature, the solar controller switches back, and the solar collector is filled up with water from the mains [2,14,15]. Severe outdoor conditions (i.e., the outside temperature is frequently below 0 °C) require the use of an antifreeze solution as thermal fluid (mostly to flow in exterior pipes of solar collector) and the installation of a heat exchanger to separate the domestic hot water circuit from the solar circuit. For these types of applications (Figure 1b), it is required to add an expansion vessel to manage the thermal fluid volume variations with the temperature and an internal heat exchanger is installed in the storage tank [9,15,16].

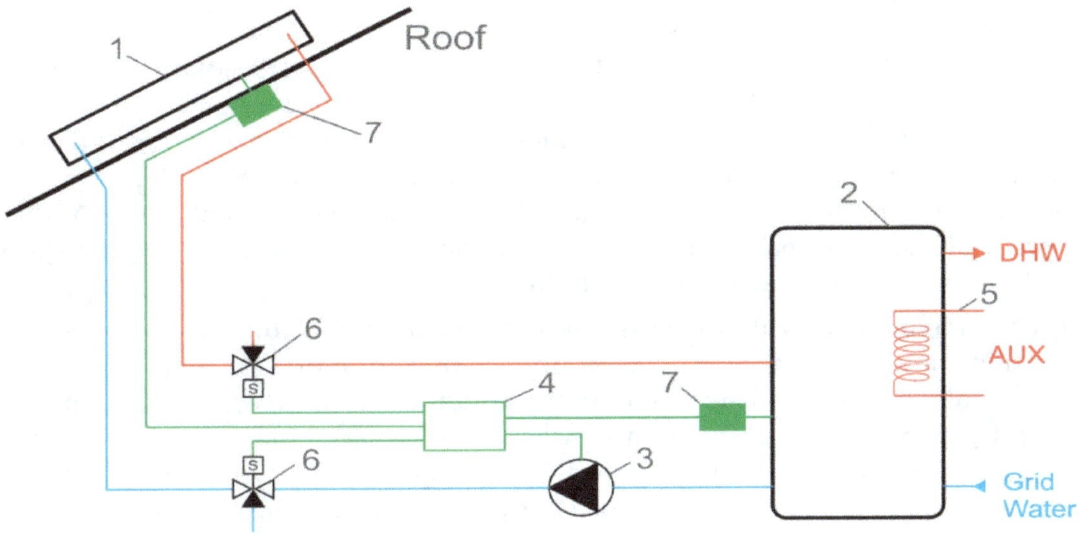

(a)

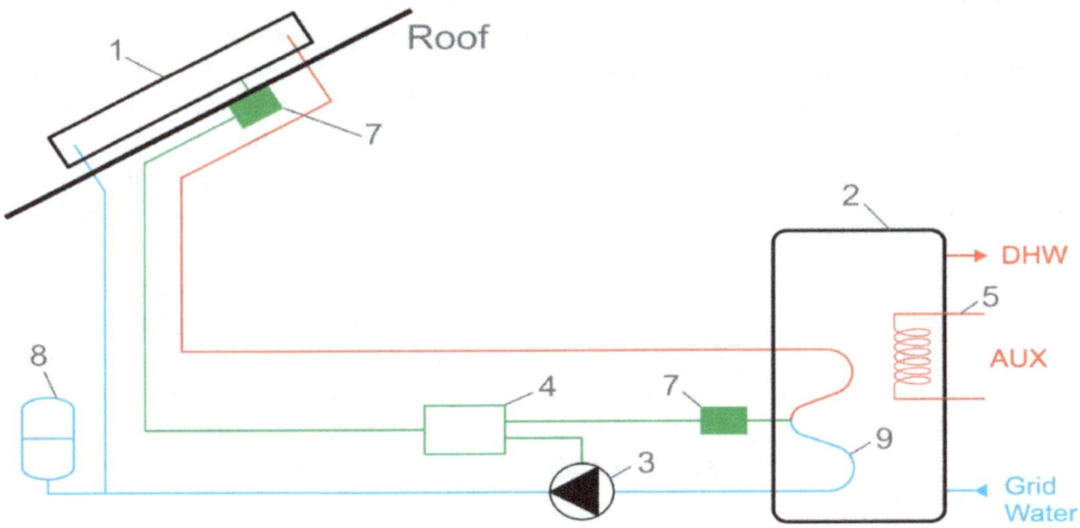

(b)

Figure 1. Solar thermal system: (**a**) with forced circulation of the water in an open-loop system with a solar controller (1), storage tank (2), hydraulic pump (3), solar controller (4), auxiliary water heater (5), valve (6), temperature sensors (7); (**b**) with forced circulation of the water with antifreeze, expansion vessel (8), and heat exchanger (9) to separate the domestic hot water circuit from the solar.

1.2. Brief Literature Review

Most of the literature concerning the analysis of solar thermal systems investigates a specific technology aiming to assess the thermal system efficiency, the system's operational conditions, or the best technological solution and appropriate size for a specific application [8,17–20]. Kalogirou [15] reviewed several types of solar thermal collectors, such as flat-plate, parabolic trough, evacuated tube, Fresnel lens, or parabolic dish, for different applications, e.g., solar water heating, integrated collector storage, space heating and cooling, air and water systems, heat pumps, and refrigeration appliances. Ong and Tong [20] studied a solar water heater with a specific design of a solar collector and storage tank. The system was sized considering distinct weather conditions (ambient temperature and solar radiation intensity). Marion et al. [21] developed a theoretical model in which a heat transfer equation was defined to predict the thermal equilibrium state of a glazed flat plate collector, considering different operational conditions. They have concluded that the optimal condition linearly depends on solar radiation. In addition, the net mechanical power is mostly affected by the fluid mass flow rate. Yongga et al. [22] studied a novel facade solar thermal system performance in summer and winter conditions. The research aimed to evaluate the efficiency of a new flat-plate collector and indoor radiation plate design, in avoiding extreme variations of the indoor thermal environment. The latest technological innovation such as the direct integration of collectors within the roofs or façades structure has shown the potentiality of being capable of regulating and optimizing the energy flows between the indoor and outdoor environments [23,24]. Islam and Morimoto [5] presented a state-of-the-art review of non-concentrating solar thermal systems, describing the design of current technology as well as the material selection criteria and engineering limitations.

The implementation of solar thermal systems has been studied in different regions, with different climatic conditions and in countries whose economic considerations for their application are very different. Artur et al. [25] applied a survey to understand the transition of current domestic hot water technology to solar thermal systems in urban areas (Maputo, Mozambique). The study demonstrated a positive outcome regarding the reduction of electricity demand and the consequent economic savings. The transition depends on the household income, governmental policies, and incentives to subsidize the technology adaptation. Košičan et al. [26] studied a solar thermal panel for space heating by determining the best design scheme for industrial and office buildings. The study was focused on the thermal energy performance of ventilated double skin façades in order to improve the thermal comfort in buildings. The authors compared different solar thermal systems considering energy, technical, economical, and commercial availability as criteria for four applications in Spain. It was concluded that there is no universal best solar thermal scheme. The choice for a particular solar thermal system is influenced by the energy consumption, technical conditions, location, and economic criteria. Sami et al. [27] published an interesting study regarding the integration of solar thermal systems (water heating) into high-energy performance housing in various regions with distinct climatic conditions in Algeria. The study considered both economic and energy criteria and allowed determining the optimal collector area that minimized the purchase and installation costs. Louvet et al. [28] used the levelized cost of energy to determine the heat cost per kWh of final energy produced by solar thermal systems in five European countries, considering different residential systems. It was concluded that it is of utmost importance to focus the research on solar assisted systems in order to reduce prices and increase the system's lifetime to cut off the economic gap when compared with the conventional heating systems. In most of the countries in study (Austria, Denmark, France, Germany, and Switzerland), the heat cost per kWh of final energy is higher for the solar-assisted heating systems, which are not subsidized. A similar study was also presented by Gautam et al. [29], who have also concluded that the high initial costs of solar thermal plants are not cost-effective when compared to the conventional technologies. Tschopp et al. [30] compared the market growth of solar thermal systems in Denmark, Germany, China, and Austria. The authors concluded that the stability of the solar thermal price is a key aspect to provide renewable heat supply, mostly because of low operation and maintenance costs and the large solar collectors' lifetime. Martinopoulos and Tsalikis [31] developed a techno-economic model of a solar system for space and water heating of remote households in Greece. The model considered different collector sizes (from 8 to 12 m^2) and storage tanks (from 0.5 to 0.65 m^3). The results showed that it is possible to minimize the energy production costs with low payback periods and a positive net present value.

Optimizing the performance of energy systems through numerical models has been an effective option for researchers [5]. Some of those models are based on thermal–economic analysis that combines thermodynamic and economic principles to assess the costs of energy production of power plant components [32,33]. Karki et al. [6] presented a simplified model to assess a solar flat-plate collector energy system performance based on a modified *f-chart* implementation that accounts the amount of energy collected. The model also includes system cost determination. The model was tested considering various geographic locations and climate conditions. The model was implemented in MS Excel, resulting in a decision support tool that allows the collectors sizing for a specific solar fraction input. Cruz-Peragon et al. [34] proposed a model of a collector by applying a 2D finite-difference method with steady and transient conditions. The model is based on Newton's method optimization to describe the critical coefficients of system, such as the emittance, plate absorbance, or convection transfer coefficient. Sharma and Diaz [35] developed a numerical model of a channel-based solar collector to evaluate thermal performance. The system was based on a flat-tube absorber enclosed with a coated glass envelope, which was used to minimize the convection losses. The steady-state model was evaluated in terms of collector efficiency. It was disclosed that the efficiency of the collector decreases with the increase in the inlet temperature because of the increase in radiative losses due to higher temperature reached at absorber surface. Similar research was conducted by Mansour [36] where the pressure drop and heat transfer inside the absorber channels were calculated by using Engineering Equation Solver EES software. Both studies have shown higher values in terms of collector efficiency with the inclusion of mini channels. Assaf and Shabani [17] presented a multi-objective optimization model of a solar–hydrogen combined heat and power system integrated with solar–thermal collectors to supply both power and hot water demand. Ust, Sahin, and Kodal [37] developed a thermo-economic analysis to maximize the power output per unit of the total cost. The model integrated physical and cost variables to optimize thermal and economic output. The costs integration is useful to disclose the most cost-effective configuration of the components and, then, improve the overall system design [38]. Lin et al. [19] applied a genetic algorithm to solar air systems to search for an optimal Pareto front of a multi-objective optimization problem. The system includes thermal energy storage using phase change materials in which the average heat transfer effectiveness and effective phase change materials charging time are defined as conflicting objective functions. Antoniadis and Martinopoulos [8] optimized a solar thermal system with storage with TRNSYS modeling software. The main objective was to assess the integration of different options of a solar collector array. The model calculates the annual domestic hot water and the space heating needs during the heating season for a typical single-family dwelling in the city of Thessaloniki (Greece). The system was optimized considering the collector area and the volume of the storage tank in order to optimize the use of the solar fraction. Yılmaz [39] proposed an original methodology for determining the optimum orientation and sizing of water-in-glass evacuated tube solar water heating systems for residential applications. The study was based on a transient model considering the system performance prediction for a year period through the System Advisor Model. Asadi et al. [40] presented multi-objective optimization of a 10 kW single-effect ammonia–water solar absorption cooling system. The model included an energy–exergy analysis, by assessing the effects of a different number of solar collector, as well as different solar collectors including a flat plate, evacuated tube, parabolic trough, and compound parabolic collectors. The system's performance was studied for different ambient temperatures. None of these revised studies considers the cost of the individual components of the solar thermal system. Authors based their studies on the total costs of the systems and payback periods from the system investment.

In this work, a simplified model for a solar thermal system was formulated with the objective to minimize the investment costs while maximizing the solar collection efficiency. The mathematical model includes the definition of physical equations that describe the operation of the three main components: the solar collector that captures the solar radiation, the pump with its controller that promotes the forced convection of the heated water, and the storage tank, which is used to accumulate heat to be used when the radiation is not available. A set of cost equations was developed considering the most important physical variables and the cost coefficients were based on real data from commercial systems. The resulting thermo-economic optimization model was applied to supress the DHW needs of a typical residential building in Portugal.

2. Materials and Methods

This section presents the characterization of solar radiation and the solar availability at a specific region of Portugal, regarding the global horizontal radiation. The DHW for a typical residential building was also estimated considering the consumption of a family of four. Afterwards, we present the mathematical model for the energy and economic optimization of the solar thermal system.

2.1. Solar Radiation in Portugal

Portugal has excellent climate conditions to implement thermal systems based on solar energy. Solar radiation data in Portugal has been collected and analyzed by the IPMA—Instituto Português do Mar e da Atmosfera (Portuguese Institute for Sea and Atmosphere) [41]. The data were collected for a period corresponding to 15 years and, in order to ensure the quality data analysis, the data were presented as annual average Global Horizontal Irradiation (GHI) values (Figure 2) considering the following assumptions: (1) for days that lack more than two hours of data, the GHI was estimated through the mean daily value of the same period for the other years; (2) for days that lack less than two hours, the values for those gaps were interpolated from adjacent hours; (3) GHI annual variability was determined through the standard deviation of the annual averages of GHI (Figure 3).

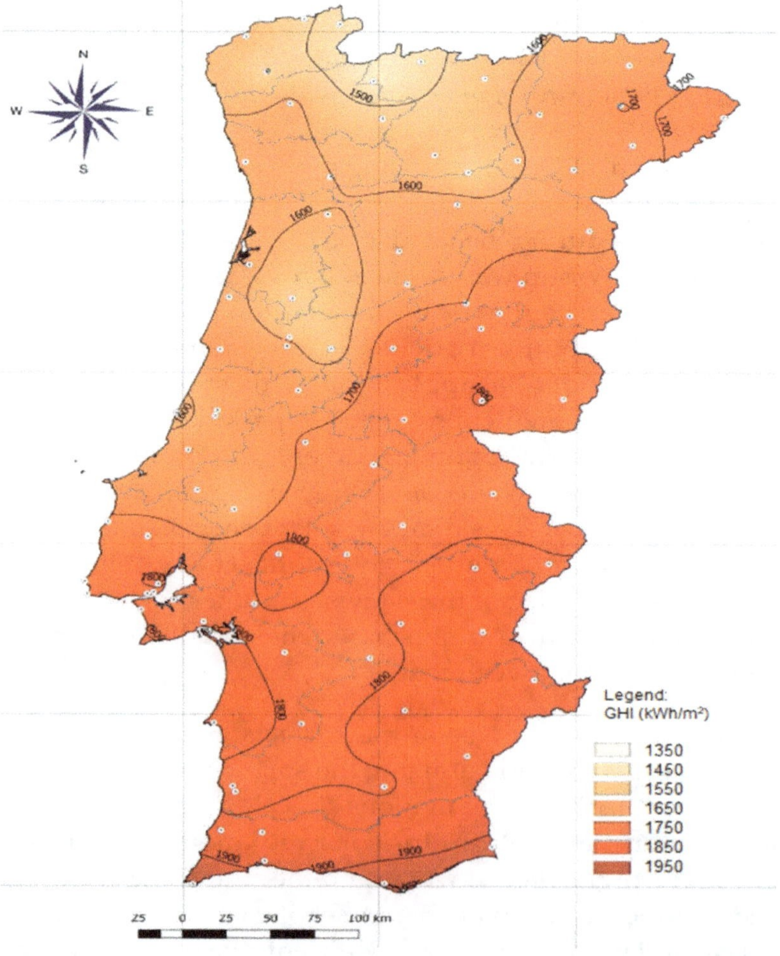

Figure 2. Annual average values of GHI (kWh/m^2) in Portugal. Data from [41].

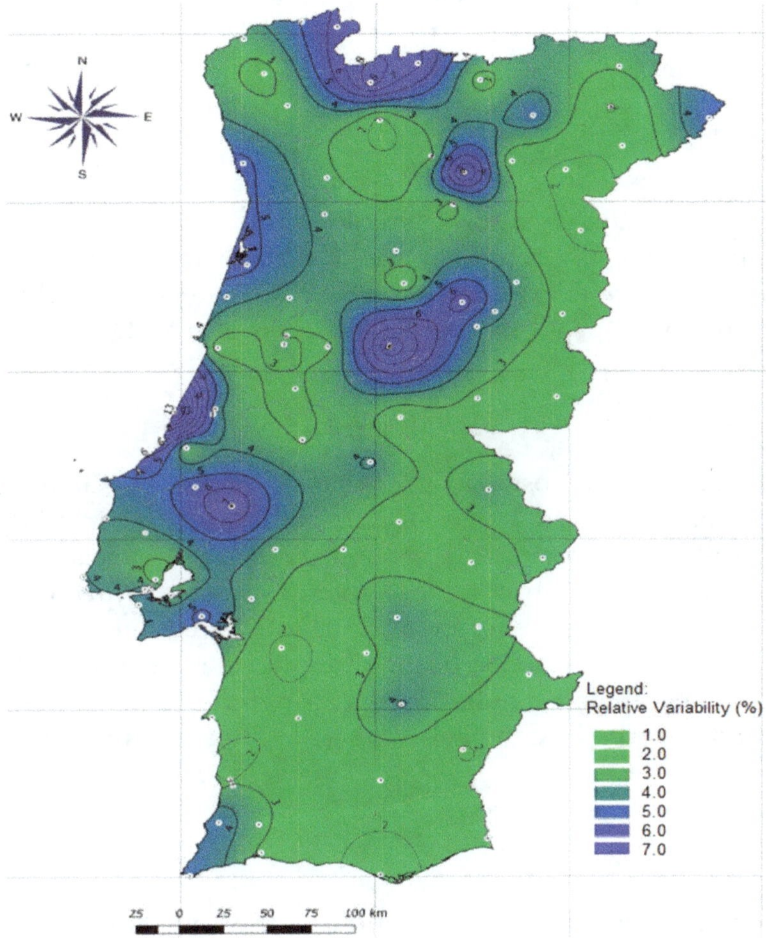

Figure 3. Annual variability of GHI values (%) in Portugal. Data from [41].

The annual availability of GHI is higher in the South of Portugal. This is due to the more favorable atmospheric conditions, resulting in a high number of sunshine hours in that region. This evidence is confirmed by the annual variability in the availability of solar radiation, which is more reduced in these regions of the Portuguese territory. The greatest relative variability occurs mainly in the coastal regions and specific areas where cloud cover is highest. These areas coincide with the locations where the average GHI values are lower. Three different locations were selected to illustrate different GHI values and its variability (Table 1).

Table 1. Annual average values of Global Horizontal Irradiation (GHI) and its variability for two different locations in Portugal.

Locations	GHI	GHI Variability
Guimarães (North)	1600 kWh/m^2	3%
Lisbon (Centre)	1890 kWh/m^2	4%
Faro (South)	1950 kWh/m^2	2%

2.2. *Typical DHW Consumption Profile of Portuguese Dwellings*

In Portugal, a typical dwelling has an average occupation of 4 people, with a consumption of 40 L of water per person and per day. The DHW needs vary during the day, depending on the period of the time and building utility. Assuming the daily domestic hot water needs, it is possible to estimate the consumption demand ($Cons_{dhw}$ in L/day). The thermal requirements for DHW (Q_{DHW}) can be calculated in accordance with the Portuguese regulation for the thermal behavior of buildings [42] as presented by Equation (1):

$$Q_{dhw} = Cons_{dhw} \times (4.187/3600) \times \Delta T \quad (\text{kWh/day}) \tag{1}$$

where ΔT is the water temperature variation from the public water supply to the usage temperature. If not calculated, the public water supply temperature can be assumed with the reference value of 15 °C, whereas the usage temperature can be assumed to be 60 °C (ΔT = 45 °C). Thus, for a dwelling with a 160 L/day of DWH consumption, the yearly thermal requirements for DHW can reach 3056.5 kWh/year.

2.3. Solar Thermal System Description and Physical Model

The solar thermal system is composed by a flat-plate solar collector able to heat up a thermal fluid to a temperature plateau of 80 °C. A transparent glass cover is used to insulate the flat-plate collector and an absorber surface allows the conversion of solar radiation, which is transferred to the transport medium that circulates in the system tubes and is then pumped for further storage. The amount of solar radiation received by the collector is affected by the total radiation on a horizontal surface (I_g) and the collector surface area (A_{solar}). A fraction of the radiation is reflected, another portion is absorbed by the gazing cover, and the remaining radiation that penetrates the transparent cover is transferred to the absorber plate and absorbed by the collector [43]. Thus, the incident heat can be defined by Equation (2):

$$Q_i = I_g \times \tau \times \alpha \times A_{solar} \tag{2}$$

where τ is the collector aperture transmittance and α is the collector absorbance. Not all incident energy is absorbed by the panel. Part of the energy is reflected, and part of it is lost through conduction–convection [44,45]. As the collector absorbs heat, the temperature inside the collector rises, but some heat is lost by convection and radiation to the surroundings [25]. The heat loss (Q_o) depends on the collector overall heat transfer coefficient (U_L) and the temperature difference between the collector temperature (T_c) and the ambient temperature (T_a), as shown in Equation (3).

$$Q_0 = U_L \times A_{solar} \times (T_c - T_a) \tag{3}$$

Thus, useful heat at the solar collector (Q_c) is then considered as the extraction heat at a steady-state condition. Therefore, this value is proportional to the useful heat that is absorbed by the solar collector as expressed by Equation (4):

$$Q_c = I_g \times \tau \cdot \sigma \times A_{solar} - U_L \times A_{solar}(T_c - T_a) \tag{4}$$

where σ represents the Stefan-Boltzmann constant (being σ equal to 5.6710^{-8} W/m^2.K^4). The most important heat losses occur on the top surface of the solar collector through convection and radiation. It is generally assumed that about 80% of the solar radiation is adsorbed by the plate, but only 45% of the heat is collected by the collector tubes [44]. A simplified Sankey diagram of the heat flow at a standard flat-plate solar collector is shown in Figure 4. This output corresponds to average values reported in the literature, assuming that a fraction of the radiation is reflected back to the sky, another fraction is absorbed by the glazed cover, and the remaining is transmitted through the glass, reaching the absorber plate as short wave radiation [46–48].

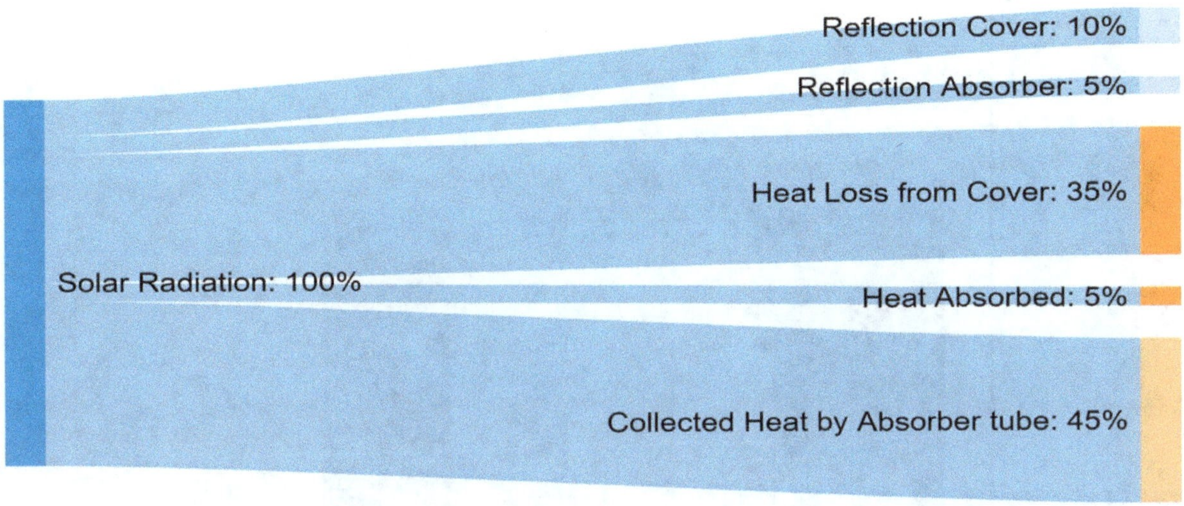

Figure 4. Sankey diagram of the heat flow at a standard flat-plate solar collector. Adapted from [46–48].

When the temperature of the solar panel is equal to the ambient temperature, i.e., when there are no convective or radiative losses to the environment, the performance of the panel is equal to the optical performance. However, the efficiency of the solar collector can be estimated by knowing its efficiency if the fluid at the entrance was at ambient temperature and removing the heat losses. Thus, the solar collection efficiency is defined by the ratio between the received and the captured energy [2]. So, the solar collector efficiency ($\eta_{collection}$) can be estimated by Equation (5):

$$\eta_{collection} = F\prime(\tau\alpha) - a_1\, T* - \, a_1 I_g\, T*^2 \tag{5}$$

where the variable T^* is the maximum temperature that can be reached by the collector, considering certain solar irradiation and ambient temperature conditions. This term can be calculated as the ratio between the temperature gradient suffered by the thermal fluid with the irradiation $\left(T^* = T_f - T_a/I_g\right)$ in °C/(W/m^2). The term F' is defined as the irrigation factor, which is included in the equation as the correction factor. The linear losses coefficient is represented by $a_1 = F'U_L$. In fact, the heat transfer factor is defined as the ratio between the heat received by the fluid and the heat absorbed by the collector.

In other words, it is the fraction of energy that is provided by the solar heating system. This solar fraction must vary between 40 and 90%. Assessing the system operation based on the hourly model, and assuming that the energy needs and losses are constant throughout the day, the energy needs and the heat exchange can be defined as schematically represented by Figure 5. The distinct areas of the scheme (A to E) represent different points of operation of the solar thermal system between its start-up and shutdown period. When solar radiation is not enough to increase the thermal fluid temperature, and therefore trigger the solar thermal system, the necessary heat for the building appliances is supplied by the auxiliary backup system, typically a conventional boiler. The auxiliary backup system is used both at early and late hours of the day, when the solar thermal system no longer has the capacity to promote heat exchange (A region). During the solar collector starting-up (first hours of the day), the solar energy available is used to heat up the system due to the low temperatures reached during the night. This process occurs until the thermal fluid temperature is enough to transfer energy to the water (B region).

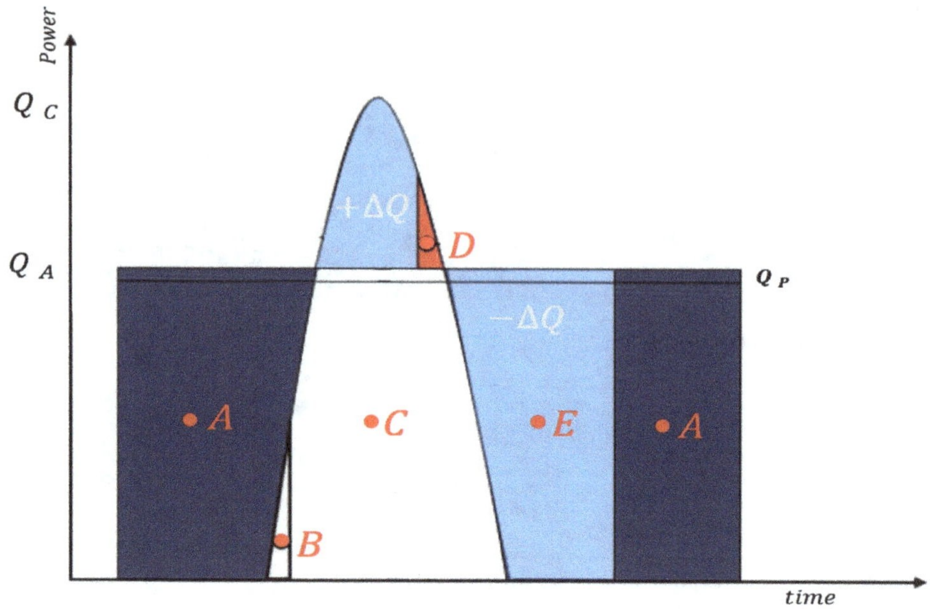

Figure 5. Schematic representation of the energy exchange of the solar collector considering its daily operation (Q_C represents the energy captured by the solar collector; Q_A represents the adsorbed; Q_P represents the thermal losses and ΔQ is the heat variation).

When solar energy is enough to increase the thermal fluid temperature, the necessary heat for the building appliances is supplied by the solar collector panel (C region) and the surplus is stored in the storage tank, which is represented by the positive heat variation (+ΔQ). The solar thermal system has a limited capacity to store the heat. From that point, all excess heat is dissipated (D region). Afterwards, when the heat captured by the solar panel is not enough to supply the useful energy requirements, the heat accumulated in the tank is then used to compensate for this difference (E region), which is represented by the negative heat variation ($-\Delta Q$). By the end of the day, the collector stops capturing solar radiation (A region).

Thus, the collector selection is mainly based on the temperature to be obtained for the thermal fluid and its cost. The heat generated in the collectors is transported to the storage tank through the hydraulic circuit. This circuit includes pipes, connecting the collectors to the remaining components and promoting the thermal fluid circulation, which is most of the time guaranteed by the pump operation (forced circulation) [2]. The mass flow pumped in the primary circuit ($\dot{V}_{pump}$) can be calculated as presented in Equation (6). This parameter mostly depends on the aperture area of the solar collector (A_{solar}) and the properties of the thermal fluid that is used in the system, such as density (ρ_{tfluid}) and the specific heat (c_p). Typically, this thermal fluid corresponds to a mixture of water with an appropriate concentration of antifreeze [15,18].

$$\dot{V}_{pump} = \left(\frac{10F'U_L}{c_{p\ tfluid}} \times A_{solar} \times 3600 \right) \times \frac{1}{\rho_{tfluid}} \tag{6}$$

Considering that the solar energy can be intermittent and it can be out of step with the hot water use, it is necessary to store the generated heat to be used when necessary. In order to avoid corrosion problems, most of the solar thermal systems use a storage tank with a heat exchanger, allowing the thermal fluid from the collector to circulate in the primary circuit and exchange heat with the secondary circuit (consumption circuit) [44]. Therefore, when choosing or sizing this component, a direct correlation can be established between the optimum volume ($V_{storage}$) of the deposit and the area of the solar collector (A_{solar}). According to [2], the optimal storage volume per unit of solar collection area is 0.070 m^3/m^2. For values above 0.090 m^3/m^2, the ratio between the efficiency of solar energy usage and the storage capacity do not lead to a greater use of incident solar energy; it only increases the storage tank volume and its cost [2,49]. This consideration can be translated into the model considering the upper and lower limit of the direct correlation between these two variables (Equation (7)).

$$V_{storage} = \lambda_{storage} A_{solar} \tag{7}$$

According to [2], the optimal storage volume per unit of solar collection area, $\lambda_{storage}$ is 0.070 m^3/m^2. Values above 0.080 m^3/m^2 do not lead to a greater use of incident solar energy but only contribute to the increase in the storage tank volume and its cost [2].

2.4. *Economic Model Definition*

The costing methodology allows the costs to be defined in a transparent manner, so that options can be validated and compared equitably. Cost estimation of the investment associated with the implementation of a particular system requires gathering and validation of the cost data. For this, one can collect cost data from the literature and technology suppliers [49,50]. Defining the economic criteria for evaluating an investment while considering the system performance allows a cost-optimal level analysis of a thermal system [49–51].

In this study, a purchase cost equation was defined for each main component of the system: solar collector, storage tank, and circulation pump. The cost of each component depends on quality and size parameters in order to scale the thermal system sizing. The mathematical expressions were based on the incorporation of energy with cost variables, from which the cost estimation is performed to evaluate the system's overall investment cost. The procedure uses the conventional chemical engineering methods for equipment sizing as well as the factorial method of cost estimation [52]. The coefficients of each cost equation were adjusted to the size of components, taking into account real market data, and the exponents were obtained by nonlinear regression. The costing methodology formulation, as well as each one of the cost equations, are briefly described because the complete model was already presented in previous works [49,53].

2.4.1. Solar Collector Cost Equation

The two most important physical variables that affect the solar collector (C_{solar}) cost are the collector area and the collection efficiency. Thus, the cost equation is given by Equation (8):

$$C_{solar} = C_{ref,solar} \times A_{ref,\ solar} \times \left(\frac{A_{solar}}{A_{ref,\ solar}} \right)^{b_{solar}} \times \eta_0^{\gamma} \times a_1^{\beta} \tag{8}$$

where $C_{ref,\,solar}$ is the reference cost coefficient, $A_{ref,\,solar}$ is the reference collector area, and A_{solar} represents the effective collector area. The equation also accounts the collector efficiency, which can be calculated as the product between the efficiency factor ($\eta_0 = 0.789$) and the linear loss coefficient (a_1). The exponents γ and β were estimated by nonlinear regression using the Advanced Process Monitor (APM) model in Matlab. The equation also includes a sizing factor (b_{solar}), which was estimated through the data collected from different commercial models with distinct specific costs and efficiencies [49].

2.4.2. Circulation Pump Cost Equation

The circulation pump (C_{pump}) cost equation was defined considering as the main sizing variable the water flow rate ($\dot{V}_{pump}$), as presented by Equation (9):

$$C_{pump} = C_{ref,pump} \times \dot{V}_{ref,pump} \left(\frac{\dot{V}_{pump}}{\dot{V}_{ref,pump}} \right)^{b_{pump}} \tag{9}$$

where the term $C_{ref, pump}$ corresponds to the reference cost for this component, $\dot{V}_{ref, pump}$ is the reference flow rate, $\dot{V}_{f, pump}$ is the effective flow rate, and b_{pump} is the component sizing exponent.

2.4.3. Storage Tank Cost Equation

The storage volume ($V_{storage}$) is the key parameter that mostly affects the storage tank cost ($C_{storage}$). Nevertheless, the equation also includes the thermal insulation correction factor of the storage deposit ($Ins_{storage}$), which is defined as a function of the insulation thickness and heat transfer coefficients based on data from commercial models. This term results from the calculation of the global heat transfer coefficient associated with the collector insulation layer $\left(Ins_{storage} = 1/U_S^{\delta}\right)$. Thus, the storage tank cost equality can be defined as presented by Equation (10):

$$C_{storage} = C_{ref,storage} \times V_{ref,storage}\left(\frac{V_{storage}}{V_{ref,storage}}\right)^{b_{storage}} \times Ins_{storage} \tag{10}$$

where $C_{ref, storage}$ refers to the reference cost coefficient, $V_{ref, storage}$ corresponds to the reference storage volume, and $b_{storage}$ is the sizing factor. All the reference coefficients were estimated considering the relative weight that it is attributed to each component in the final purchase cost of the solar thermal system. The sizing factors values were obtained after a sensitivity analysis, which was based on technical information from commercial systems [53]. The final values of the reference coefficients used in the economic model are presented in Table 2.

Table 2. Reference coefficients assumed in the definition of cost equations.

Component	Parameter	Value
Solar collector	Reference cost coefficient, $C_{ref, solar}$	298.5 €/m^2
	Reference collector area, $A_{ref, solar}$	2.50 m^2
	Sizing factor, b_{solar}	0.5
Circulation Pump	Reference cost coefficient, $C_{ref, pump}$	322.5 €/(m^3.h)
	Reference flow rate, $\dot{V}_{ref, pump}$	0.98 m^3/h
	Sizing factor, b_{pump}	0.3
Storage tank	Reference cost coefficient, $C_{ref, storage}$	3647.5 €/m^3
	Reference storage volume, $V_{ref, storage}$	0.32 m^3
	Sizing factor, $b_{storage}$	0.3
	Thermal insulation correction factor, $Ins_{storage}$	0.985

3. Optimization Model

The studies available in the literature are mostly focused on the analysis of either a solar thermal system with a single fixed technology, where the goal is to optimize a single component and its operational conditions or find the best size for a specific application [9,12,54,55]. Multi-objective optimization has been defined as finding a vector of decision variables while minimizing or maximizing different objective functions, given several constraints [56].

3.1. Definition of Objective Function, Decision Variables and Constraints

In the present work, two objectives namely maximizing the solar collector efficiency, *Max* f_1, and minimizing the annualized investment cost of the solar thermal system, *Min* f_2, are presented in Equation (11):

$$Objective\ functions \begin{cases} Maximize\ f_1 \\ f_1 = -\eta_{collection} \\ \\ Minimize\ f_2 \\ f_2 = C_{inv} \times \left[\frac{i_e(1+i_e)^n}{(1+i_e)^n - 1} \right] \end{cases} \quad (11)$$

where C_{inv} corresponds to the sum of three costs equations with the installation costs. The expenses with the installation costs ($C_{installation}$) typically represent a percentage of the investment with the solar collector panel (20% of the purchase cost). Thus, the total investment costs C_{inv} can be estimated by Equation (12):

$$C_{inv} = \left(1.2C_{solar} + C_{pump} + C_{storage}\right). \quad (12)$$

In order to determine the equal amounts of annual cash flows of the investment, the C_{inv} value is multiplied by the capital recovery factor, considering an effective rate of return, i_e, over n years of the solar thermal system operational time.

Three decision variables were selected for the numerical model: linear loss coefficient of the solar collector (a_1), the solar collection area (A_{solar}), and the tank storage volume ($V_{storage}$). Upper and lower bounds were defined in order to limit the so-called decision variable space. Those simple limits are presented by Equations (13)–(15).

$$3.5 \le a_1 \le 5.8 \left(W/(m^2K)\right) \quad (13)$$

$$2.0 \le A_{solar} \le 10.0 \left(m^2\right) \quad (14)$$

$$0.060 \le V_{Storage} \le 0.740 \left(m^3\right) \quad (15)$$

The linear constraints were assumed in the model. These constraints aim to limit the operational parameters that give significance to the thermal model, including the guarantee that there is a temperature difference that promotes the heat transfer between the thermal fluid and the water (Equation (16)); the temperature limits for the thermal fluid should be limited considering their physical properties and guarantee the minimum temperature differential to promote the heat transfer process (Equation (17)).

$$T_f - T_a > 10\ (°C) \quad (16)$$

$$15 < T_f < 180\ (°C) \quad (17)$$

3.2. Optimization Algorithm

Different algorithms can be used to solve constrained optimization engineering problems. Some methods are based on gradient calculations, whereas others are derivative-free methods, such as the search methods or the genetic and evolutionary algorithms [38]. The use of gradient-based methods requires that the objective function is a continuous and the derivatives exist. The derivative-free methods are suitable for solving optimization problems in practical contexts, with a considerable numbers of decision variables (less than 100) and without requiring a feasible initial solution or high computational effort [57–59].

Engineering problems have been solved by applying genetic algorithms [60,61]. While a classical algorithm generates a single optimal solution point at each iteration, a genetic algorithm generates a population of points, and the best point in the population approaches an optimal solution. It is a method for solving both constrained and unconstrained optimization problems based on a natural selection process that mimics biological evolution. The algorithm repeatedly modifies a population of individual solutions. At each step, the genetic algorithm randomly selects individuals from the current population and uses them as parents to produce the children for the next generation. Over successive generations, the population "evolves" toward an optimal solution. The population size depends on the nature of the problem but typically contains several hundreds or thousands of possible solutions. Usually, the population is generated randomly, allowing the entire range of possible solutions (the search space). The solutions may be "seeded" in areas where optimal solutions are likely to be found. A genetic algorithm requires a representation of the solution domain and a fitness function to evaluate it. The fitness function is a particular objective function that is used to give the merit of a given solution. After each simulation step, the worst solutions are excluded. Thus, fitness functions indicate how close a certain solution came to meeting the overall specification of the optimization problem. These algorithms are usually applied to find global minima for highly nonlinear problems [56].

Non-dominated Sorting Genetic Algorithm-II (NASGA-II) is a non-dominated and sorting-based multi-objective evolutionary algorithm used to solve problems that require constraint-handling [62]. This algorithm is a non-dominant and elitist genetic algorithm that picks individuals with better rank (fitness function value), which contributes to the population diversity [63]. This method has a reduced convergence time and it can handle mathematical models with nonlinear functions when compared with gradient-based methods [59]. The elitist selection mechanisms choose the best solutions of each generation without applying any operators. However, since a controlled elitist genetic process is used, the method allows a balance between the exploration and exploitation of the objective functions' space and creates a feasible population regarding the defined linear constraints. Each solution is denoted by the vector of the decision variables. After the objective function and constraints evaluation, the obtained values for the population are scored through their comparison with the fitness function ones. The population and the offspring are combined into one matrix to extend the population, by computing the rank and crowding distance. In this study, a variant of NSGA-II was used as the optimization algorithm. The flow chart of the NSGA-II employment at MatLab is presented in Figure 6.

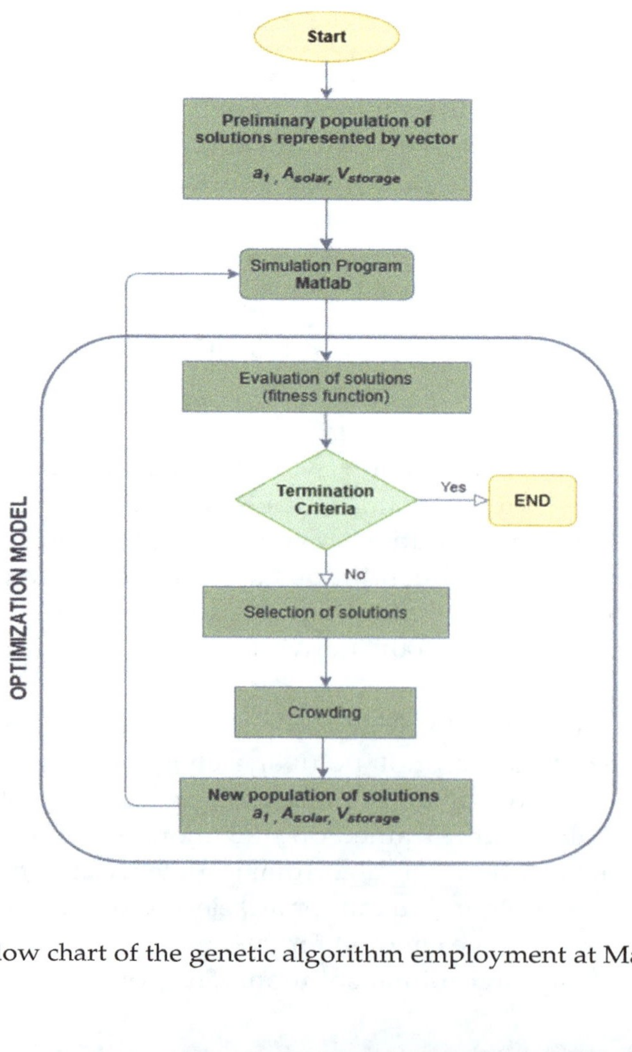

Figure 6. Flow chart of the genetic algorithm employment at MatLab [59].

Then, the Pareto optimal solutions can be obtained, considering the trade-off between competing objectives. The optimization problem was solved with "gamultiobj" MatLab routine, considering a population size of 60 and a Pareto fraction of 0.35.

4. Results and Discussion

The numerical simulations were carried out considering a base-case scenario with a solar panel optimally oriented, 45° south, and set with an inclination of 51° angle. The thermal fluid was defined as a solution of water plus 22.5% of propilenoglycol ($\rho_{tfluid} = 1190$ kg/m^3, $c_{p,tfluid} = 4.045$ kJ/(kg.K)) [64]. For simulation purposes, it was assumed the same temperature at the absorbing plate and cover. Thus, the heat losses from the absorbing surface are assumed to be proportional to the gradient temperature between the absorbing surface and its surroundings.

The useful lifetime of solar thermal systems varies between 15 and 20 years [49]. For the base-case scenario, it was considered a period of 15 years.

4.1. Main Results

The optimization converged when the average change in the spread of Pareto solutions was lower than the minimum distance between objective function values. The objective function was evaluated a total of 29,881 times, and 497 generations were generated by minimizing the annualized investment cost and maximizing the solar collection efficiency. Figure 7 presents the multi-objective optimization results, which are expressed as a two-dimensional Pareto curve. Each point of the Pareto front represents the combination of the decision variables for which it is not possible to improve one of the objective functions without worsening the other one, accounting for the model constraints. This simulation was carried out considering the incident irradiation equal to 950 W/m^2 (base-case scenario).

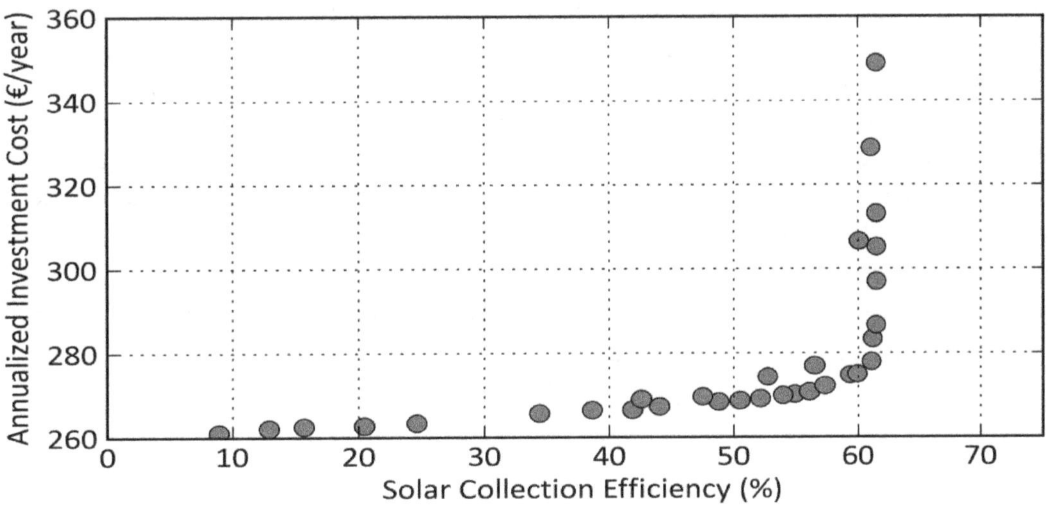

Figure 7. Pareto optimal frontier in objectives' solution space for an incident radiation of 950 W/m^2.

Regarding the optimal solution space, the solar collection efficiency varies between 9.0% and 61.6%, and the annualized investment costs vary from 261.0 €/year to 358.24 €/year. From the analysis of the Pareto curve, it is shown that gains in overall efficiency above 60% result in a relative increase of about 16.3% in the annualized investment costs of the solar thermal system. On the other hand, for investments below 270 €/year in the system, the solar collector efficiency is extremely low (less than 50%). A possible explanation for these results might be related to the correctness of the sizing exponents of the solar collector, which should be adjusted after elaborating a more in-depth sensitivity analysis. Thus, based on the best trade-off between both objectives, an optimal solution point near 60% of efficiency and a system whose optimal physical solution allows an investment cost between 270 and 280 €/year should be selected.

The maximum and minimum value of the decision variables within the optimal frontiers in the objectives' solution space is presented in Table 3. According to the results, the solar collector linear coefficient loss is a parameter that does not undergo a great variation. The optimal solar collector area value varies between 4.00 and 6.25 m^2. These results are in agreement with the literature because, according to [2], an equivalent area of 1 m^2 of solar panel is required to suppress, on average, a demand of 500 kWh/year. With these values, the solar thermal system is able to meet an annual thermal requirement varying between 2000 and 3120 kWh/year, which are within the DHW consumption of the typical Portuguese dwelling (estimated to reach a DHW thermal requirement of 3056.5 kWh).

Table 3. Decision variables within the optimal frontiers in the objectives' solution space.

Decision Variable	Minimum Value	Maximum Value
Solar collector linear coefficient loss, a_1	3.684 W/m^2.K	3.692 W/m^2.K
Solar collector area, A_{solar}	4.00 m^2	6.25 m^2
Storage tank volume, $V_{storage}$	0.203 m^3	0.439 m^3

Regarding the storage tank, the results depict a system with a volume between 0.203 m^3 (203 L) and 0.439 m^3 (439 L). This result shows that the storage volume per unit of solar collection area is within the expected values (0.07 m^3/m^2) for which the storage capacity is adjusted with the efficiency of solar energy usage [2].

The point on the Pareto front nearest to the equilibrium between the objectives is the "solution point" that considers the best trade-off between the objective functions. Thus, a point from the Pareto curve was selected in order to evaluate and analyze the physical and economic parameters for an optimal solution combination (Table 4). We selected a point that corresponds to a panel with a solar area of 4.17 m^2 and a linear coefficient loss of 3.684 W/m^2.K, a storage volume of 0.275 m^3, and a pump flow rate of 0.1364 m^3/h. This combination allows obtaining a system with a solar collection efficiency of 61.2% and with an annualized investment cost of 283.3 €/year (for a system with 15 years of lifetime).

Table 4. Physical parameters of the "solution point" nearest to the trade-off equilibrium.

Physical Parameter	Value
Solar collector linear coefficient loss, a_1	3.684 W/m^2.K
Solar collector area, A_{solar}	4.17 m^2
Storage tank volume, $V_{storage}$	0.275 m^3
Pump flow rate, $\dot{V}_{pump}$	0.1364 m^3/h

Considering the selected characteristics, the purchase cost of the thermal system components is presented in Table 5. The analysis shows that the solar collector and the storage tanks are the most expensive components of the solar thermal system, representing 42% and 43%, respectively. The installation costs represent about 8% of the system total cost, which corresponds to a value of 2545.0 €. The results show that for a solar–thermal system collector with 4.17 m^2, the system's specific cost represents 610.3 €/m^2 of solar area. This value is relatively close to the commercial data, which ranges between 400 and 600 €/m^2 [49,65].

Table 5. Purchase cost considering the optimal solution for the trade-off between objective functions and cost structure considering the relative share of each component.

Component	Purchase Cost	Relative Cost
C_{solar}	1059.6 €	
$C_{storage}$	1098.6 €	
C_{pump}	174.9 €	
$C_{installation}$	211.9 €	
Total purchasing cost	2545.0 €	

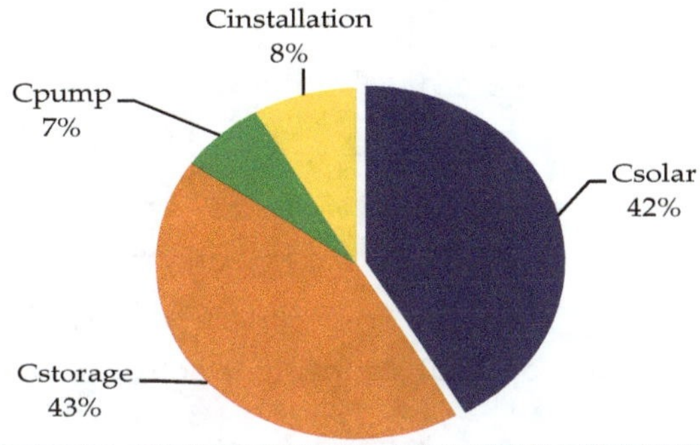

4.2. Sensitivity Analysis

As previously stated, the efficiency of a solar collector depends upon the irradiation, area exposed to solar radiation, wind velocity, diffuse particles in the atmosphere, fluid flow rate in the system, the material used in the absorber plate, and the heat transfer coefficient. However, one of the crucial parameters is the incident radiation, which affects the amount of energy that is transferred to the working fluid. It is desirable to make the difference between the absorbing surface temperature and the fluid as small as practicable in order to maximize the collection efficiency.

Thus, a sensitivity analysis was also performed in order to understand the influence of different solar irradiation values on the optimization model performance. The numerical simulations were carried out considering a lower and an upper value of incident radiation, 900 and 1000 W/m^2, as presented by Figures 8 and 9, respectively. When comparing the Pareto fronts, the simulations disclosed an interesting trade-off between both objective functions. At higher values of irradiance (1000 W/m^2), the solution space shows a set of optimal points with a lower dispersion of efficiency values for a more constrained variation of annualized investment costs. The solar collection efficiency varies between 62.8% and 73.2%, and the annualized investment costs vary from 272.4 €/year to 326.0 €/year. At lower values of irradiance (900 W/m^2), the solar collection efficiency varies between 38.1% and 72.5% and the annualized investment costs vary from 265.5 to 350.6 €/year. The greater the incidence of radiation, the higher the collection efficiency of the panel, which requires smaller solar collecting areas, which reduces the investment cost for the system purchase. This outcome can be evaluated by estimating the annual investment cost for the optimal solution of storage tank volume and solar collector area values.

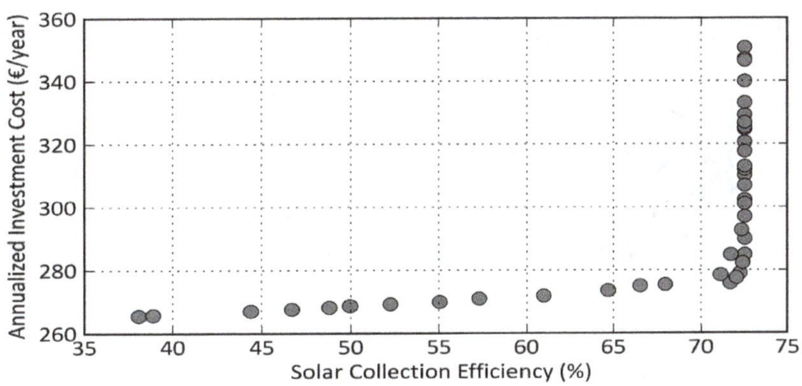

Figure 8. Pareto optimal frontier in the objectives' solution space for an incident radiation of 900 W/m^2.

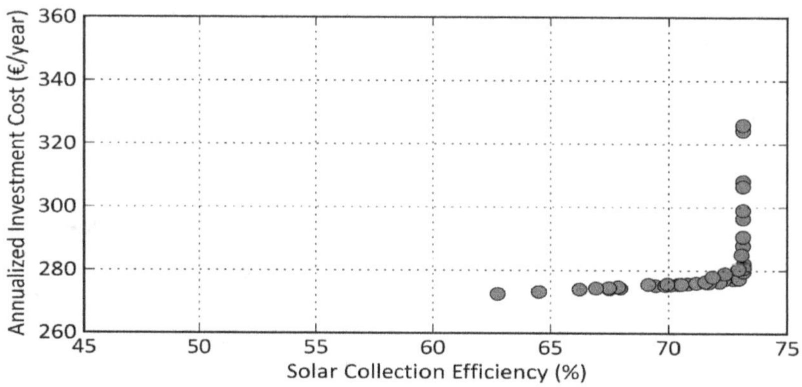

Figure 9. Pareto optimal frontier in the objectives' solution space for an incident radiation of 1000 W/m^2.

Two-dimensional contour maps (Figure 10) were generated to represent the annualized investment cost as a function of the storage tank volume and solar collector area, considering different values of incident radiation. The results were obtained by applying the Kriging method, which is an estimation technique that provides a minimum error-variance of the solution data.

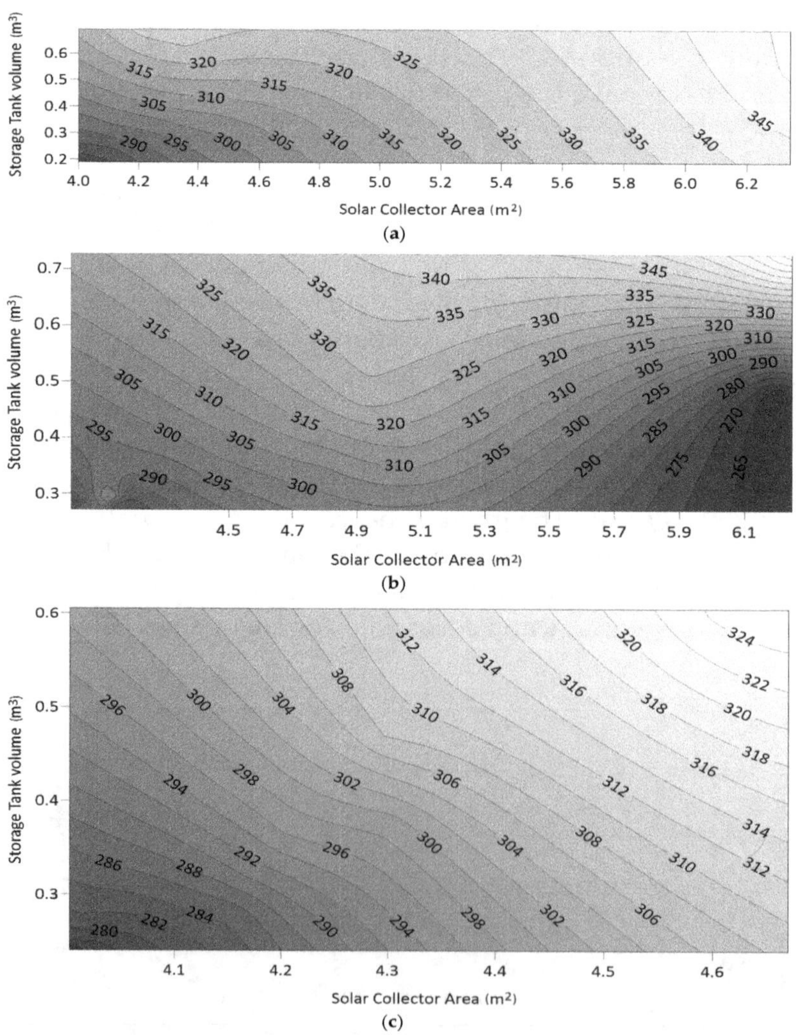

Figure 10. Annualized investment cost as a function of storage tank volume and solar collector area: (**a**) assuming an incident radiation of 900 W/m^2; (**b**) assuming an incident radiation of 950 W/m^2; and (**c**) assuming an incident radiation of 1000 W/m^2.

5. Main Conclusions

In this paper, a multi-objective optimization model was applied to numerically simulate a solar thermal system. The mathematical modelling comprises the three main components: the solar collector that captures the solar radiation, the pump with its controller that promotes the forced convection of the heated water; and the storage tank, which is responsible for accumulating heat to be used when the radiation is not available. A set of cost equations was developed considering the most important physical variables, and the cost coefficients were based on real data from commercial systems. The resulting model was applied to supress the DHW needs of a typical residential building in Portugal. A typical dwelling with an occupation of four people has a daily consumption of about 160 L, which represents a DHW thermal requirement of 3056.5 kWh per year.

The costing methodology was defined considering the quality and sizing variables that most affect the cost of the system components. Regarding the solar collector, the solar area and the collecting efficiency are the most important physical parameters. The flow rate and the storage volume are the variables that affect the cost of the pump and storage tank, respectively. The coefficients of the cost equations were based on real data from commercial systems.

The multi-objective approach aimed to both minimize the system investment costs while maximizing the solar collection efficiency. The simulations were performed through a MatLab code by using an elitist variant of NASGA-II. The solar collection area and the linear loss coefficient as well as the tank storage volume were defined as decision variables. Upper and lower limits were attributed to the decision variables in order to give significance to the mathematical model. The numerical simulations were carried out considering that the solar panel was optimally oriented (45° south, with an inclination of 51° angle). The incident radiation was fixed with the value of 950 W/m^2.

The analysis of the Pareto curve led to the conclusion that a system with an annualized investment cost between 270 and 280 €/year allows reaching a collection efficiency of 60%. Considering the trade-off between the objective functions, a point from the Pareto curve was selected in order to evaluate the physical and economic parameters for an optimal solution: a panel with a solar area of 4.17 m^2 and with a linear coefficient loss of 3.684 W/m^2.K; a storage volume of 0.275 m^3, and a pump flow rate of about 0.1364 m^3/h. This combination allows obtaining a system with a solar collection efficiency of 61.2% and the acquisition of a system with an annualized investment cost of 283.3 €/year (for a system with 15 years of lifetime). For this optimal solution, the total purchasing cost is 2545.0 €, whereas the solar collector and the storage tank are the most expensive components, representing 42% and 43% of the investment costs, respectively. These results represents a specific cost of 610.3 €/m^2 per unit of solar collection area, which is relatively low when compared with the commercial data, which ranges between 400 and 600 €/m^2.

A sensitivity analysis was also performed in order to understand the influence of different solar irradiation values on the optimization model performance. Higher values of incident radiation result in systems that optimally have higher efficiencies, requiring smaller areas for solar collecting, which reduces the investment cost for the system purchase.

Regarding the manuscript limitations, a profile of solar radiation over the day should be added to the model in order to study more realistically the limitations of this type of renewable energy systems. Due to the intermittency of this energy source, it should also include the needs and costs regarding the backup system. As future work, it is intended to perform a sensitivity analysis in order understand the influence of systems lifetime and the interest rate in the annualized investment cost of the solar thermal system.

Author Contributions: Conceptualization, A.C.F.; Formal analysis, J.C.T. and S.T.; Supervision, J.C.T. and S.T.; Validation, A.C.F.; Writing—original draft, A.C.F.; Writing—review and editing, A.S. All authors have read and agreed to the published version of the manuscript.

Acknowledgments: Authors acknowledge the support given by FCT through R&D Units Project Scope UIDB/00319/2020 (ALGORITMI) and R&D Units Project Scope UIDP/04077/2020 (METRICS).

References

1. Visa, I.; Duta, A.; Moldovan, M.; Burduhos, B.; Neagoe, M. *Solar Energy Conversion Systems in the Built Environment*; Green Energy and Technology; Springer International Publishing: Cham, Germany, 2020; pp. 327–339.
2. Lebeña, E.P.; Costa, J.C. Instaladores de Equipamentos Solares térmicos. In *Conversão Térmica da Energia Solar*; Sociedade Portuguesa de Energia Solar: Lisbon, Portugal, 2008; pp. 1–89.
3. Paulescu, M.; Paulescu, E.; Gravila, P.; Badescu, V. *Weather Modeling and Forecasting of PV Systems Operation*; Springer: London, UK, 2013; Volume 103, pp. 127–179.
4. Tsalikis, G.; Martinopoulos, G. Solar energy systems potential for nearly net zero energy residential buildings. *Sol. Energy* **2015**, *115*, 743–756. [CrossRef]
5. Islam, M.; Morimoto, T. Advances in low to medium temperature non-concentrating solar thermal technology. *Renew. Sustain. Energy Rev.* **2018**, *82*, 2066–2093. [CrossRef]
6. Karki, S.; Haapala, K.R.; Fronk, B.M. Technical and economic feasibility of solar flat-plate collector thermal energy systems for small and medium manufacturers. *Appl. Energy* **2019**, *254*, 113649. [CrossRef]
7. Ramos, A.; Guarracino, I.; Mellor, A.; Alonso-álvarez, D.; Childs, P.; Ekins-daukes, N.J.; Markides, C.N. *Solar-Thermal and Hybrid Photovoltaic-Thermal Systems for Renewable Heating*; Imperial College: London, UK, 2017.
8. Antoniadis, C.N.; Martinopoulos, G. Optimization of a building integrated solar thermal system with seasonal storage using TRNSYS. *Renew. Energy* **2019**, *137*, 56–66. [CrossRef]
9. Araújo, A.; Silva, R.; Pereira, V. Solar thermal modeling for rapid estimation of auxiliary energy requirements in domestic hot water production: On-off versus proportional flow rate control. *Sol. Energy* **2019**, *177*, 68–79. [CrossRef]
10. Martinopoulos, G.; Tsalikis, G. Diffusion and adoption of solar energy conversion systems—The case of Greece. *Energy* **2018**, *144*, 800–807. [CrossRef]
11. Klein, S.; Beckman, W. A general design method for closed-loop solar energy systems. *Sol. Energy* **1979**, *22*, 269–282. [CrossRef]
12. Chen, C.; Diao, Y.; Zhao, Y.; Wang, Z.; Liang, L.; Wang, T.; Zhu, T.; Ma, C. Thermal performance of a closed collector–storage solar air heating system with latent thermal storage: An experimental study. *Energy* **2020**, *202*, 117764. [CrossRef]
13. Xiao, Q.; Cao, J.; Zhang, Y.; Li, L.; Xu, T.; Yuan, W. The application of solar-to-thermal conversion phase change material in novel solar water heating system. *Sol. Energy* **2020**, *199*, 484–490. [CrossRef]
14. Kalogirou, S.A. Designing and Modeling Solar Energy Systems. In *Solar Energy Engineering*; Elsevier BV: Amsterdam, The Netherlands, 2014; pp. 583–699.
15. Kalogirou, S.A. Solar thermal collectors and applications. *Prog. Energy Combust. Sci.* **2004**, *30*, 231–295. [CrossRef]
16. Zeng, R.; Zhang, X.; Deng, Y.; Li, H.; Zhang, G. Optimization and performance comparison of combined cooling, heating and power/ground source heat pump/photovoltaic/solar thermal system under different load ratio for two operation strategies. *Energy Convers. Manag.* **2020**, *208*, 112579. [CrossRef]
17. Assaf, J.; Shabani, B. Multi-objective sizing optimisation of a solar-thermal system integrated with a solar-hydrogen combined heat and power system, using genetic algorithm. *Energy Convers. Manag.* **2018**, *164*, 518–532. [CrossRef]
18. Tian, Y.; Zhao, C. A review of solar collectors and thermal energy storage in solar thermal applications. *Appl. Energy* **2013**, *104*, 538–553. [CrossRef]
19. Lin, W.; Ma, Z.; Ren, H.; Gschwander, S.; Wang, S. Multi-objective optimisation of thermal energy storage using phase change materials for solar air systems. *Renew. Energy* **2019**, *130*, 1116–1129. [CrossRef]
20. Ong, K.S.; Tong, W.L. System Performance of U-Tube and Heat Pipe Solar Water Heaters. *J. Appl. Sci. Eng.* **2012**, *15*, 105–110.
21. Marion, M.; Voicu, I.; Tiffonnet, A.-L. Study and optimization of a solar subcritical organic Rankine cycle. *Renew. Energy* **2012**, *48*, 100–109. [CrossRef]
22. Yongga, A.; Li, N.; Long, J.; He, Y. Thermal performance of a novel solar thermal facade system in a hot-summer and cold-winter zone. *Sol. Energy* **2020**, *204*, 106–114. [CrossRef]

23. Frattolillo, A.; Canale, L.; Ficco, G.; Mastino, C.C.; Dell'Isola, M. Potential for Building Façade-Integrated Solar Thermal Collectors in a Highly Urbanized Context. *Energies* **2020**, *13*, 5801. [CrossRef]
24. Katsaprakakis, D.A.; Zidianakis, G.; Yiannakoudakis, Y.; Manioudakis, E.; Dakanali, I.; Kanouras, S. Working on Buildings' Energy Performance Upgrade in Mediterranean Climate. *Energies* **2020**, *13*, 2159. [CrossRef]
25. Artur, C.; Neves, D.; Cuamba, B.C.; Leão, A.J. Domestic hot water technology transition for solar thermal systems: An assessment for the urban areas of Maputo city, Mozambique. *J. Clean. Prod.* **2020**, *260*, 121043. [CrossRef]
26. Košičan, J.; Picazo, M.; Ángel, P.; Vilcekova, S. A Multicriteria Methodology to Select the Best Installation of Solar Thermal Power in a Family House. *Energies* **2020**, *13*, 1047. [CrossRef]
27. Sami, S.; Semmar, D.; Hamid, A.; Mecheri, R.; Yaiche, M. Viability of integrating Solar Water Heating systems into High Energy Performance housing in Algeria. *Energy* **2018**, *149*, 354–363. [CrossRef]
28. Louvet, Y.; Fischer, S.; Furbo, S.; Giovannetti, F.; Helbig, S.; Köhl, M.; Mugnier, D.; Philippen, D.; Veynandt, F.; Vajen, K. Economic comparison of reference solar thermal systems for households in five European countries. *Sol. Energy* **2019**, *193*, 85–94. [CrossRef]
29. Gautam, A.; Chamoli, S.; Kumar, A.; Singh, S. A review on technical improvements, economic feasibility and world scenario of solar water heating system. *Renew. Sustain. Energy Rev.* **2017**, *68*, 541–562. [CrossRef]
30. Tschopp, D.; Tian, Z.; Berberich, M.; Fan, J.; Perers, B.; Furbo, S. Large-scale solar thermal systems in leading countries: A review and comparative study of Denmark, China, Germany and Austria. *Appl. Energy* **2020**, *270*, 114997. [CrossRef]
31. Tsilingiridis, G.; Martinopoulos, G. Thirty years of domestic solar hot water systems use in Greece—Energy and environmental benefits—Future perspectives. *Renew. Energy* **2010**, *35*, 490–497. [CrossRef]
32. Ferreira, A.C.; Nunes, M.; Teixeira, S.F.C.F.; Leão, C.P.; Silva, Â.M.; Teixeira, J.C.F.; Martins, L. An economic perspective on the optimisation of a small-scale cogeneration system for the Portuguese scenario. *Energy* **2012**, *45*, 436–444. [CrossRef]
33. Ferreira, A.C.; Nunes, M.; Martins, L.; Teixeira, S.F.C.F. Maximum Profit of a Cogeneration System Based on Stirling Thermodynamic Cycle. In Proceedings of the 2014 14th International Conference on Computational Science and Its Applications, Guimaraes, Portugal, 30 June–3 July 2014; Institute of Electrical and Electronics Engineers (IEEE): Piscataway, NJ, USA, 2014; pp. 156–160.
34. Cruz-Peragón, F.; Palomar, J.M.; Casanova, P.; Dorado, M.; Manzano-Agugliaro, F. Characterization of solar flat plate collectors. *Renew. Sustain. Energy Rev.* **2012**, *16*, 1709–1720. [CrossRef]
35. Sharma, N.; Diaz, G. Performance model of a novel evacuated-tube solar collector based on minichannels. *Sol. Energy* **2011**, *85*, 881–890. [CrossRef]
36. Mansour, M.K. Thermal analysis of novel minichannel-based solar flat-plate collector. *Energy* **2013**, *60*, 333–343. [CrossRef]
37. Sahin, B.; Kodal, A. Performance analysis of an endoreversible heat engine based on a new thermoeconomic optimization criterion. *Energy Convers. Manag.* **2001**, *42*, 1085–1093. [CrossRef]
38. Bejan, A.; Tsatsaronis, G.; Moran, M. *Thermal Design and Optimization*; Wiley: New York, NY, USA, 1996.
39. Yılmaz, I.H. Residential use of solar water heating in Turkey: A novel thermo-economic optimization for energy savings, cost benefit and ecology. *J. Clean. Prod.* **2018**, *204*, 511–524. [CrossRef]
40. Asadi, J.; Amani, P.; Amani, M.; Kasaeian, A.; Bahiraei, M. Thermo-economic analysis and multi-objective optimization of absorption cooling system driven by various solar collectors. *Energy Convers. Manag.* **2018**, *173*, 715–727. [CrossRef]
41. Cavaco, A.; Silva, H.; Canhoto, P.; Neves, S.; Neto, J.; Collares Pereira, M. *Global Solar Radiation in Portugal and Its Variability, Monthly and Yearly*; Institute of Earth Sciences: Evora, Portugal, 2016.
42. *Desempenho Energético dos Edifícios*; DL 118/2013 Decreto-Lei n.o 118/2013; Diário da República: Portugal, 2013; Volume 159, pp. 4988–5005.
43. Kumar, L.; Hasanuzzaman, M.; Rahim, N. Global advancement of solar thermal energy technologies for industrial process heat and its future prospects: A review. *Energy Convers. Manag.* **2019**, *195*, 885–908. [CrossRef]
44. El Mkadmi, C.; Wahed, M.A. Optimization of a solar thermal system for low temperature industrial heating process. In Proceedings of the 2016 International Renewable and Sustainable Energy Conference (IRSEC), Marrakech, Morocco, 14–17 November 2016; Institute of Electrical and Electronics Engineers (IEEE): Piscataway, NJ, USA, 2016; pp. 313–319.

45. Menke, C.; Gebauer, M.; Jonas, D. Optimization of Large-scale Solar Thermal Systems for Student Dormitories in Germany. In Proceedings of the 49th Annual Conference of the Australian Solar Energy Society, Sydney, Australia, 1 December 2011.
46. Belyakov, N. *Solar energy. Sustainable Power Generation*, 1st ed.; Elsevier: Amsterdam, The Netherlands, 2019; pp. 417–438.
47. Rao, P.; Francis, J.; Love, J.T. Two dimensional analysis of a flat-plate solar collector. In Proceedings of the 11th Thermophysics Conference, Atlanta, GA, USA, 16–20 June 2014; American Institute of Aeronautics and Astronautics (AIAA): Reston, VA, USA, 1976; pp. 1–4.
48. Islam, M.; Morimoto, T. Performance Prediction of Solar Collector Adsorber Tube Temperature using a Nonlinear Autoregressive Model with eXogenous Input. *Int. J. Comput. Appl.* **2015**, *114*, 24–32. [CrossRef]
49. Ferreira, A.C.; Silva, A. Application of a costing methodology to estimate capital costs of solar thermal systems in residential Portuguese context. *Int. J. Sustain. Energy Plan. Manag.* **2020**, *26*, 33–46. [CrossRef]
50. Lazzaretto, A.; Tsatsaronis, G. SPECO: A systematic and general methodology for calculating efficiencies and costs in thermal systems. *Energy* **2006**, *31*, 1257–1289. [CrossRef]
51. Tronchin, L.; Tommasino, M.C.; Fabbri, K. On the cost-optimal levels of energy-performance requirements for buildings: A case study with economic evaluation in Italy. *Int. J. Sustain. Energy Plan. Manag.* **2014**, *3*, 49–62. [CrossRef]
52. Biegler, L.T.; Grossmann, I.E.; Westerberg, A.W. *The Factorial Method of Cost Estimation*; Physical and Chemical Engineering Sciences; Prentice Hall International: Upper Saddle River, NJ, USA, 1997.
53. Ferreira, A.C.; Silva, A.M. Development of a costing methodology for solar thermal systems: Application to Portuguese scenario. In Proceedings of the 4th International Conference on Energy & Environment: Bringing together Engineering and Economics, Guimarães, Portugal, 16 May 2019; Paula Ferreira, U., Isabel Soares, F., Eds.; University of Minho: Guimarães, Portugal, 2019; pp. 64–70.
54. Hachem, H.; Gheith, R.; Aloui, F.; Ben Nasrallah, S. Technological challenges and optimization efforts of the Stirling machine: A review. *Energy Convers. Manag.* **2018**, *171*, 1365–1387. [CrossRef]
55. Attonaty, K.; Pouvreau, J.; Deydier, A.; Oriol, J.; Stouffs, P. Thermodynamic and economic evaluation of an innovative electricity storage system based on thermal energy storage. *Renew. Energy* **2020**, *150*, 1030–1036. [CrossRef]
56. Rao, R.V.; Patel, V. Multi-objective optimization of heat exchangers using a modified teaching-learning-based optimization algorithm. *Appl. Math. Model.* **2013**, *37*, 1147–1162. [CrossRef]
57. Ferreira, A.C.; Vaz, A.I.F.; Teixeira, S.F.; Teixeira, J.C. Thermal-economic design of a micro-gas turbine CHP system using a multi-objective optimization approach. In Proceedings of the PANACM 2015—1st Pan-American Congress on Computational Mechanics, in Conjunction with the 11th Argentine Congress on Computational Mechanics, MECOM 2015, Buenos Aires, Argentina, 27–29 April 2015.
58. Ferreira, A.C.; Rocha, A.M.A.C.; Teixeira, S.F.C.F.; Nunes, M.; Martins, L. On Solving the Profit Maximization of Small Cogeneration Systems. In Proceedings of the Computer Vision—ECCV 2020; Springer Science and Business Media LLC: Berlin/Heidelberg, Germany, 2012; Volume 7335, pp. 147–158.
59. Ferreira, A.C.; Silva, Â.M.; Teixeira, S.F.C.F. Multi-objective Optimization of Solar Thermal Systems Applied to Residential Building in Portugal. In Proceedings of the Computer Vision—ECCV 2020; Springer Science and Business Media LLC: Berlin/Heidelberg, Germany, 2019; pp. 26–39.
60. Valdés, M.; Durán, M.D.; Rovira, A. Thermoeconomic optimization of combined cycle gas turbine power plants using genetic algorithms. *Appl. Therm. Eng.* **2003**, *23*, 2169–2182. [CrossRef]
61. Ahmadi, M.H.; Sayyaadi, H.; Mohammadi, A.H.; Barranco-Jimenez, M.A. Thermo-economic multi-objective optimization of solar dish-Stirling engine by implementing evolutionary algorithm. *Energy Convers. Manag.* **2013**, *73*, 370–380. [CrossRef]
62. Deb, K.; Pratap, A.; Agarwal, S.; Meyarivan, T. A fast and elitist multiobjective genetic algorithm: NSGA-II. *IEEE Trans. Evolut. Comput.* **2002**, *6*, 182–197. [CrossRef]
63. MathWorks. Create genetic algorithm options structure 2018, 2014a edition. Available online: https://www.mathworks.com/help/gads/gaoptimset.html (accessed on 12 September 2018).
64. Greenstream Publishing, Solar Electricity Handbook. Available online: http://www.solarelectricityhandbook.com/solar-irradiance.html (accessed on 1 January 2019).
65. Vulcano Thermal Solar Systems Catalog Prices—Vulcano. 2017. Available online: www.vulcano.pt (accessed on 1 February 2017).

2

Optimization of an Inter-Plant Hydrogen Network: A Simultaneous Approach to Solving Multi-Period Optimization Problems

Rusong Han [1], Lixia Kang [1,2], Yinghua Jiang [1], Jing Wang [1] and Yongzhong Liu [1,2,3,*]

1 Department of Chemical Engineering, Xi'an Jiaotong University, Xi'an 710049, China; hanrsg@stu.xjtu.edu.cn (R.H.); lx_kang@mail.xjtu.edu.cn (L.K.); jiangyinghua@stu.xjtu.edu.cn (Y.J.); wjjycg0916@stu.xjtu.edu.cn (J.W.)
2 Shaanxi Key Laboratory of Energy Chemical Process Intensification, Xi'an 710049, China
3 Key Laboratory of Thermo-Fluid Science and Engineering, Ministry of Education, Xi'an 710049, China
* Correspondence: yzliu@mail.xjtu.edu.cn

Abstract: Inter-plant hydrogen integration can reduce the consumption of hydrogen utility in petrochemical parks. However, the fluctuation of operating conditions will lead to complex multi-period problems of hydrogen network integration. This work develops a simultaneous optimization approach to solving multi-period optimization problems for the inter-plant hydrogen network. To do this, we consider the inter-plant hydrogen integration and the fluctuation of operating conditions in each plant at the same time, and aim to minimize the total annualized cost of the entire hydrogen system of all plants involved. An industrial case study of a three-plant hydrogen network with seven subperiods was adopted to verify the effectiveness of the proposed method. Results show that the optimal structure and the corresponding scheduling scheme can be obtained when the lowest cost of the system is targeted. Compared with the stepwise methods, the proposed approach features taking the characteristics of all subperiods into account simultaneously and making the structure of the hydrogen network much more effective and economical. For the scheduling schemes, the utilization efficiency of the internal hydrogen sources is increased by hydrogen exchange among the plants.

Keywords: multi-period hydrogen network; inter-plant; superstructure; simultaneous optimization approach

---

1. Introduction

Chemical industry parks generally consist of companies that produce a variety of chemical products, where hydrogen is usually used as a utility [1]. The integration of the hydrogen network is an effective approach to optimize the allocation of hydrogen utility and reduce operation costs [2]. In these parks, the quantity and quality of hydrogen supply and demand in each company are usually different, which makes it possible to integrate the hydrogen networks for all plants [3]. Comparing to integrating the hydrogen network in a single plant, integrating hydrogen systems for multiple plants could further reduce utility consumption and improve the utilization efficiency of hydrogen by allowing the transportation of hydrogen streams across plants [4]. Nevertheless, the supply and demand of hydrogen in the plants change over time, and the flowrates fluctuation in one plant does not synchronize with the other one. Therefore, to make the best use of the hydrogen utilities in the chemical industry parks, multi-period methods are often adopted to optimize the holistic hydrogen network by considering the varying operating conditions of the hydrogen streams in the whole park.

For the multi-period method applied in the optimization of the hydrogen network, the operating duration is usually divided into several subperiods to approximate the fluctuations of the operating conditions, and the operational parameters are kept constant in each subperiod. Liang et al. [5] presented a multi-period optimization model for the hydrogen network and investigated the influences of the number and the duration of subperiods on the optimal design of the hydrogen network. The solution strategies for the optimization of multi-period hydrogen network can be categorized into two kinds of methods, i.e., stepwise methods and simultaneous methods. For the stepwise strategies, the problem is decomposed into subproblems to reduce the number of model variables, and the final hydrogen network can be obtained by synthesizing the design scheme of each subperiod [6–8]. Although the scale of the model is reduced, it is difficult to appraise the performance of the hydrogen network in each subperiod in a comprehensive way. Thus, the hydrogen network structures obtained by the stepwise methods are usually local optima. On the contrary, the simultaneous methods fully consider the trade-off between each subperiod, and the configurations and schedule schemes of the hydrogen network with much better performance can be achieved. Nevertheless, the solution process could be time-consuming [9] when the scale of the model is enlarged [10].

For the inter-plant integration of the hydrogen network, there are basically three ways. The first way is to share the hydrogen utilities by pipelines to directly meet the demands of hydrogen sinks in all plants. The second one is that the purifier is used to increase the purities of some internal hydrogen sources in one plant and deliver them to other plants. The last one is that intermediate hydrogen headers are set up among these plants. Deng et al. [11] realized that integrating the hydrogen systems through hydrogen pipelines and centralized purifiers. In their work, all possible connections of the inter-plant hydrogen network were considered in the superstructure, and the influences of design conditions of the purifiers on the performances were studied comprehensively. Kang et al. [12] integrated the inter-plant hydrogen network through intermediate headers, of which the hydrogen purity and pressure were optimized in turn. Lou et al. [13] synthesize the inter-plant hydrogen network by the cross-plant pipelines and purifiers. In their method, the hydrogen allocation among the plants was optimized by using a transshipment model, and then the intra-plant structure of individual hydrogen networks was designed. Although the abovementioned studies can be used to optimize integrating the inter-plant hydrogen network, the influences of the operating condition fluctuations on the structure and scheduling schemes are somehow ignored. To consider the fluctuations of the operating conditions, Shehata et al. [14] used the results of a single period model to synthesize the multi-period hydrogen network and achieved a significant reduction in operating costs. However, this stepwise approach fails to consider the relationships and trade-offs among subperiods, and the design scheme of the hydrogen system is greatly affected by the worst case, which may cause an unnecessary increase of investment cost in the plants. Therefore, it is necessary to develop a multi-period simultaneous method to realize the optimal design of the inter-plant hydrogen system considering the fluctuations of the operating conditions in all plants.

To address the abovementioned issues, a multi-period simultaneous optimization method for the inter-plant hydrogen system, which features a mixed-integer linear programming (MILP) problem, is proposed to coordinate the supply and demand of hydrogen and realize the lowest total annual cost (TAC) of the hydrogen network in a chemical industry park. In this method, the hydrogen networks in different plants are connected by the hydrogen supply pipeline across the plants and the centralized purifier in each plant. A case study is used to illustrate the effectiveness and advantages of the proposed method.

The rest of this paper is arranged as follows. The problem statement is given in Section 2, followed by introducing the multi-period optimization model of the inter-plant hydrogen network in Section 3. The case study is given in Section 4, and the conclusions are drawn in Section 5.

2. Problem Statement

Figure 1 presents the superstructure of the hydrogen system with N plants in a chemical industry park, which is composed of hydrogen sources, hydrogen sinks, purifiers, compressors, intra-plant pipelines, cross-plant pipelines, and a fuel gas system.

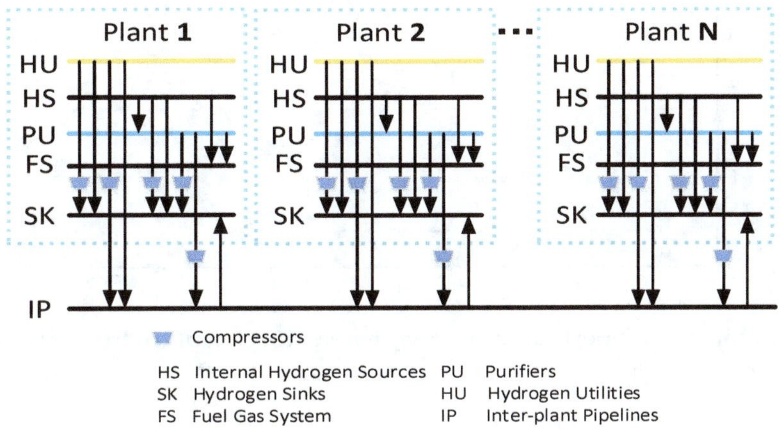

Figure 1. The superstructure of the inter-plant hydrogen network.

The hydrogen sources consist of internal hydrogen sources ($i \in \mathbf{HS}$) and hydrogen utilities ($j \in \mathbf{HU}$), and the set of all the hydrogen sources in the inter-plant hydrogen network can be expressed as $s \in \mathbf{SC}$, $\mathbf{SC} = \mathbf{HU} \cup \mathbf{HS}$. The set of hydrogen–consuming units are denoted as hydrogen sinks ($k \in \mathbf{SK}$). The hydrogen utilities can be delivered to the hydrogen sinks within a plant via the intra-plant pipelines, or to hydrogen sinks of other plants via the cross-plant pipelines. The internal hydrogen source can only be delivered to the hydrogen sinks within a plant, or to the purifier as a feed stream, or directly discharged to the fuel gas system. The centralized purifier ($m \in \mathbf{PU}$) of each plant only deals with the internal hydrogen source streams from the same plant, and the product stream can be allocated to all hydrogen sinks, and the residual stream is fed to the fuel gas system. The compressors are used to elevate the pressure of the hydrogen streams to meet the inlet requirements of the hydrogen sinks and the purifiers. In the hydrogen network of plant n ($n \in \mathbf{N}$), the flowrates, purities, and pressures of hydrogen sources and sinks are known. In addition, the distances among the plants are also known.

The main assumptions are listed as follows.

1. A dedicated hydrogen pipeline is set for each cross-plant match;
2. In each plant, at most one centralized purifier can be set up;
3. A dedicated compressor is placed for each match that needs elevating pressure.

The fluctuations of the hydrogen sources and hydrogen sinks in different plants may change out of synchronization. Hence, for the plants involved, to integrate the hydrogen networks in different plants, the partitioning of subperiod should be carried out in terms of fluctuation of hydrogen sources and sinks throughout the plants, where the operational parameters are kept constant in each subperiod, as shown in Figure 2. In the figure, F_s denotes the flowrate of hydrogen source, and F_k denotes the flowrate of hydrogen sink. The operating time of interest can be divided into p subperiods, in which the flowrates and purities of the hydrogen streams are constant. Then the problem of integrating hydrogen networks in multiple plants is transformed into a multi-period optimization problem of the inter-plant hydrogen network. Moreover, the mathematical programming model needs to be established and solved to obtain the structure of the inter-plant hydrogen network that can accommodate the fluctuating of the hydrogen sources and hydrogen sinks in different plants.

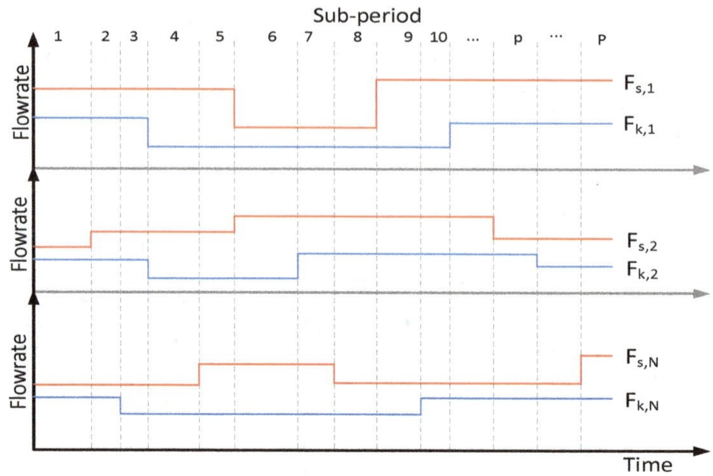

Figure 2. The partitioning of subperiods and flowrates fluctuation of hydrogen sources and hydrogen sinks in different plants. F_S: the flowrate of hydrogen source; F_K: the flowrate of hydrogen sink.

3. Mathematical Programming Model for Multi-Period Optimization of Inter-Plant Hydrogen Networks

3.1. Objective Function

In the model, the hydrogen networks of the three plants are integrated as a whole, where hydrogen utilities are shared among plants and can be transported through cross-plant pipelines to implement the industrial symbiosis. The objective is to minimize the total annual cost (TAC) of the entire hydrogen network in a chemical industry park, which can be formulated as

$$\min TAC = C_{H_2} - C_{fuel} + C_{ele} + Af \cdot \left(C_{pur} + C_{pipe} + C_{com}\right) \tag{1}$$

where TAC is the total annual cost, which is the summation of the operating cost and the capital cost. The operating cost includes the cost of hydrogen utility, C_{H_2}, the revenue of fuel gas, C_{fuel}, electricity purchasing cost, C_{ele}; the capital cost includes the purifier cost, C_{pur}, the pipeline cost, C_{pipe} and the compressor cost, C_{com}.

Af denotes the annualized factor, which is calculated by

$$Af = \frac{fi(1+fi)^q}{(1+fi)^q - 1} \tag{2}$$

where fi is the annual interest rate, and q is the number of years for depreciation. In this work, fi is 0.05 and q is 5.

The cost of hydrogen utility, C_{H_2}, is calculated by

$$C_{H_2} = \sum_{p\in \mathbf{P}} \left[t_p \cdot \sum_{j\in \mathbf{HU}} \sum_{k\in \mathbf{SK}} \left(3600 \times F^p_{j,k} \cdot e_j\right) \right] \tag{3}$$

where t_p is the duration of subperiod p, and e_j represents the price of the hydrogen utility. $F^p_{j,k}$ stands for the flowrate supplied from hydrogen utility j to hydrogen sink k under subperiod p.

The revenue of fuel gas, C_{fuel}, is calculated by

$$C_{\text{fuel}} = e_{\text{heat}} \cdot \sum_{p \in \mathbf{P}} 3600 \times t_p \left\{ \begin{array}{l} \sum\limits_{m \in \mathbf{PU}} \left[(1 - r_m) \cdot \sum\limits_{s \in \mathbf{SC}} (F^p_{s,m} \cdot y^p_s) \right] \cdot \Delta H^\ominus_{\text{c,H}_2} + \\ \sum\limits_{i \in \mathbf{HS}} (F^p_{i,SUR} \cdot y^p_i) \cdot \Delta H^\ominus_{\text{c,H}_2} + \\ \sum\limits_{i \in \mathbf{HS}} \left[F^p_{i,SUR} \cdot (1 - y^p_i) \right] \cdot \Delta H^\ominus_{\text{c,CH}_4} + \\ \sum\limits_{m \in \mathbf{PU}} \left[F^p_{m,resd} - (1 - r_m) \cdot \sum\limits_{s \in \mathbf{SC}} (F^p_{s,m} \cdot y^p_s) \right] \cdot \Delta H^\ominus_{\text{c,CH}_4} \end{array} \right\} \tag{4}$$

where e_{heat} is the price of heat energy. r_m is the ratio of the flowrate of pure hydrogen in the product stream to that in the feed stream. $F^p_{s,m}$ stands for the flowrate supplied from hydrogen source s to purifier m, and y^p_s denotes the hydrogen purity of hydrogen source s. For the internal hydrogen source i ($i \in \mathbf{HS}$), $F^p_{i,SUR}$ is the flowrate allocated to the fuel gas system and y^p_i denotes the hydrogen purity. $F^p_{m,resd}$ denotes the flowrates of the residual stream for purifier m. $\Delta H^\ominus_{\text{c,H}_2}$ and $\Delta H^\ominus_{\text{c,CH}_4}$ represent the standard combustion heat of hydrogen and methane.

The electricity cost of the compressors can be calculated by

$$C_{\text{ele}} = \sum_{p \in \mathbf{P}} t_p \cdot e_{\text{ele}} \cdot \left(\sum_{k \in \mathbf{SK}} \sum_{s \in \mathbf{SC}} W^p_{s,k} + \sum_{m \in \mathbf{PU}} \sum_{s \in \mathbf{SC}} W^p_{s,m} + \sum_{k \in \mathbf{SK}} \sum_{m \in \mathbf{PU}} W^p_{m,k} \right) \tag{5}$$

where e_{ele} denotes the unit price of electricity.

$W^p_{a,b}$ represents the power of the compressor between supplier a and receiver b in subperiod p [15], which is calculated by

$$\begin{gathered} W^p_{a,b} = z^{p,\text{P}}_{a,b} \cdot F^p_{a,b} \frac{c_p T}{\eta} \cdot \left[\left(\frac{P^p_b}{P^p_a} \right)^{\frac{\gamma - 1}{\gamma}} - 1 \right] \\ z^{p,\text{P}}_{a,b} \in \left\{ z^{\text{P}}_{s,k}, z^{\text{P}}_{s,m}, z^{\text{P}}_{m,k} \right\} \forall s \in \mathbf{SC}, m \in \mathbf{PU}, k \in \mathbf{SK}, p \in \mathbf{P} \end{gathered} \tag{6}$$

where $z^{p,\text{P}}_{a,b}$ is used to judge whether a compressor is necessary to be installed or not; $F^p_{a,b}$ denotes the flowrate from supplier a to receiver b; c_p is the specific heat at constant pressure; T is the inlet temperature and η is the efficiency of the compressor; P^p_a and P^p_b are the inlet and outlet pressures of the compressor; γ represents the ratio of specific heat.

The capital cost of the pipelines can be estimated by

$$C_{\text{pipe}} = \sum_{k \in \mathbf{SK}} \sum_{s \in \mathbf{SC}} C_{s,k} + \sum_{m \in \mathbf{PU}} \sum_{s \in \mathbf{SC}} C_{s,m} + \sum_{k \in \mathbf{SK}} \sum_{m \in \mathbf{PU}} C_{m,k} + \sum_{i \in \mathbf{HS}} C_{i,\text{fuel}} \tag{7}$$

The capital cost of each pipeline meets the following inequality.

$$\begin{gathered} C_{a,b} \geq \left\{ z^p_{a,b} \cdot a_{\text{pipe}} + b_{\text{pipe}} \left[\frac{F^p_{a,b}}{\max(P^p_a, P^p_b)} \right] \right\} \cdot d_{a,b} \\ C_{a,b} \in \left\{ C_{s,k}, C_{s,m}, C_{m,k} \right\} \forall s \in \mathbf{SC}, m \in \mathbf{PU}, k \in \mathbf{SK} \\ z^p_{a,b} \in \left\{ z^p_{s,k}, z^p_{s,m}, z^p_{m,k} \right\} \forall s \in \mathbf{SC}, m \in \mathbf{PU}, k \in \mathbf{SK}, p \in \mathbf{P} \end{gathered} \tag{8}$$

where the binary variable $z^p_{a,b}$ is used to indicate whether the connection between supplier a and receiver b exists in subperiod p. If the flowrate is non-zero, $z^p_{a,b}$ is equal to one, otherwise $z^p_{a,b}$ is zero which means that the connection does not exist. $C_{a,b}$ is the capital cost of the pipeline between supplier a and receiver b; a_{pipe} and b_{pipe} are the capital cost coefficients of the pipelines [12]; $d_{a,b}$ represents the distance between supplier a and receiver b.

The capital cost of the pipeline from hydrogen utility i to the fuel gas system is denoted as $C_{i,\text{fuel}}$, which meets the following inequality.

$$C_{i,\text{fuel}} \geq \left[a_{\text{pipe}} \cdot z_{i,\text{fuel}} + b_{\text{pipe}} \cdot \left(\frac{F^{p}_{i,SUR}}{P^{p}_{i}} \right) \right] \cdot d_{i,\text{fuel}} \qquad \forall i \in \mathbf{HS}, p \in \mathbf{P} \tag{9}$$

where the binary variable $z_{i,\text{fuel}}$ is used to indicate whether the connection from the internal hydrogen source i to the fuel gas system exists or not; $d_{i,\text{fuel}}$ denotes the distance between internal hydrogen source i and the fuel gas system.

The capital cost of the compressors can be expressed as

$$C_{\text{com}} = \sum_{k \in \mathbf{SK}} \sum_{s \in \mathbf{SC}} C^{\text{P}}_{s,k} + \sum_{m \in \mathbf{PU}} \sum_{s \in \mathbf{SC}} C^{\text{P}}_{s,m} + \sum_{k \in \mathbf{SK}} \sum_{m \in \mathbf{PU}} C^{\text{P}}_{m,k} \tag{10}$$

The capital cost of each compressor is denoted as $C^{\text{P}}_{a,b}$, which meets the following inequality.

$$\begin{gathered} C^{\text{P}}_{a,b} \geq C^{p,\text{P}}_{a,b} \\ C^{\text{P}}_{a,b} \in \left\{ C^{\text{P}}_{s,k}, C^{\text{P}}_{s,m}, C^{\text{P}}_{m,k} \right\} \quad \forall s \in \mathbf{SC}, m \in \mathbf{PU}, k \in \mathbf{SK} \\ C^{p,\text{P}}_{a,b} = \left(a_{\text{com}} \cdot z^{p}_{a,b} \cdot z^{p,\text{P}}_{a,b} + b_{\text{com}} \cdot W^{p}_{a,b} \right) \times 10^4 \\ z^{p}_{a,b} \in \left\{ z^{p}_{s,k}, z^{p}_{s,m}, z^{p}_{m,k} \right\} \quad \forall s \in \mathbf{SC}, m \in \mathbf{PU}, k \in \mathbf{SK}, p \in \mathbf{P} \end{gathered} \tag{11}$$

where $C^{p,\text{P}}_{a,b}$ denotes the capital cost of the compressor, which is calculated by its power in subperiod p; a_{com} and b_{com} are the capital cost coefficients of the compressors [16].

The required power of the compressor in each subperiod is less than the rated power, given by

$$\begin{gathered} W_{a,b} \geq W^{p}_{a,b} \quad \forall p \in \mathbf{P} \\ W_{a,b} \in \left\{ W_{s,k}, W_{s,m}, W_{m,k} \right\} \quad \forall s \in \mathbf{SC}, m \in \mathbf{PU}, k \in \mathbf{SK} \end{gathered} \tag{12}$$

where $W_{a,b}$ denotes the rated power of the compressor.

The capital cost of the purifiers can be expressed as

$$C_{\text{pur}} = \sum_{m \in \mathbf{PU}} C_m \tag{13}$$

C_m denotes the capital cost of centralized purifier m, which meets the following inequality.

$$\begin{gathered} C_m \geq C^{p}_{m} \ \forall m \in \mathbf{PU}, p \in \mathbf{P} \\ C^{p}_{m} = \left(a_{\text{pur}} \cdot z_m + b_{\text{pur}} \cdot F^{p}_{m} \right) \times 10^4 \quad \forall m \in \mathbf{PU}, p \in \mathbf{P} \end{gathered} \tag{14}$$

where C^{p}_{m} stands for the capital cost of centralized purifier m under subperiod p; a_{pur} and b_{pur} are the capital cost coefficients of the purifiers [16]; the binary variable z_m is used to indicate whether the purifier m exists or not; F^{p}_{m} is the feed flowrate for purifier m in subperiod p.

3.2. Constraints

3.2.1. Constraints of Hydrogen Sources and Hydrogen Sinks

In a certain subperiod, each hydrogen source can be allocated to hydrogen sinks, purifiers, and the fuel gas system. The flowrate balance of the hydrogen source can be formulated as

$$F^{p}_{s} = \sum_{k \in \mathbf{SK}} F^{p}_{s,k} + \sum_{m \in \mathbf{PU}} F^{p}_{s,m} + F^{p}_{s,SUR} \quad \forall s \in \mathbf{SC}, p \in \mathbf{P} \tag{15}$$

where F_s^p is the flowrate of hydrogen source s in subperiod p; $F_{s,k}^p$ stands for the flowrate supplied from hydrogen source s to hydrogen sink k. For the hydrogen utility j ($j \in \mathbf{HU}$), $F_{j,SUR}^p$ represents the excessive production capacity.

In subperiod p, hydrogen sources and purifiers should meet the requirement of a hydrogen sink. The flowrate balance of the hydrogen sink can, thus, be expressed as

$$F_k^p = \sum_{s \in \mathbf{SC}} F_{s,k}^p + \sum_{m \in \mathbf{PU}} F_{m,k}^p \quad \forall k \in \mathbf{SK}, p \in \mathbf{P} \tag{16}$$

where F_k^p represents the flowrate supplied to hydrogen sink k in subperiod p; $F_{m,k}^p$ denotes the flowrate supplied from purifier m to hydrogen sink k.

The flowrate supplied to hydrogen sink k should not be less than the required flowrate, $F_{k,req}^p$, given by

$$F_k^p \geq F_{k,req}^p \quad \forall k \in \mathbf{SK}, p \in \mathbf{P} \tag{17}$$

The hydrogen purity limit of each hydrogen sink in subperiod p should satisfy

$$\sum_{s \in \mathbf{SC}} F_{s,k}^p y_s^p + \sum_{m \in \mathbf{PU}} F_{m,k}^p y_{m,prod}^p \geq F_k^p y_{k,req}^p \quad \forall k \in \mathbf{SK}, p \in \mathbf{P} \tag{18}$$

where $y_{m,prod}^p$ represents the hydrogen purity of the product stream of purifier m; $y_{k,req}^p$ is the minimum purity required by hydrogen sink k. This inequality constraint is linear because these are three parameters.

3.2.2. Constraints of Purifiers

In any subperiod p, the flowrate balance of purifier m can be expressed as

$$F_m^p = \sum_{s \in \mathbf{SC}} F_{s,m}^p \quad \forall m \in \mathbf{PU}, p \in \mathbf{P} \tag{19}$$

$$F_m^p = F_{m,prod}^p + F_{m,resd}^p \quad \forall m \in \mathbf{PU}, p \in \mathbf{P} \tag{20}$$

where $F_{m,prod}^p$ denotes the flowrates of product stream for purifier m.

The product stream of the purifier is all allocated to hydrogen sinks, i.e.,

$$F_{m,prod}^p = \sum_{k \in \mathbf{SK}} F_{m,k}^p \quad \forall m \in \mathbf{PU}, p \in \mathbf{P} \tag{21}$$

For the purifier m, the hydrogen recovery ratio, r_m, is defined as

$$r_m \cdot \sum_{s \in \mathbf{SC}} \left(F_{s,m}^p \cdot y_s^p\right) = F_{m,prod}^p \cdot y_{m,prod}^p \quad \forall m \in \mathbf{PU}, p \in \mathbf{P} \tag{22}$$

3.2.3. Constraints of Connections

If a connection between supplier a (e.g., hydrogen utility, internal hydrogen source, and product stream of purifier) and receiver b (e.g., hydrogen sink and inlet stream of purifier) exists, its flowrate should be within the lower bound $F_{a,b}^{p,\mathrm{L}}$ and the upper bound $F_{a,b}^{p,\mathrm{U}}$, i.e.,

$$\begin{gathered} z_{a,b}^p F_{a,b}^{p,\mathrm{L}} \leq F_{a,b}^p \leq z_{a,b}^p F_{a,b}^{p,\mathrm{U}} \\ F_{a,b}^p \in \left\{F_{s,k}^p, F_{s,m}^p, F_{m,k}^p\right\} \forall s \in \mathbf{SC}, m \in \mathbf{PU}, k \in \mathbf{SK}, p \in \mathbf{P} \end{gathered} \tag{23}$$

The maximum flowrate of the pipeline connected to the fuel gas system, F^{U}_{fuel}, should satisfy

$$F^{p}_{i,SUR} \leq z_{i,fuel} \cdot F^{U}_{fuel} \quad \forall i \in \mathbf{HS}, p \in \mathbf{P} \tag{24}$$

The maximum feed flowrate for purifier m is F^{U}_{m}, and it should satisfy

$$F^{p}_{m} \leq F^{U}_{m} \cdot z_{m} \quad \forall m \in \mathbf{PU}, p \in \mathbf{P} \tag{25}$$

3.2.4. Constraints of Inter-Plant Matches

The hydrogen utility cannot be sent to the purifier, i.e.,

$$z^{p}_{j,m} = 0 \qquad \forall j \in \mathbf{HU}, m \in \mathbf{PU}, p \in \mathbf{P} \tag{26}$$

In this work, the internal hydrogen source is solely allocated to the purifier or hydrogen sinks in the same plant, i.e.,

$$z^{p}_{i,m} = \delta z'^{p}_{i,m} \; \forall i \in \mathbf{HS}, m \in \mathbf{PU}, p \in \mathbf{P} \tag{27}$$

where

$$z'^{p}_{i,m} = \begin{cases} 0 \\ 1 \end{cases} \forall i \in \mathbf{HS}_{n}, m \in \mathbf{PU}_{n'}, p \in \mathbf{P}$$

$$\delta = \begin{cases} 0 & n \neq n' \\ 1 & n = n' \end{cases}$$

$$z^{p}_{i,k} = \delta z'^{p}_{i,k} \quad \forall i \in \mathbf{HS}, k \in \mathbf{K}, p \in \mathbf{P} \tag{28}$$

where

$$z'^{p}_{i,k} = \begin{cases} 0 \\ 1 \end{cases} \forall i \in \mathbf{HS}_{n}, k \in \mathbf{K}_{n'}, p \in \mathbf{P}$$

where n denotes the serial number of the plant ($n \in \mathbf{N}$); $z'^{p}_{i,m}$ and $z'^{p}_{i,k}$ are both binary variables.

3.2.5. The Necessity of Setting Up a Compressor

If the pressure of a hydrogen source is lower than that of a hydrogen sink or purifier to be sent, its pressure should be raised by a compressor. In this work, $z^{p,P}_{a,b}$ is used to judge whether a compressor is necessary to be set or not, i.e.,

$$\begin{gathered} z^{p,P}_{a,b} = \begin{cases} 0 \; P^{p}_{b} - P^{p}_{a} \leq 0 \\ 1 \; P^{p}_{b} - P^{p}_{a} > 0 \end{cases} \\ z^{p,P}_{a,b} \in \left\{ z^{p,P}_{s,k}, z^{p,P}_{s,m}, z^{p,P}_{m,k} \right\} \forall s \in \mathbf{SC}, m \in \mathbf{PU}, k \in \mathbf{SK}, p \in \mathbf{P} \end{gathered} \tag{29}$$

Equations (1)–(29) constitute the mathematical programming model of the multi-period hydrogen network for integration of inter-plant hydrogen network, which features a MILP problem. When only one subperiod is considered, this model is reduced to a single period model, which is presented in Appendix A.

4. Case Study

4.1. Fundamental Data and Subperiod Partitioning

In this case study, there are three plants, denoted by Plant A [17], Plant B [18], and Plant C [5], in a chemical industry park. The distance between Plant A and Plant B is 10 km, and the distance between Plant B and Plant C is also 10 km. Plant A is 20 km away from Plant C.

Hydrogen streams can be delivered among the three plants through cross-plant hydrogen pipelines. The annual operating time of the plants is assumed to be 8000 h. According to Figure 2 and the fluctuating characteristics of hydrogen sources and sinks, the operating time of the inter-plant hydrogen network is divided into seven subperiods, as listed in Table 1, in which the division of the subperiod and the flowrates of each subperiod are also presented. In this case study, the hydrogen utilities include S1, S8, S9, S10, and S19, and their prices are listed in Table 2.

Table 1. Subperiod partitioning of the inter-plant hydrogen network and parameters of hydrogen sources and hydrogen sinks.

Subperiod		**1**	**2**	**3**	**4**	**5**	**6**	**7**		
	Time [h]	**0**	**1456**	**2388**	**3046**	**4177**	**5496**	**7022**	**8000**	
	$\Delta t_{subperiod}$ [h]	**1456**	**932**	**658**	**1131**	**1319**	**1526**	**978**		
Plant	**S/K**	**Flowrate [mol·s^{-1}]**							**H_2 Purity [mol%]**	**Pressure [MPa]**
A	S1	1115.0	1115.0	1115.0	1115.0	1115.0	1115.0	1115.0	0.95	2.1
	S2	216.0	216.0	216.0	237.6	237.6	224.7	224.7	0.80	2.1
	S3	113.6	113.6	113.6	126.6	126.6	124.4	124.4	0.80	8.3
	S4	144.8	144.8	144.8	109.5	109.5	180.3	180.3	0.75	2.4
	S5	80.8	80.8	80.8	100.1	100.1	67.9	67.9	0.75	2.8
	S6	20.0	20.0	20.0	14.4	14.4	18.3	18.3	0.70	2.4
	S7	31.2	31.2	31.2	36.5	36.5	25.9	25.9	0.65	1.4
	K1	521.0	521.0	521.0	580.4	580.4	570.3	570.3	0.95	13.8
	K2	486.7	486.7	486.7	367.9	367.9	605.6	605.6	0.93	3.4
	K3	246.7	246.7	246.7	305.8	305.8	207.3	207.3	0.90	4.1
	K4	75.8	75.8	75.8	54.4	54.4	69.5	69.5	0.80	3.4
	K5	54.7	54.7	54.7	64.2	64.2	45.4	45.4	0.75	2.1
B	S8	881.5	881.5	881.5	881.5	881.5	881.5	881.5	0.97	7
	S9	312.1	312.1	312.1	312.1	312.1	312.1	312.1	0.95	1.2
	S10	312.1	312.1	312.1	312.1	312.1	312.1	312.1	0.95	1.2
	S11	124.9	124.9	133.6	133.6	133.6	133.6	131.3	0.90	1.4
	S12	536.9	536.9	509.2	509.2	509.2	509.2	589.3	0.92	1.1
	S13	5.4	5.4	7.1	7.1	7.1	7.1	6.1	0.60	1.4
	S14	20.9	20.9	24.0	24.0	24.0	24.0	16.5	0.46	1.6
	S15	13.2	13.2	10.9	10.9	10.9	10.9	17.3	0.71	1.5
	S16	24.8	24.8	30.9	30.9	30.9	30.9	19.2	0.83	1.5
	S17	90.5	90.5	55.4	55.4	55.4	55.4	81.6	0.76	1.2
	S18	61.2	61.2	52.8	52.8	52.8	52.8	69.3	0.65	1.2
	K6	74.4	74.4	98.6	98.6	98.6	98.6	85.0	0.90	7
	K7	47.4	47.4	54.7	54.7	54.7	54.7	37.5	0.87	5
	K8	156.1	156.1	129.0	129.0	129.0	129.0	203.6	0.92	7
	K9	299.7	299.7	372.8	372.8	372.8	372.8	232.1	0.96	7
	K10	287.2	287.2	175.7	175.7	175.7	175.7	259.0	0.98	10
	K11	549.4	549.4	474.5	474.5	474.5	474.5	622.1	0.94	20
C	S19	496.0	496.0	496.0	496.0	496.0	496.0	496.0	0.999	3.5
	S20	437.1	480.8	480.8	480.8	363.0	363.0	363.0	0.92	3.5
	S21	142.3	103.0	103.0	103.0	137.9	137.9	137.9	0.71	2.5
	S22	21.4	27.0	27.0	27.0	31.4	31.4	31.4	0.70	2.5
	K12	292.6	236.0	236.0	236.0	286.3	286.3	286.3	0.93	13.0
	K13	281.3	332.5	332.5	332.5	373.2	373.2	373.2	0.87	8.7
	K14	28.7	34.8	34.8	34.8	29.6	29.6	29.6	0.87	5.0
	K15	34.1	40.6	40.6	40.6	33.8	33.8	33.8	0.85	3.5
	K16	34.3	29.5	29.5	29.5	26.8	26.8	26.8	0.85	3.0

Table 2. Prices of hydrogen utilities [5,17,18].

Hydrogen Source	S1 [17]	S8 [18]	S9 [18]	S10 [18]	S19 [5]
e_{hu} [CNY·mol^{-1}]	0.01	0.009	0.011	0.011	0.015

In this case, the pressure swing adsorption (PSA) is selected as the centralized purifier. The pressures of PSA for feed stream and product stream are both 1.2 MPa, the hydrogen recovery ratio of PSA is assumed to be 0.9, and the hydrogen purity of the product stream is 99%. The residual stream of PSA is discharged to the fuel gas system, of which the pressure is 0.06 MPa. In the calculation, the fixed cost coefficient a and the variable cost coefficient b for the compressor are 69 and 1.164 [16], and those for the pipeline are 32 and 28.12 [12], and those for the purifier are 302.3 and 14.25 [16]. The electricity price is 0.8 CNY/kW·h, and the price of heat energy is 0.025 CNY/MJ. In the calculations, the models were coded on GAMS 24.1 platform running on a PC with 2.93 GHz CPU and 6 GB RAM, and CPLEX was used as the solver.

4.2. Optimal Design of Multi-Period Inter-Plant Hydrogen Network

Figure 3 shows the topological structure of the optimal design of the multi-period inter-plant hydrogen network, and the red lines represent the cross-plant pipelines. Results indicate that the centralized purifier is set up for each plant, and the product stream of the purifier only supplies the hydrogen sinks within the same plant. The total number of connections in the hydrogen network is 50, and only three matches across the three plants. Furthermore, it can be seen from Figure 3 and Table 1 that the hydrogen in plant B is relatively surplus in the inter-plant hydrogen network. Since the hydrogen utility S8 in plant B has higher pressure and purity, it is delivered to the hydrogen sinks with high-pressure requirements in Plant A and Plant C through the cross-plant pipelines. In addition, Plant A also supplies hydrogen utility to Plant C.

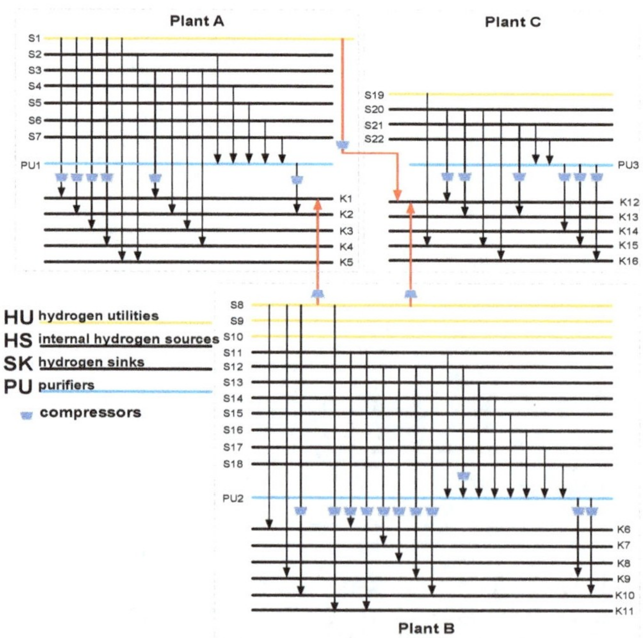

Figure 3. The structure of the inter-plant hydrogen network, obtained by the proposed method.

Table 3 presents the comparison of the effects of integrating the inter-plant hydrogen network with integrating the individual plant hydrogen network. The results show that the utilization of hydrogen is more efficient when the hydrogen streams can be transported among the plants via inter-plant pipelines and compressors, since the consumption of hydrogen utility is reduced by 14.3% compared with the scheme of individual plant integration. The total annual cost decreases by 4.22%. Therefore, the inter-plant integration can effectively achieve the reduction of the operating cost and TAC as well. However, the number of connections increases from 40 to 50, and the investment cost increases by 19.5%, indicating that a more complex network structure is reached.

Table 3. Comparison of the inter-plant integration and individual plant integration.

Item	Inter-Plant Network	Individual Hydrogen Network			
	A + B + C	A	B	C	Total
Number of matches	50	15	13	12	40
Consumption of hydrogen in utility [10^9 mol]	48.424	25.832	24.626	4.872	55.330
Hydrogen cost [10^8 CNY·y^{-1}]	4.940	2.937	2.234	0.731	5.902
Investment cost [10^8 CNY·y^{-1}]	0.974	0.377	0.292	0.146	0.815
Operating cost [10^8 CNY·y^{-1}]	4.429	2.579	1.482	0.756	4.817
TAC [10^8 CNY·y^{-1}]	5.403	2.956	1.774	0.901	5.631

It is also worth mentioning that different values of the annualized factor Af have a big impact on the optimal structure and the scheduling scheme, and according to Equation (2), Af is influenced by the annual interest rate (*fi*) and the number of years for depreciation (q). Therefore, we analyze the influence of *fi* and q on *TAC*, as is shown in Figure 4. *TAC* increases with the growth of the annual interest rate (*fi*) or the decrease of the years for depreciation (q), and meanwhile, the investment cost of the system is reduced, and the network structure is simplified, but this is at the expense of operation cost.

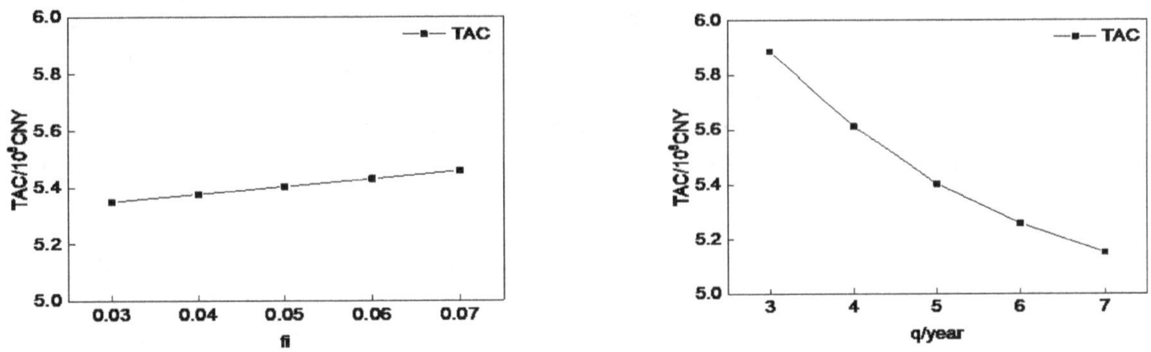

Figure 4. The influence of fi and q on the total annual cost (*TAC*).

4.3. Comparison of the Proposed Method and Other Design Methods

To further clarify the advantages and effectiveness of the proposed method, we compared the simultaneous method and stepwise methods, such as the structure-merged method and the structure-fixed method in this section.

4.3.1. The Results of Structure-Merged Method and Structure-Fixed Method

1. The structure-merged method

In this method, the multi-period hydrogen network is synthesized in two steps. In the first step, the single-period optimization model of the inter-plant hydrogen network is solved, and the optimal hydrogen network structure of each subperiod can be obtained. In the second step, the hydrogen network structures of the seven subperiods are merged. The final structure of the inter-plant hydrogen network accommodates all the matches that have ever occurred in all subperiods, and the capacity of pipeline and compressors are assigned to the maximum capacity in all subperiods.

Table 4 shows the number of matches, investment costs, and operation costs of the single period structure and the merged final structure. It can be seen that the investment cost does not increase significantly after the structure merged. The reason is that the structure obtained under each subperiod is similar, and only a few devices are newly constructed. The total matching number of hydrogen networks is 48, which is less than the one obtained by the simultaneous method. However, its investment cost is 16% higher than that of the simultaneous method.

Table 4. Characteristics of the multi-period hydrogen network, obtained by the structure-merged method.

Item	Position	Step1 The Sequence Number of the Optimized Subperiod							Step2 Multi-Period Hydrogen Network
		1	2	3	4	5	6	7	
Number of matches	intra-plant	35	35	39	38	36	37	38	45
	inter-plant	2	2	2	2	1	1	2	3
Investment cost [10^8 CNY·y^{-1}]	intra-plant	0.885	0.864	0.806	0.818	0.844	0.852	0.939	1.056
	inter-plant	0.029	0.029	0.039	0.039	0.035	0.035	0.042	0.071
	total	0.914	0.893	0.845	0.857	0.879	0.887	0.981	1.127
Operation cost [10^8 CNY·y^{-1}]	total	4.232	4.218	3.981	3.909	4.451	4.793	4.956	4.396

2. The structure-fixed method

In this method, the first step is to integrate the hydrogen network with the data under the pth subperiod. The obtained intra-plant structure of the hydrogen network is fixed as the initial structure, whereas the inter-plant structure of the hydrogen network is allowed to extend. Then the optimal design and operation scheme under the other six subperiods are optimized on the basis of the initial structure. It is noted that all of the intra-plant connections newly installed are forbidden. Finally, the obtained structures are merged into the final hydrogen network, and the total annual cost is denoted as TAC_p. Set p as 1~7 and perform the above steps, respectively. Seven design schemes can be obtained. When the number of subperiod is p, the model of the single period hydrogen network is solved p^2 times.

Table 5 shows the number of inter-plant and intra-plant matches, investment cost, operating cost, and the TAC of the seven design schemes. It can be seen that compared with the simultaneous method, the number of intra-plant connections decreases, and the number of inter-plant connections increases. The newly added inter-plant pipelines are used to adjust the hydrogen exchange among the plants to meet the hydrogen demand in other subperiods. The selection of the initial structure also affects the structure and cost of the final hydrogen network.

Table 5. Costs and the number of matches for the scheme obtained by the structure-fixed method.

Item	The Sequence Number of the Fixed Subperiod						
	1	2	3	4	5	6	7
Number of intra-plant matches	35	35	39	38	36	37	38
Number of inter-plant matches	9	8	5	8	7	4	5
Number of matches	44	43	44	46	43	41	43
Investment cost [10^8 CNY·y^{-1}]	1.13	1.11	1.10	1.17	1.14	1.06	1.08
Operation cost [10^8 CNY·y^{-1}]	4.39	4.40	4.39	4.38	4.47	4.47	4.39
Total annual cost [10^8 CNY·y^{-1}]	5.53	5.50	5.49	5.55	5.61	5.53	5.48

As shown in Table 5, when the structure of the single period hydrogen network scheme obtained under the subperiod 7 is fixed as the initial structure, the total annual cost of the inter-plant hydrogen network is 5.478×10^8 CNY·y^{-1}, which is lower than that of other schemes. This scheme is chosen as the result of the structure-fixed method in the following analysis.

It is worthy of noting that there are seven subperiods in this case. For the structure-merged method and structure-fixed method, we need to solve the model of the single period hydrogen network for 7 times and 49 times, respectively, whereas we simply solved the hydrogen integration model once by using the simultaneous method. Thus, the computation load is reduced.

4.3.2. The Hydrogen Exchange among the Plants

The configuration of cross-plant connections is a significant feature of the inter-plant hydrogen network. In this section, we will discuss the results related to the cross-plant hydrogen matches. According to the abovementioned results, the numbers of the cross-plant matches obtained by the proposed method, the structure-merged method, and the structure-fixed method are 3, 3, and 5, respectively, and the average flowrates of hydrogen exchange in the whole year are 363 mol·s^{-1}, 257 mol·s^{-1}, and 283 mol·s^{-1}, respectively. It appears that, in the proposed method, more hydrogen transportation across the plants is realized by fewer inter-plant connections.

In specific, to reflect the connections among the plants, the amounts of hydrogen exchange via the inter-plant pipelines are listed in Tables 6–8. It can be seen from these tables that Plant B always outputs hydrogen to plant A and plant C, and Plant A always exports hydrogen to Plant C. In this case, Plant B supplies surplus hydrogen to the other two plants owing to its utility S8 with higher pressure and lower price, whereas Plant C usually imports hydrogen from other plants because of its own relatively expensive utility.

Table 6. The amount of hydrogen exchange among the plants in the proposed method.

Subperiod		Amount of Pure Hydrogen [mol·s^{-1}]		
		A	B	C
1	A			131
	B	189		13
	C			
2	A			91
	B	189		13
	C			
3	A			
	B	189		105
	C			
4	A			147
	B	189		105
	C			
5	A			147
	B	189		105
	C			
6	A			147
	B	126		99
	C			
7	A			131
	B	189		13
	C			

Table 7. The amount of hydrogen exchange among the plants in the structure-merged method.

Subperiod		Amount of Pure Hydrogen [mol·s^{-1}]		
		A	B	C
	A			
1	B	80		124
	C			
	A			
2	B	99		105
	C			
	A			
3	B	184		105
	C			
	A			
4	B	184		105
	C			
	A			
5	B			278
	C			
	A			
6	B			278
	C			
	A			26
7	B			242
	C			

Table 8. The amount of hydrogen exchange among the plants in the structure-fixed method.

Subperiod		Amount of Pure Hydrogen [mol·s^{-1}]		
		A	B	C
	A			
1	B	90		124
	C			
	A			
2	B	108		106
	C			
	A		19	
3	B	196		106
	C			
	A		19	
4	B	196		106
	C			
	A		19	
5	B	34		269
	C			
	A		19	
6	B	34		269
	C			
	A			26
7	B			242
	C			

It is worthy of noting that the cross-plant pipelines are fully used to transport hydrogen among the plants in each subperiod in the scheme obtained by the proposed method, whereas the inter-plant pipelines are idle sometimes in the scheme obtained by the stepwise method, from Plant B to Plant A during the subperiod 5 to the subperiod 7, for example. It implies that in the simultaneous method proposed in this work, the interactions among the subperiods are considered to coordinate the hydrogen demands of each plant. Nevertheless, the interaction among the subperiods is somehow neglected in the stepwise methods.

4.3.3. Economic Analysis of the Three Methods

In this section, we will further discuss the economic performances of the design schemes obtained by the three methods. Table 9 shows the comparison of the costs of design schemes obtained by the stepwise methods and the proposed method. As shown in the table, a more economical design scheme can be obtained by the proposed method. Although the operating cost of the proposed method is slightly higher, the reduction of the investment costs is remarkable.

Table 9. Comparison of costs of design schemes obtained by the three methods.

Method	Cost [10^8 CNY·y^{-1}]								
	Investment				Operation				Total
	Compressor	PSA	Pipeline	Sum	Utility	Electricity	Fuel Profit	Sum	
This work	0.581	0.345	0.048	0.974	4.940	1.112	−1.623	4.429	5.403
The structure-merged method	0.688	0.393	0.046	1.127	4.968	1.113	−1.685	4.396	5.523
The structure-fixed method	0.641	0.393	0.050	1.084	4.948	1.120	−1.673	4.395	5.478

It can also be seen from Table 9 that the operating cost accounts for about 80% of the total annual cost, which is composed mostly of the hydrogen utility cost. In the proposed method, the relation among each subperiod is taken into account, leading to a reasonable allocation of hydrogen sources with different pressures and purities. Thus, the utility cost and the electricity expense are both the lowest. Meanwhile, a lower fuel profit is reached by the reduction of fuel gas streams owing to the internal hydrogen sources being directly matched with the sinks.

As shown in Table 9, the investment cost obtained by the proposed method is 0.973×10^8 CNY·y^{-1}, which is 10–14% lower than those obtained by the two stepwise methods. The cost of the compressors and the purifiers are both lower too. The capacities of compressors and purifiers show the same trend, as shown in Table 10.

Table 10. Comparison of the device capacity of the three methods.

Method	Sum of the Compressor Power [kW]	Capacities of PSA Units [mol·s^{-1}]			
		A	B	C	Sum
This work	20,051	493	361	130	984
The structure-merged method	23,928	496	465	169	1130
The structure-fixed method	22,372	496	465	169	1130

To further analyze the reduction of investment costs, we will revisit the performances of single period networks listed in Table 4. It indicates that the cost and the equipment capacities of subperiod 7 are the largest among all the subperiods. Since the structure obtained by the stepwise method is based on the structure of the subperiod 7, its investment cost is also larger. In contrast, the design scheme obtained by the simultaneous method is a trade-off among all of the subperiods, and its investment cost is less than the costs for subperiod 7. It can be inferred that the design scheme by using the stepwise methods are a little sensitive to the situation in subperiods. The worst case may be avoided by the simultaneous method, due to considering the interactions among the subperiods.

5. Conclusions

When the hydrogen networks in chemical industry parks are integrated, it is necessary to coordinate the temporal and spatial demand and supply of hydrogen in both the intra-plant and inter-plant. To address the optimization problem of integrating multi-period hydrogen network in multiple plants, a multi-period simultaneous optimization method for the inter-plant hydrogen network integration was proposed, in which the hydrogen supply across the plants is implemented through cross-plant pipelines, and a MILP model is developed for the optimization of inter-plant hydrogen network. A case study of a three-plant hydrogen network with seven subperiods is utilized to demonstrate the effectiveness and advantages of the proposed method. Moreover, the optimal design scheme was compared with those of the stepwise methods, including the structure-merged method and the structure-fixed method.

The results show that the proposed method can give a design and scheduling scheme for the inter-plant hydrogen system with lower TAC. Compared with the stepwise methods, the proposed method simplifies the solving steps and is able to provide a better network structure and scheduling scheme. The investment cost obtained by the proposed method is lower than those obtained by the two stepwise methods. In the proposed method, the interactions among the subperiods are considered to coordinate the fluctuating demand and supply of hydrogen in each plant. The cross-plant pipelines are fully used to transport hydrogen among the plants in each subperiod in the scheme obtained by the proposed method, whereas the inter-plant pipelines are idle sometimes in the scheme obtained by the stepwise methods. The proposed method is suitable for integrating the inter-plant hydrogen system with predictable operational parameters and can also be used to study the influence of inter-plant integration and multi-period operation on the design of the hydrogen system. However, the fluctuations in operating conditions may change randomly in practice. Therefore, the uncertainties should be considered during integration. Of course, it deserves further work in the future.

Author Contributions: Conceptualization, Y.L. and L.K.; methodology, L.K. and Y.L.; software, R.H., Y.J., and L.K.; validation, R.H., Y.J., and J.W.; formal analysis, R.H., Y.J., and J.W.; investigation, Y.L. and L.K.; data curation, R.H., J.W. and Y.J.; writing—original draft preparation, R.H.; writing—review and editing, R.H., Y.J., Y.L. and L.K.; visualization, R.H. and Y.L.; supervision, Y.L. and L.K.; project administration, Y.L.; funding acquisition, Y.L. and L.K. All authors have read and agreed to the published version of the manuscript.

Acknowledgments: Our deepest gratitude goes to the anonymous reviewers for their careful work and thoughtful suggestions that have helped improve this paper substantially.

Nomenclature

Parameters

a	fixed cost coefficient of devices or pipelines
b	variable cost coefficient of devices or pipelines
c_p	specific heat at constant pressure
$d_{a,b}$	distance between supplier a and receiver b
$d_{i,\text{fuel}}$	distance of the pipeline between internal hydrogen source i and fuel gas system
e	price of energy or hydrogen utility
F_s^p	flowrate of source s in subperiod p
$F_{k,req}^p$	required flowrate for hydrogen sink k in subperiod p
F_m^{U}	maximum flowrate for purifier m
$F_{\text{fuel}}^{\text{U}}$	maximum flowrate for pipeline to fuel gas system
P_a^p	pressure level of device a in subperiod p
r_m	hydrogen recovery ratio of purifier m
T	inlet temperature of compressor
t_p	operating time of subperiod p
y_s^p	hydrogen purity of hydrogen source s in subperiod p

$y^p_{m,prod}$	product hydrogen purity for purifier m in subperiod p
$y^p_{k,req}$	required hydrogen purity of hydrogen sink k in subperiod p
$\Delta H^\ominus_c$	standard combustion heat
δ	binary parameters to judge whether two devices exist in the same plant
γ	ratio of specific heat
η	efficiency of compressor
Continuous variables	
C_{H_2}	cost of hydrogen utility
C_{fuel}	revenue of fuel gas
C_{ele}	electricity purchasing cost
C_{pur}	investment cost for the purifiers
C_{pipe}	investment cost for the pipelines
C_{com}	investment cost for the compressors
$C_{a,b}$	cost of the pipeline between supplier a and receiver b
$C_{i,fuel}$	cost of the pipeline between internal hydrogen source i and fuel gas system
$C^P_{a,b}$	cost of the compressor between supplier a and receiver b
$C^{p,P}_{a,b}$	cost of the compressor between supplier a and receiver b in subperiod p
C_m	cost of purifier m
C^p_m	cost of purifier m calculated by its capacity in subperiod p
$F^p_{a,b}$	flowrate of the connection between supplier a and receiver b in subperiod p
$F^p_{i,SUR}$	flowrate from internal hydrogen source i to fuel gas system in subperiod p
$F^p_{j,SUR}$	surplus flowrate of hydrogen utility j in subperiod p
F^p_k	flowrate of hydrogen sink k in subperiod p
F^p_m	inlet flowrate for purifier m in subperiod p
$F^p_{m,prod}$	product flowrate for purifier m in subperiod p
$F^p_{m,resd}$	residue flowrate for purifier m in subperiod p
TAC	total annual cost
$W^p_{a,b}$	power of the dedicated compressor in subperiod p
$W_{a,b}$	the rated power of the dedicated compressor
Binary variables	
$z^p_{a,b}$	the existence of the connection between supplier a and receiver b in subperiod p
$z^{p,P}_{a,b}$	the existence of the dedicated compressor in subperiod p
z_m	the existence of purifier m
$z_{i,fuel}$	the existence of the connection between internal hydrogen source i to fuel gas system
$z'^p_{i,a}$	binary variable for internal hydrogen source i toward receiver a in subperiod p
Sets	
SC	hydrogen source
HS	internal hydrogen source
HU	hydrogen utility
SK	hydrogen sink
PU	purifier
FS	fuel gas system
N	plant
P	subperiod
Superscripts	
L	lower bound
U	upper bound
Subscripts	
pipe	pipeline
com	compressor
ele	electricity
pur	purifier

fuel	fuel gas
req	required
prod	product
resd	residue
SUR	surplus

Appendix A Single Period Optimization Model for the Inter-Plant Hydrogen Network

The objective TAC_p is given in Equation (A1), which can be calculated by Equations (2), (6), (29), and (A4)–(A12). And the constraints are Equations (15)–(23), (26)–(28), (A2) and (A3).

$$\min TAC_p = C_{\mathrm{H_2},p} - C_{\mathrm{fuel},p} + C_{\mathrm{power},p} + Af \cdot (C_{\mathrm{psa},p} + C_{\mathrm{pipe},p} + C_{\mathrm{com},p}) \tag{A1}$$

$$\text{St. } F^p_{i,SUR} \le z^p_{i,\mathrm{fuel}} \cdot F^{\mathrm{U}}_{\mathrm{fuel}} \quad \forall i \in \mathbf{HS}, p \in \mathbf{P} \tag{A2}$$

$$F^p_m \le F^U_m \cdot z^p_m \quad \forall m \in \mathbf{PU}, p \in \mathbf{P} \tag{A3}$$

where

$$C_{\mathrm{H_2},p} = 3600 \times 8000 \cdot \sum_{j \in \mathbf{HU}} \sum_{k \in \mathbf{SK}} \left(F^p_{j,k} \cdot e_j\right) \tag{A4}$$

where 8000 is the annual operating time.

$$C_{\mathrm{fuel},p} = e_{\mathrm{heat}} \times 3600 \times 8000 \left\{ \begin{array}{l} \sum\limits_{m \in \mathbf{PU}} \left[(1 - r^p_m) \cdot \sum\limits_{s \in \mathbf{SC}} (F^p_{s,m} \cdot y^p_s)\right] \cdot \Delta H^{\ominus}_{\mathrm{c,H_2}} + \\ \sum\limits_{i \in \mathbf{HS}} (F^p_{i,SUR} \cdot y^p_i) \cdot \Delta H^{\ominus}_{\mathrm{c,H_2}} + \\ \sum\limits_{i \in \mathbf{HS}} \left[F^p_{i,SUR} \cdot (1 - y^p_i)\right] \cdot \Delta H^{\ominus}_{\mathrm{c,CH_4}} + \\ \sum\limits_{m \in \mathbf{PU}} \left[F^p_{m,resd} - (1 - r^p_m) \cdot \sum\limits_{s \in \mathbf{SC}} (F^p_{s,m} \cdot y^p_s)\right] \cdot \Delta H^{\ominus}_{\mathrm{c,CH_4}} \end{array} \right\} \tag{A5}$$

$$C_{\mathrm{ele},p} = e_{\mathrm{ele}} \times 8000 \left(\sum_{k \in \mathbf{SK}} \sum_{s \in \mathbf{SC}} W^p_{s,k} + \sum_{m \in \mathbf{PU}} \sum_{s \in \mathbf{SC}} W^p_{s,m} + \sum_{k \in \mathbf{SK}} \sum_{m \in \mathbf{PU}} W^p_{m,k} \right) \tag{A6}$$

$$C_{\mathrm{pipe},p} = \sum_{k \in \mathbf{SK}} \sum_{s \in \mathbf{SC}} C^p_{s,k} + \sum_{m \in \mathbf{PU}} \sum_{s \in \mathbf{SC}} C^p_{s,m} + \sum_{k \in \mathbf{SK}} \sum_{m \in \mathbf{PU}} C^p_{m,k} + \sum_{i \in \mathbf{HS}} C^p_{i,\mathrm{fuel}} \tag{A7}$$

$$\begin{array}{l} C^p_{a,b} = \left\{ a_{\mathrm{pipe}} \cdot z^p_{a,b} + b_{\mathrm{pipe}} \cdot \left[\dfrac{F^p_{a,b}}{\max(P_a, P_b)} \right] \right\} \cdot d_{a,b} \\ z^p_{a,b} \in \left\{ z^p_{s,k}, z^p_{s,m}, z^p_{m,k} \right\} \forall s \in \mathbf{SC}, m \in \mathbf{PU}, k \in \mathbf{SK}, p \in \mathbf{P} \end{array} \tag{A8}$$

$$C^p_{i,\mathrm{fuel}} = \left[a_{\mathrm{pipe}} \cdot z^p_{i,\mathrm{fuel}} + b_{\mathrm{pipe}} \cdot \left(\frac{F^p_{i,SUR}}{P_i} \right) \right] \cdot d_{i,\mathrm{fuel}} \quad \forall i \in \mathbf{HS}, p \in \mathbf{P} \tag{A9}$$

$$\begin{array}{l} C_{\mathrm{com},p} = \sum\limits_{a,b} \left(a_{\mathrm{com}} \cdot z^p_{a,b} \cdot z^{p,\mathrm{P}}_{a,b} + b_{\mathrm{com}} \cdot W^p_{a,b} \right) \times 10^4 \\ z^p_{a,b} \in \left\{ z^p_{s,k}, z^p_{s,m}, z^p_{m,k} \right\} \forall s \in \mathbf{SC}, m \in \mathbf{PU}, k \in \mathbf{SK}, p \in \mathbf{P} \end{array} \tag{A10}$$

$$C_{\mathrm{psa},p} = \sum_{m \in \mathbf{PU}} C^p_m \tag{A11}$$

$$C^p_m = \left(a_{\mathrm{pur}} \cdot z^p_m + b_{\mathrm{pur}} \cdot F^p_m \right) \times 10^4 \quad \forall m \in \mathbf{PU}, p \in \mathbf{P} \tag{A12}$$

References

1. Hwangbo, S.; Lee, I.-B.; Han, J. Multi-period stochastic mathematical model for the optimal design of integrated utility and hydrogen supply network under uncertainty in raw material prices. *Energy* **2016**, *114*, 418–430. [CrossRef]
2. Jagannath, A.; Elkamel, A.; Karimi, I.A. Improved Synthesis of Hydrogen Networks for Refineries. *Ind. Eng. Chem. Res.* **2014**, *53*, 16948–16963. [CrossRef]

3. Deng, C.; Zhou, Y.; Chen, C.-L.; Feng, X. Systematic approach for targeting interplant hydrogen networks. *Energy* **2015**, *90*, 68–88. [CrossRef]
4. Lou, J.; Liao, Z.; Jiang, B.; Wang, J.; Yang, Y. Robust optimization of hydrogen network. *Int. J. Hydrog. Energy* **2014**, *39*, 1210–1219. [CrossRef]
5. Liang, X.; Kang, L.; Liu, Y. Impacts of Subperiod Partitioning on Optimization of Multiperiod Hydrogen Networks. *Ind. Eng. Chem. Res.* **2017**, *56*, 10733–10742. [CrossRef]
6. Liang, X.; Kang, L.; Liu, Y. The Flexible Design for Optimization and Debottlenecking of Multiperiod Hydrogen Networks. *Ind. Eng. Chem. Res.* **2016**, *55*, 2574–2583. [CrossRef]
7. Kang, L.; Liang, X.; Liu, Y. Design of multiperiod hydrogen network with flexibilities in subperiods and redundancy control. *Int. J. Hydrogen Energy* **2018**, *43*, 861–871. [CrossRef]
8. Kuo, C.-C.; Chang, C.-T. Improved Model Formulations for Multiperiod Hydrogen Network Designs. *Ind. Eng. Chem. Res.* **2014**, *53*, 20204–20222. [CrossRef]
9. Jiao, Y.; Su, H.; Hou, W.; Li, P. Design and Optimization of Flexible Hydrogen Systems in Refineries. *Ind. Eng. Chem. Res.* **2013**, *52*, 4113–4131. [CrossRef]
10. Liu, F.; Zhang, N. Strategy of Purifier Selection and Integration in Hydrogen Networks. *Chem. Eng. Res. Design* **2004**, *82*, 1315–1330. [CrossRef]
11. Deng, C.; Zhou, Y.Y.; Jiang, W.; Feng, X. Optimal design of inter-plant hydrogen network with purification reuse/recycle. *Int. J. Hydrogen Energy* **2017**, *42*, 19984–20002. [CrossRef]
12. Kang, L.; Liang, X.; Liu, Y. Optimal design of inter-plant hydrogen networks with intermediate headers of purity and pressure. *Int. J. Hydrogen Energy* **2018**, *43*, 16638–16651. [CrossRef]
13. Lou, Y.; Liao, Z.; Sun, J.; Jiang, B.; Wang, J.; Yang, Y. A novel two-step method to design inter-plant hydrogen network. *Int. J. Hydrogen Energy* **2019**, *44*, 5686–5695. [CrossRef]
14. Shehata, W.M.; Shoaib, A.M.; Gad, F.K. Inter-plant hydrogen integration with regeneration placement and multi-period consideration. *Egypt. J. Pet.* **2018**, *27*, 553–565. [CrossRef]
15. Wu, S.; Liu, G.; Yu, Z.; Feng, X.; Liu, Y.; Deng, C. Optimization of hydrogen networks with constraints on hydrogen concentration and pure hydrogen load considered. *Chem. Eng. Res. Des.* **2012**, *90*, 1208–1220. [CrossRef]
16. Hallale, N.; Liu, F. Refinery hydrogen management for clean fuels production. *Adv. Environ. Res.* **2001**, *6*, 81–98. [CrossRef]
17. Elkamel, A.; Alhajri, I.; Almansoori, A.; Saif, Y. Integration of hydrogen management in refinery planning with rigorous process models and product quality specifications. *Int. J. Process Syst. Eng.* **2011**, *1*, 302–330. [CrossRef]
18. Liao, Z.; Wang, J.; Yang, Y.; Rong, G. Integrating purifiers in refinery hydrogen networks: A retrofit case study. *J. Clean. Prod.* **2010**, *18*, 233–241. [CrossRef]

3

Numerical and Experimental Study of a Vortex Structure and Energy Loss in a Novel Self-Priming Pump

Hao Chang [1,2], Ramesh K. Agarwal [2,*], Wei Li [1,2], Ling Zhou [1,2] and Weidong Shi [3,*]

1 Research Center of Fluid Machinery Engineering and Technology, Jiangsu University, Zhenjiang 212013, China; haochang@wustl.edu (H.C.); 1000003263@ujs.edu.cn (W.L.); lingzhoo@hotmail.com (L.Z.)

2 Department of Mechanical Engineering and Materials Science, Washington University in St. Louis, St. Louis, MO 63130, USA

3 School of Mechanical Engineering, Nantong University, Nantong 226019, China

* Correspondence: rka@wustl.edu (R.K.A.); wdshi@ujs.edu.cn (W.S.)

Abstract: The self-priming pump as an essential energy conversion equipment is widely used in hydropower and thermal power plants. The energy losses in the internal flow passage of the pump directly affect its work efficiency. Therefore, it is important to improve the internal flow characteristic of the pump. In the present work, a novel self-priming pump which starts without water is proposed; this pump can reduce the energy consumption as well as the time needed to start its operation. The spatial structure of the vortices in the pump is investigated by employing the Q criterion with the numerical solution of the vorticity transport equation. Based on the morphology, the vortices can be separated into three categories: Trailing Edge Vortex (TEV), Leading Edge Vortex (LEV) and Gap Leakage Vortex (GLV). Generally, the morphology of the TEV is more disorderly than that of LEV and GLV, and the intensity of TEV is significantly higher than that of the other two vortices. To determine the magnitude and distribution of energy loss in the pump, entropy production analysis is employed to study the influence of blade thickness on energy characteristics of the pump. It is found that with an increase in the flow rate, the location of energy loss transfers from the trailing edge to the leading edge of the blade, and viscous entropy production (VEP) and turbulence entropy production (TEP) are the dominant factors which influence the energy conversion in the pump. More importantly, employing the blade with a thin leading edge and a thick trailing edge can not only significantly reduce the impact of incoming flow under over-load condition (flow rate higher than the design condition) but can also increase the efficiency of the pump. Thus, an increase in thickness of the blade from the leading edge to the trailing edge is beneficial for improving the pump performance. The results of this paper can be helpful in providing guidelines for reducing the energy loss and in improving the performance of a self-priming pump.

Keywords: vortex structure; energy loss; entropy production; self-priming pump

1. Introduction

The self-priming pump as an essential form of energy conversion equipment is widely used in hydropower and thermal power plants. The energy loss in the internal flow passage of the pump directly affects its work efficiency. Therefore, it is important to improve its internal flow characteristics by improving the design of its blades [1]. In recent years, considerable research has been conducted on improving the energy characteristics of the pump by modifying its blade profile. Wang et al. [2] proposed an innovative impeller with forward-curving blade by numerically and experimentally investigating the relationship between the energy loss in the impeller and the blade inlet angle, and obtained an optimal range for the inlet angles. Elyamin et al. [3] investigated the influence of the number of blades on the energy performance by using numerical simulations and concluded that the increase in number of blades can weaken the secondary flow, resulting in a reduction in the energy loss of the impeller. Shao et al. [4] used high-speed photography to study the internal characteristics of the pump with an impeller with different geometrical structures; they found that the complex vortex structures at low flow rates (partial-load conditions) can be eliminated by decreasing the outlet angle and width of the blade. Nejad and Riasi [5] conducted an experimental test to analyze the relationship between the blade profile and the energy performance; the blades were manufactured from straight to curved profiles with uniform inlet and outlet angles. They showed that reducing the blade angle is beneficial for improving the pump efficiency. Han et al. [6] selected the blade wrap angle and blade exit angle as an analysis variable in their numerical and experimental investigation to optimize the pump performance; they considered the effect of blade parameters on the distribution of turbulent kinetic energy, which affects the performance of the pump. Jeon et al. [7] designed a novel S-shape blade by employing the design of experiment (DOE) and the response surface method (RSM), and showed that the blade's efficiency was enhanced by 3% compared to under the original model. Nejad at el. [8] modified the blade profile into a bucket shape and studied the influence of blade angle, chord, and other parameters on the hydraulic performance; they concluded that the maximum efficiency of the pump can be improved by employing an optimal combination of these parameters. Wang et al. [9–11] pointed out the factors influencing the energy loss in a multistage centrifugal pump and analyzed the hydraulic loss caused by surface roughness of the impeller shroud experimentally.

However, while most of the investigations noted above studied the effect of varying the number of blades, inlet angle, outlet width, wrap angle, and riding position on the energy efficiency of a pump; on the effect of modification of blade thickness on pump performance has hardly been studied. Furthermore, the influence of unstable vortex structures on the internal flow passage of the pump and their contribution to energy loss has not been studied. In addition, the overall energy loss analysis has only been conducted using the conventional approach by determining the hydraulic efficiency or turbulent kinetic energy, which does not provide the main details of energy loss, namely its location, sources, and type. In the present work, entropy production analysis is introduced to identify the sources, locations, and reasons for energy loss/dissipation. Wang [12] investigated the cavitation performance of the LNG cryogenic pump; they showed that the evolution and degradation that occur due to cavitation can be determined by applying the entropy production model, and then verified the results of their analysis through an experiment. Hou et al. [13] combined the orthogonal design with entropy production analysis to optimize the geometric parameters of the pump; the optimal design was obtained which improved the energy performance of the pump. Gu et al. [14] analyzed the influence of the clocking positions on the hydraulic loss and obtained the energy loss distribution in the pump by employing entropy production analysis. Li et al. [15] introduced the concept of wall entropy production in their investigation; they pointed out that a large deviation in energy loss occurs with respect to experiments without considering the losses that occur due to the wall effect. Thus, entropy production analysis is used in this paper to accurately evaluate the energy loss that occurs in the pump.

In this paper, a novel self-priming pump start without water has been proposed and studied by using the Q criterion with vorticity transport equation and employing the entropy production analysis for quantification of energy losses. The vortex structures inside and energy characteristics of the pump with different blade thicknesses are systematically investigated by using numerical simulations. Numerical simulations are validated for one of the most promising blade thickness distribution in terms of pump head and efficiency by conducting an experiment. Finally, guidelines for obtaining a good thickness distribution for the blades are provided for improving the flow characteristics and performance of the pump.

2. Structure of Novel Self-Priming Pump

Traditional centrifugal pumps must fill the pump with water before starting, which takes lot of time and results in energy loss. Therefore, a novel self-priming pump is proposed in this paper which can start without water. Figure 1 shows the structure of the proposed novel self-priming pump; it shows the inlet, self-priming system, compressor, pump, outlet and outlet valve of the novel self-priming pump. The specifications of the pump are listed in Table 1.

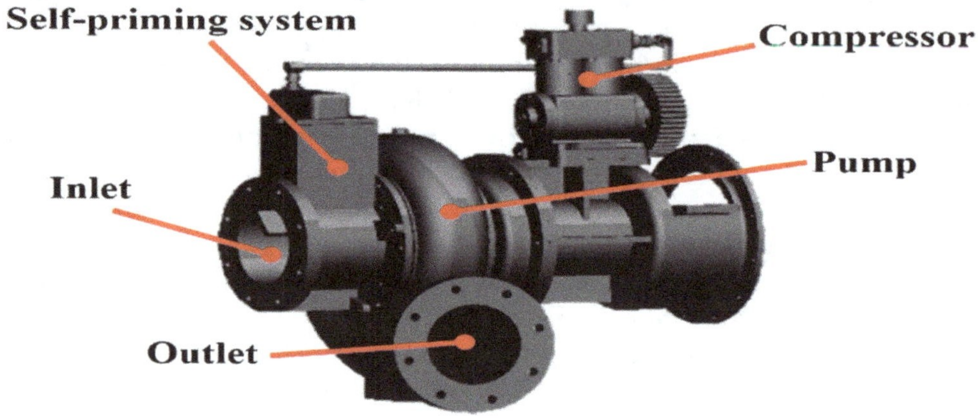

Figure 1. The structure and components of the pump.

Table 1. Specifications of the pump.

Parameter	Symbol	Value
Head	H	45 m
Design flow condition	Q_d	500 m^3/h
Rotation speed	n	2200 r/min
Specific speed	n_s	172
Impeller inlet diameter	D_1	200 mm
Impeller outlet diameter	D_2	280 mm
Blade numbers	Z	6
Blade wrap angles	ϕ	120°
Volute inlet width	b_1	80 mm
Volute outlet diameter	D_4	200 mm

As shown in Figure 1, the self-priming system is installed immediately after the inlet and consists of an induction chamber, filter screen, valve seat, ball valve, plate, valve chamber, intake tube, nozzle 1, joint, and nozzle 2, as shown in Figure 2. The inlet tube extends into the water before the pump starts, and then the compressor starts running with the operation of the pump. The jet-stream created in the compressor is discharged into nozzle 1 through the intake tube. The inner structure of nozzle 1 first shrinks and then smoothly expands, which results in a high-velocity jet-stream being created through nozzle 1.

A joint is installed between the nozzle 1 and nozzle 2 and is connected with the valve chamber. Hence, the high-velocity jet-stream passes through the joint and nozzle 2 exhausts directly into the atmosphere. At the same time, the internal pressure of the joint is continuously decreasing which leads to pushing the air in the valve box into the joint and eventually discharging it from nozzle 2. Accordingly, the pressure in the valve box gradually reduces. When the gravity of the ball valve is less than the pressure difference experienced by it, the ball valve is lifted from the valve seat and the air in the induction chamber flows into the valve box and eventually drains into the atmosphere. Furthermore, since the induction chamber is installed at the inlet, the air in the pump exhausts into the atmosphere. Ultimately, water flows into the pump along the inlet tube and the pump is filled with water, which achieves the pump start without water. When the process of self-priming is finished, the ball valve is merged in the water. Under the effect of gravity, it returns to the valve seat and closes the connection between the induction chamber and the valve box, avoiding the energy loss resulting from the water discharging into the atmosphere.

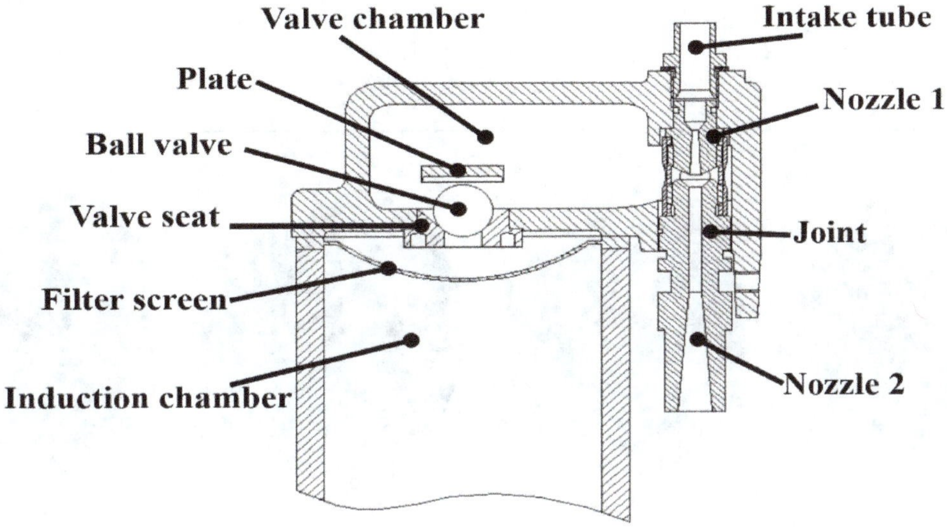

Figure 2. Structure of the self-priming system.

The outlet valve is installed on the outlet and is connected to the atmosphere in order to prevent the air flow into the pump. A rubber plate is pressed against the inclined exit of the outlet tube. When the air inside the pump is drained, the pump is filled with water and under the effect of pressure, the rubber plate separates from the exit of the outlet tube, and thus the self-priming of the pump is completed. The pump thus accomplishes the waterless start and can deliver water normally. The structure of the outlet valve as shown in the Figure 3.

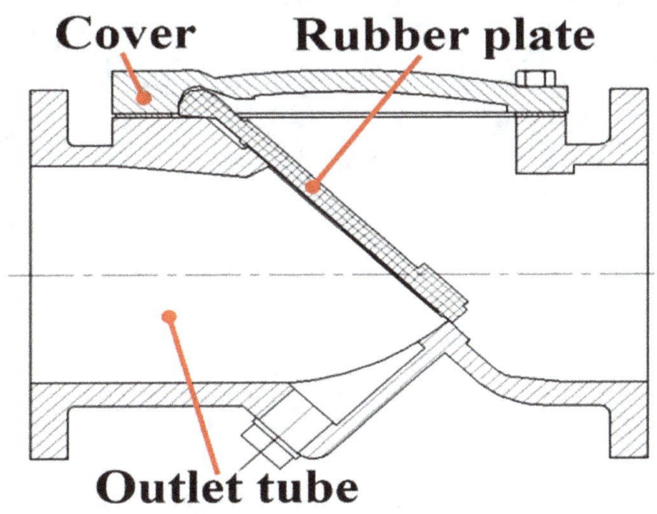

Figure 3. Structure of the outlet valve.

The geometric parameters of the blade cannot only decide the work capability of the pump, but can also affect the internal flow characteristic In line with our previous investigation, four different blade thickness distributions are proposed [16]; these are shown in Table 2. The 3D models of four blade thickness distribution schemes is shown in the Figure 4. Scheme 1 and scheme 2 employ the uniform thin and thick blade respectively, while the scheme 3 and scheme 4 employ the non-uniform increasing and decreasing blade thickness distribution respectively.

Table 2. Blades with different thickness distributions.

Scheme	Thickness of Leading Edge (mm)	Thickness of Middle Part (mm)	Thickness of Trailing Edge (mm)
1	3	3	3
2	6	6	6
3	3	6	6
4	6	6	3

Scheme 1 Scheme 2 Scheme 3 Scheme 4

Figure 4. The 3D models based on four blade thickness distribution schemes given in Table 2.

3. Computational Model and Numerical Method

3.1. Computational Domain

Figure 5 shows the computational domains of s novel self-priming pump including the outlet, volute, impeller, self-priming system, and inlet. In addition, the stable inflow and outflow can be obtained by extending the inlet and outlet, respectively.

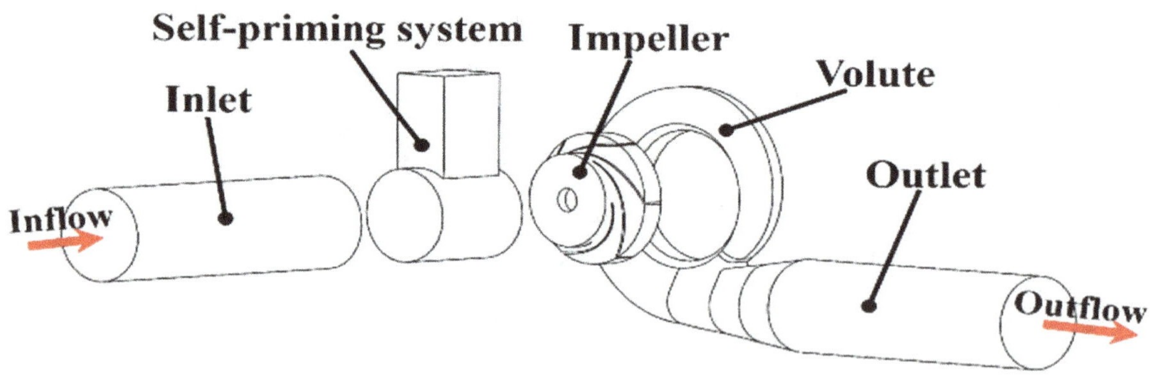

Figure 5. The computational domains.

3.2. Grid Generation

The computational domains of the self-priming pump are discretized into a hexahedral structure by employing the ICEM-CFD (The Integrated Computer Engineering and Manufacturing code for Computational Fluid Dynamics). The leading edge of the blade and tongue of the volute are refined to capture the details of the internal flow characters, as shown in Figures 6 and 7.

Since the grid resolution can significantly affect the accuracy of the simulation, grid sensitivity analysis was conducted. As shown in Figure 8, the number of grid elements was varied from 1.2×10^6 to 5.3×10^6 to evaluate the grid independence of the solution. When the number of grid elements exceeds 2.6×10^6, the deviation in pump efficiency and pump head are reduced to less than 1% by further increasing the number of grid elements. Therefore, considering the computational efficiency of the simulation, the grid elements are set as 2.6×10^6 in all simulations reported in this paper and y + is less than 60. The number of grid elements for the impeller, volute, outlet, inlet, and self-priming system are 478,998, 720,480 504,000, 696,000 and 272,232, respectively.

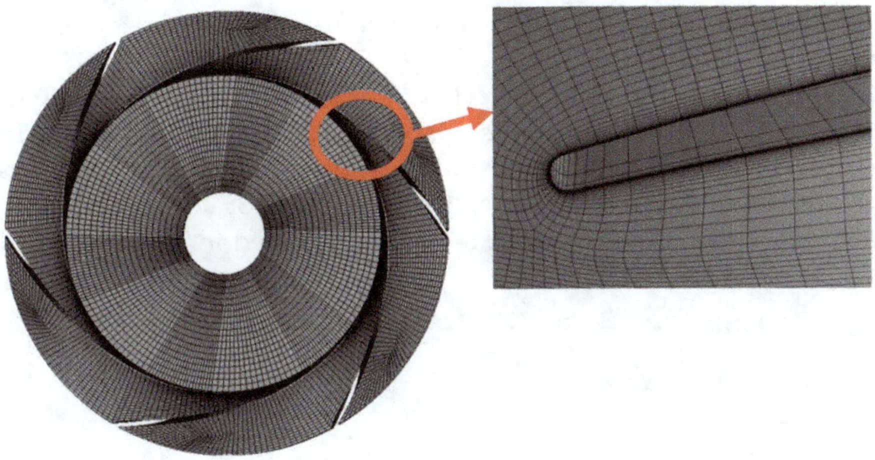

Figure 6. Grid distribution in an impeller.

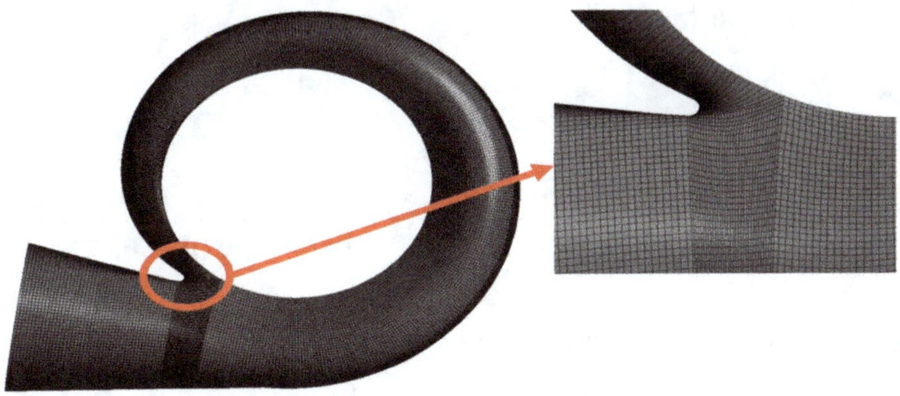

Figure 7. Grid distribution in the volute.

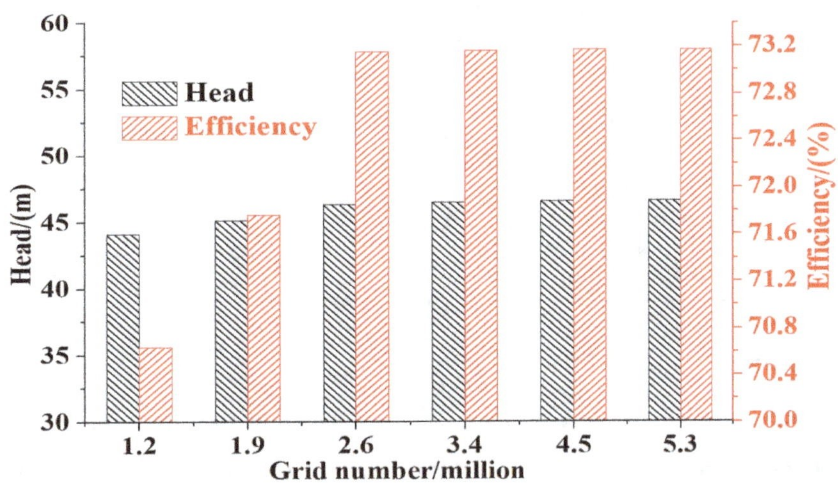

Figure 8. Sensitivity analysis of the number of grid elements used for the solution.

3.3. Calculation Method

In this paper, the SST k–ω turbulence model with Reynolds-Averaged Navier-Stokes (RANS) equations is employed for developing the numerical solution using the commercial CFD solver ANSYS CFX 17.1. The fluid medium is incompressible water at 25 °C. Considering the impeller as a rotating domain, the surfaces at the inlet and outlet of the impeller are set as the frozen rotor interfaces. Figure 9 shows the sensitivity of the calculations to time steps. The step sizes of 1/120 T (2.27×10^{-4} s), 1/60 T (4.54×10^{-4} s) and 1/30 T (9.08×10^{-4} s) are selected as time steps in the investigation. By comparing the pressure at the outlet with different time steps, it was found that small variations exist among the results when using different time steps. Based on this study on the influence of time steps on the solution, 1/60 T (4.54×10^{-4} s) is selected as the time step for computational efficiency without compromising accuracy [17,18]. The numerical setup is described in Table 3.

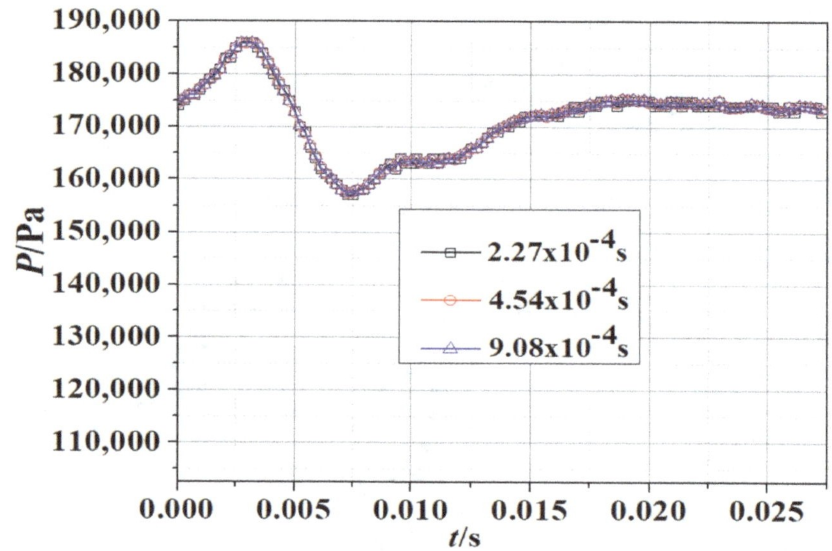

Figure 9. Analysis of the sensitivity of time steps in the numerical simulation.

Table 3. The numerical setup.

CFD Software	ANSYS CFX 17.1
Turbulence model	SST k–ω
Flow medium	Water at 25 °C
Inlet boundary condition	Total pressure (1 atm)
Outlet boundary condition	Mass flow
Wall roughness	50 μm
Steady state	Frozen rotor
Transient state	Transient rotor-stator
Advection scheme	High resolution
RMS residual	0.00001
Transient calculation time	0.2 s
Time step	4.54×10^{-4} s
Turbulence intensity	5%

4. Results and Discussion

4.1. Analysis of the Vortex Structure

Due to the complex structure of the novel self-priming pump, several types of vortices are generated in the internal flow field, namely the secondary flow vortex, reflux vortex, and flow separation vortex. Therefore, to reduce the influence of the vortices on the hydraulic performance of the pump, the *Q* criterion is employed to investigate the morphology and structure of vortices. Figure 10 shows the vortices in the impeller and the pressure distributions on the impeller for the four different blade thickness distributions given in Table 2. The equations of the Q criterion can be written as follows [19]:

$$\lambda^3 + P\lambda^2 + Q\lambda + R = 0. \tag{1}$$

Considering water as an incompressible medium, the equivalent equations are:

$$P = S_{ii} = 0 \tag{2}$$

$$Q = (\Omega_{ij}\Omega_{ji} - S_{ij}S_{ji})/2 \tag{3}$$

$$R = -\left(S_{ij}S_{jk}S_{ki} + 3\Omega_{ij}\Omega_{jk}S_{ki}\right)/3 \tag{4}$$

and

$$S_{ij} = \left(\partial u_i/\partial x_j + \partial u_j/\partial x_i\right)/2 \tag{5}$$

$$\Omega_{ij} = \left(\partial u_i/\partial x_j - \partial u_j/\partial x_i\right)/2. \tag{6}$$

In Equation (1), P, Q, and R represent the invariants of velocity gradient tensor. The Q criterion was proposed based on the second invariant of velocity gradient tensor. S_{ij} and Ω_{ij} represent tensors of strain rate and vorticity, respectively and reflect the deformation and rotational motion. Figure 10 depicts the three-dimensional (3D) vortex structures with blade thickness distribution given in Table 2 under various flow conditions. According to the morphology and structure of vortices, they can be separated into three categories, namely the leading edge vortex (LEV), trailing edge vortex (TEV), and gap leakage vortex (GLV).

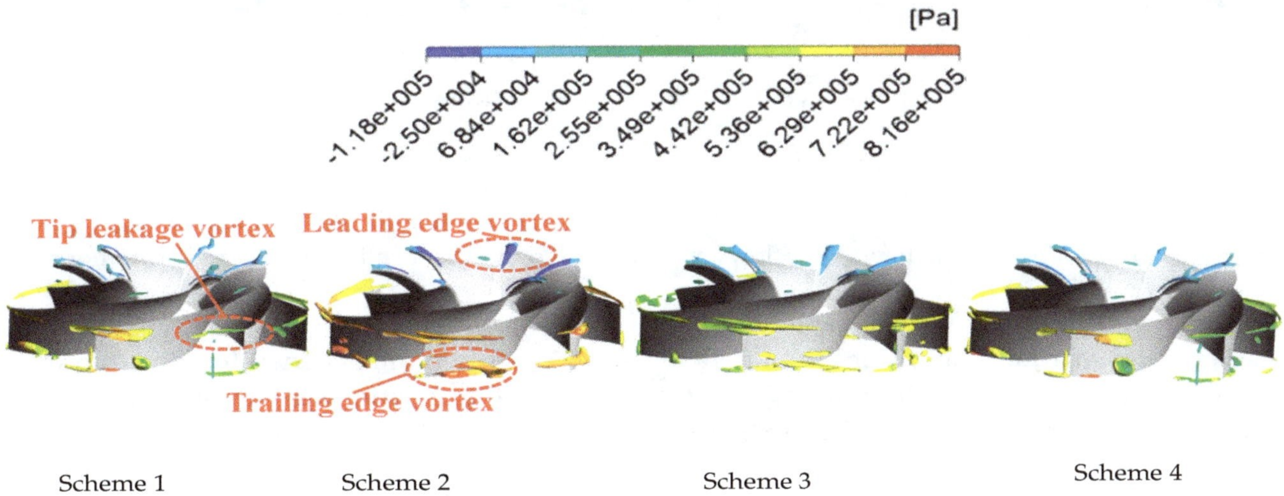

(**a**) Vortex spatial structure at 0.2 Q_d

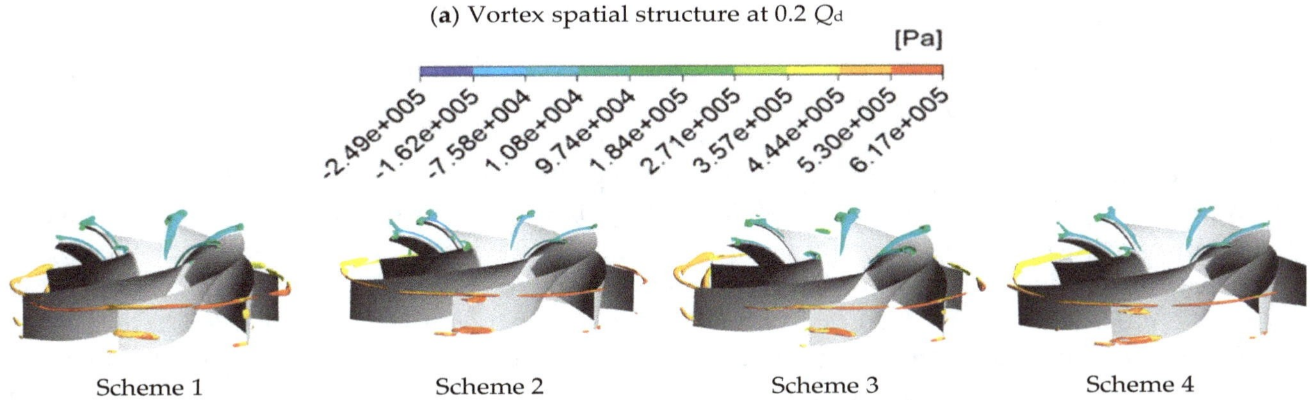

(**b**) Vortex spatial structure at 0.6 Q_d

Figure 10. *Cont.*

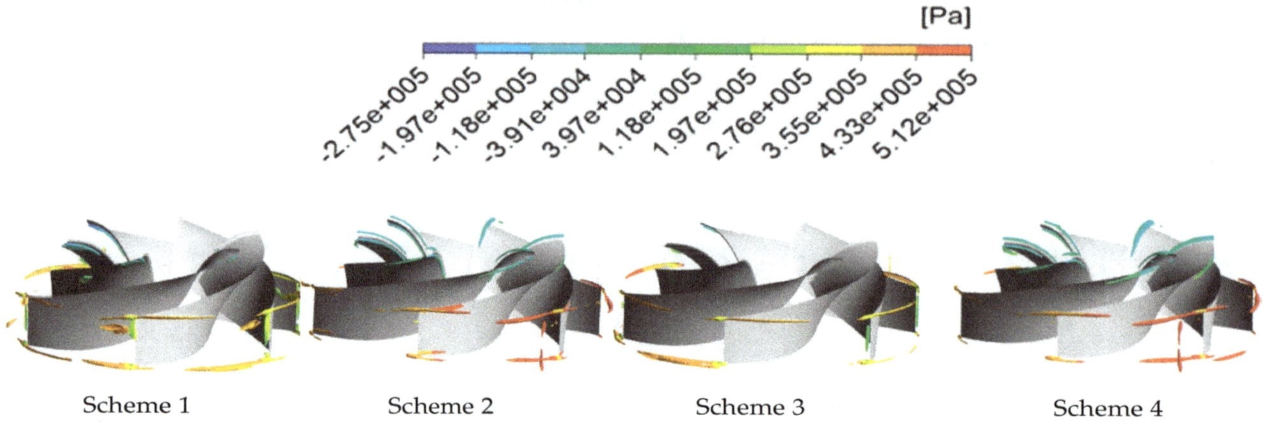

(**c**) Vortex spatial structure at 1.0 Q_d

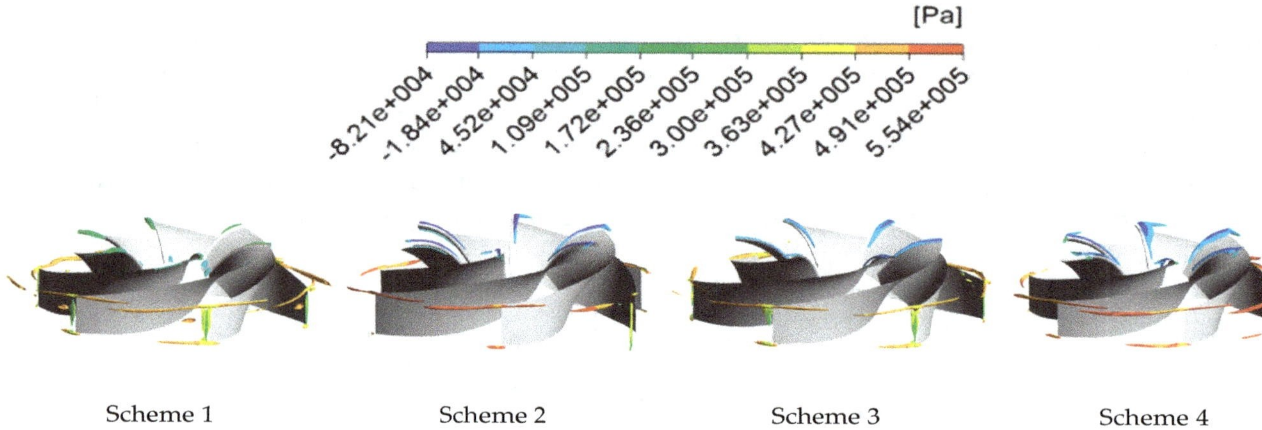

(**d**) Vortex spatial structure at 1.4 Q_d

Figure 10. 3D vortex structure for various flow conditions for four different blade thickness distributions.

As shown in Figure 10, GLV is generated at the gap between the blade and volute, which looks like a silk ribbon surrounding the entire impeller. This vortex structure is generated due to pressure difference between the pressure surface of the blade (PSB) and the suction surface of the blade (SSB), which results in water flowing from PSB to SSB through the gap and interacting with the mainstream in the impeller passage. The rotation direction of the vortex is the same as the impeller. It can also be noted that with the increase in the flow condition, the pressure difference between the PSB and SSB gradually decreases, which results in a corresponding decline in the size and strength of GLV.

Meanwhile, due to the impact of the incoming flow on the impeller, LEV can be observed at the leading edge of the blade. By comparing the LEV at various flow conditions, it can be concluded that the size and strength of LEV has a negative correlation with the flow condition; in other words, the size and strength of LEV reaches the maximum at the partial-load flow condition ($0.2Q_d$) and reaches the minimum at $1.0Q_d$. The incoming flow on the impeller at the design flow condition is considered as the non-impact flow, that is, the internal flow field is smooth without any vortices. However, when the flow condition decreases from $1.0Q_d$ to the $0.2Q_d$, the axial velocity remains the same, but the meridional velocity gradually decreases, which results in the increase of the impeller incidence angle. Therefore, the impact of incoming flow is felt at the leading edge of the blade. When the flow condition increases from $1.0Q_d$ to the $1.4Q_d$, the incidence angle is larger than the design incidence angle which also leads to generation of LEV at leading edge of the blade. In addition, it can be noted that the negative pressure is created at the leading edges of the blades which may induce cavitation that can deteriorate the impeller. The incoming flow at the inlet of the impeller is squeezed, which a causes large change in the velocity direction and thus the pressure reduces sharply at the leading edge of the blade. In addition, since the expelling coefficient of the blade with thicker leading edge is larger than that with thinner leading edge, the magnitude and area of negative pressure on the thicker leading edge blade (Schemes 2 and 4) is significantly larger than that on the thinner leading edge blade (Schemes 1 and 3). Furthermore, as the flow condition increases, the area of the negative pressure gradually decreases.

At the partial-load conditions ($0.2Q_d$–$0.6Q_d$), the TEV can be seen at the trailing edge of the blade, as shown in Figure 10a,b, where the shape of the TEV is like a bubble attached to the PSB. Generally, the morphology of the TEV is more disordered than that of the LEV and GLV, and the pressure and intensity of TEV are significantly higher than for the other two vortices. As the flow condition continues to increase, the intensity of the TEV gradually declines. Furthermore, it can be observed that the TEV of the Scheme 2 is the most disorderly one, which has a twisted shape. This not only effects the flow movement of the impeller passage, but also decreases the hydraulic performance. Therefore, based on these observations on vortex structures, it can be concluded that the characteristics of the TEV are more unstable than those of LEV and GLV.

To further investigate the formation principles of the unsteady vortex in detail, the vorticity transport equation [20,21] is used in this research:

$$\frac{D\vec{\omega}}{Dt} = \left(\vec{\omega}\cdot\nabla\right)\vec{V} - \vec{\omega}(\nabla\cdot\vec{V}) + \frac{\nabla\rho_m \times \nabla p}{\rho_m^2} + \upsilon\nabla^2\vec{\omega}. \quad (7)$$

In Equation (7), $D\vec{\omega}/Dt$ is the rate of the vorticity variation, $(\vec{\omega}\cdot\nabla)\vec{V}$ is the vortex stretching variable (VSV) caused by the velocity gradients, and $\vec{\omega}(\nabla\cdot\vec{V})$ represents the vortex dilation, which reveals the influence of the medium compressibility on the vorticity. The term $\nabla\rho_m \times \nabla p/\rho_m^2$ is the baroclinic torque, which is induced by the gradient of both the pressure and density. $\upsilon\nabla^2\vec{\omega}$ is the viscidity diffusion variable (VDV), which determines the effect of viscidity diffusion on vorticity. Considering the fact that the medium is water which is incompressible, the terms $\vec{\omega}(\nabla\cdot\vec{V})$ and $\nabla\rho_m \times \nabla p/\rho_m^2$ are neglected in this paper.

Figure 11 shows the distribution of the viscidity diffusion variable (VDV) on the PSB for various flow conditions. It can be observed that an oval shape VDV is generated near the trailing edge of the blade, and the magnitude of the VDV gradually decreases with an increase in the flow condition. The VDV distribution reaches maximum at the trailing edge at a flow condition of 100 m^3/h and it moves towards the leading edge for high flow condition of 700 m^3/h. Since the blade passage is full of uneven flow and axial eddies at the partial-load conditions, the induced relative velocity on the PSB is smaller than the ideal design relative velocity. The separation of the flow in the boundary layer occurs due to the effect of the positive incidence angle. The conclusion can be drawn that the thicker the blade is, the greater the flow separation becomes. Furthermore, when the flow condition reaches the design value, the VDV distribution reduces to its minimum. And at an over-load condition (700 m^3/h), a larger VDV is generated at the leading edge.

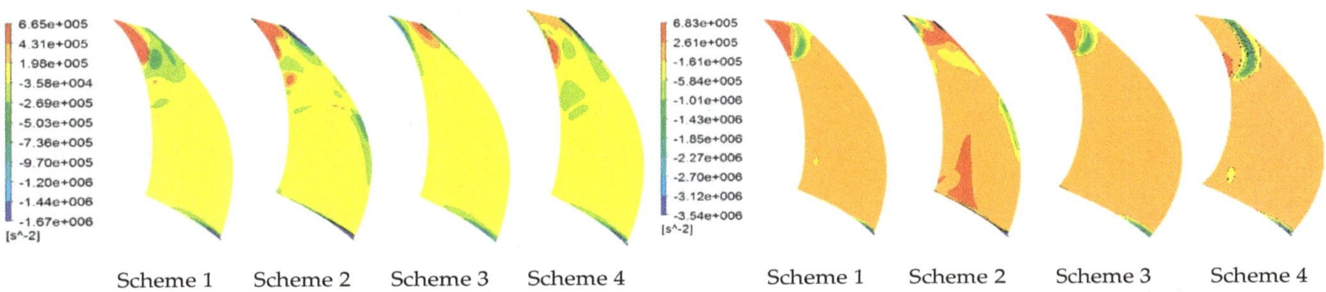

(a) The viscidity diffusion variable (VDV)distribution at 100 m^3/h

(b) The VDV distribution at 300 m^3/h

Figure 11. *Cont.*

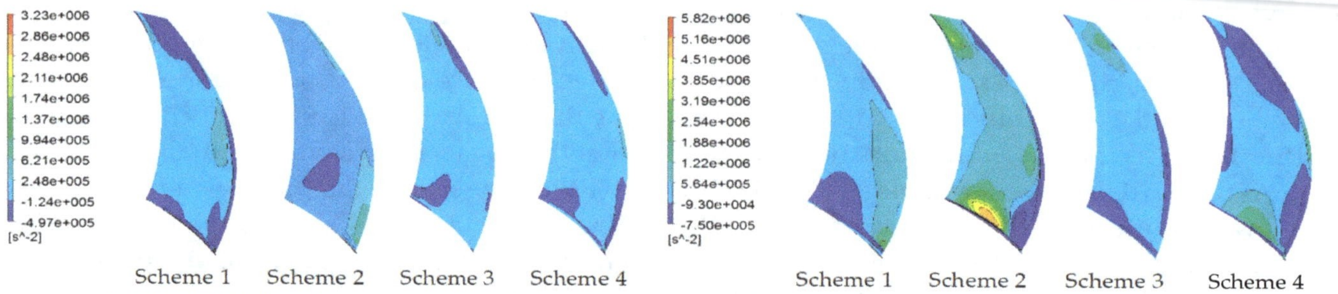

(**c**) The VDV distribution at 500 m³/h (**d**) The VDV distribution at 700 m³/h

Figure 11. The VDV distribution at various flow conditions.

The vortex Stretching Variable (VSV) distribution is similar to the VDV distribution, as shown in Figure 12. The VSV distribution is based on the angular momentum conservation principle [22]. VSV first reduces and then gradually increases; the lowest value of VSV is obtained at the design condition. Based on this discussion, it can be concluded that the VSV plays an essential role in the formation of the vortex, and more importantly that increasing the thickness of the blade is an effective way to decrease the VSV.

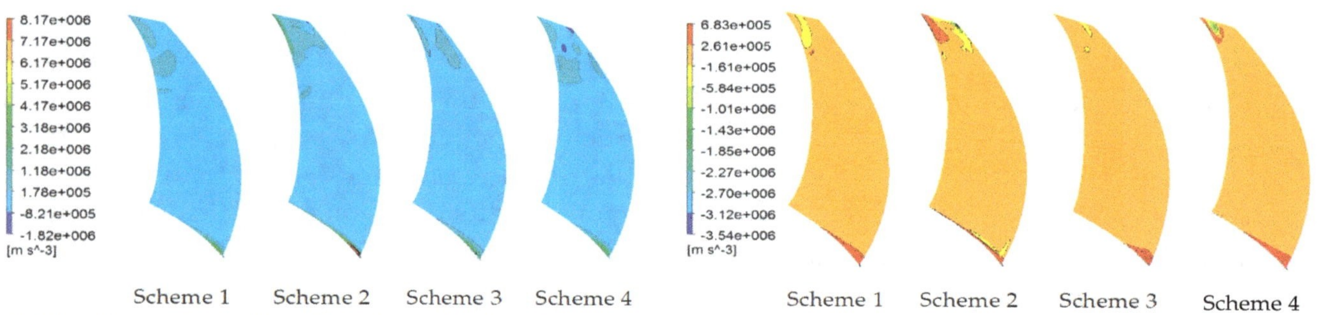

(**a**) The Vortex Stretching Variable (VSV)distribution at 100 m³/h (**b**) The VSV distribution at 300 m³/h

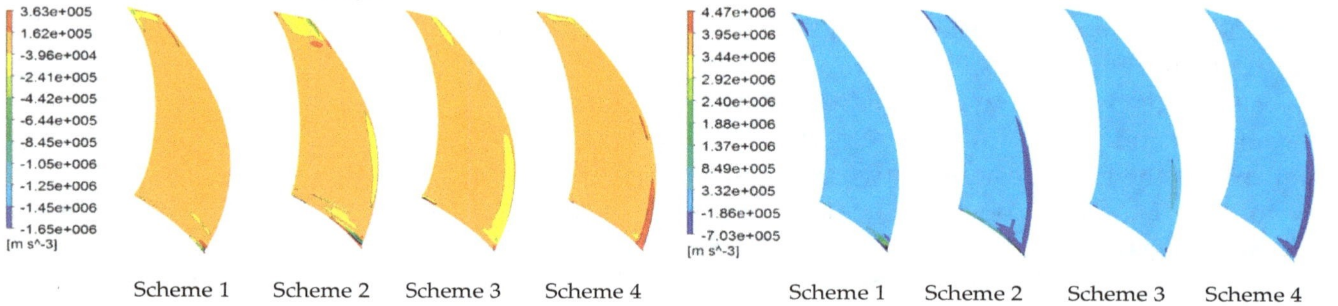

(**c**) The VSV distribution at 500 m³/h (**d**) The VSV distribution at 700 m³/h

Figure 12. The VSV distribution at various flow conditions.

4.2. Energy Characteristic of the Pump

The self-priming pump consumes and converts energy during its operation. Noting that the entropy always increases in the irreversible thermal processes, the entropy production is calculated to evaluate the energy loss of the pump. The total entropy production can be obtained as follows [15]:

$$S_{\text{Total}} = S_H + S_V + S_T + S_W \tag{8}$$

In Equation (8), S_H represents the entropy production caused by the heat conversion. Since the self-priming pump operates under the condition of a constant temperature, this term is neglected in this study. S_W denotes the entropy production due to the dissipation of kinetic energy and potential energy on the wall (WEP). S_V is the entropy production from the viscidity diffusion (VEP) and S_T is the entropy production generated by the turbulence which can determined by the loss in turbulent kinetic energy (TEP). The rate of the entropy production is obtained as follows: [12]:

The rate of entropy production from VEP can be expressed as:

$$s_v = \frac{\mu}{T}\left[2\left\{\left(\frac{\partial \overline{u}}{\partial x}\right)^2 + \left(\frac{\partial \overline{v}}{\partial y}\right)^2 + \left(\frac{\partial \overline{w}}{\partial z}\right)^2\right\} + \left(\frac{\partial \overline{u}}{\partial y} + \frac{\partial \overline{v}}{\partial x}\right)^2 + \left(\frac{\partial \overline{u}}{\partial z} + \frac{\partial \overline{w}}{\partial x}\right)^2 + \left(\frac{\partial \overline{v}}{\partial z} + \frac{\partial \overline{w}}{\partial y}\right)^2\right] \tag{9}$$

where μ is the dynamic viscosity and T is the temperature.

The rate of entropy production from VEP can be expressed as:

$$s_t = \frac{\mu}{T}\left[2\left\{\overline{\left(\frac{\partial u'}{\partial x}\right)^2} + \overline{\left(\frac{\partial v'}{\partial y}\right)^2} + \overline{\left(\frac{\partial w'}{\partial z}\right)^2}\right\} + \overline{\left(\frac{\partial u'}{\partial y} + \frac{\partial v'}{\partial x}\right)^2} + \overline{\left(\frac{\partial u'}{\partial z} + \frac{\partial w'}{\partial x}\right)^2} + \overline{\left(\frac{\partial v'}{\partial z} + \frac{\partial w'}{\partial y}\right)^2}\right] \tag{10}$$

Since S_T cannot be solved directly, the rate of the TEP can obtained as follows [13]:

$$s_t = \frac{\rho \varepsilon}{T} \tag{11}$$

where ρ is the density of water (1000 kg/m^3) and ε is the turbulence dissipation.

It should be noted that due to the velocity gradient on the wall, and the dissipation of kinetic and potential energy occurs during the water flow between the turbulent layer and the laminar layer. Most of the previous work in this area has neglected the energy loss of the WEP. In this paper, the rate of WEP is calculated by the expression [23]:

$$s_w = \frac{\tau_w v_w}{T} \tag{12}$$

The WEP on the wall is obtained as:

$$S_W = \int_A \frac{\tau_w v_w}{T} dA \tag{13}$$

where τ_w represents the shear stress of the wall, while v_w is the speed of the first fluid layer near the wall. Thus, the VEP and TEP generated in the fluid region can be obtained as follows:

$$S_T = \int_V s_t dV \tag{14}$$

$$S_V = \int_V s_v dV. \tag{15}$$

Figure 13 shows the entropy production of the impeller for various flow conditions. It can be observed that the TEP of the blades with a thick leading edge (Schemes 2 and 4) is larger than that of the blades with a thin leading edge (Schemes 1 and 3). Furthermore, the TEP of the blade with increasing thickness from the leading edge to the middle section and trailing edge (Scheme 3) is lower than that of the uniformly thin blade (Scheme 1). As the flow condition gradually increases, the TEP first decreases drastically but then smoothly increases at the design flow condition. In addition, the decrement rate of TEP at partial-load condition is larger than the growth rate under the over-load condition, and the TEP at design flow condition drops down to the minimum value. By comparing the VEP and WEP for different blade thicknesses, it can be observed that their variation trends are consistent with TEP; however, both the VEP and WEP of Scheme 2 are higher than those of other schemes. Furthermore, as the flow condition increases, both the VEP and WEP continue to rise but the growth rate of VEP is larger than that of WEP. More importantly, comparing the magnitudes of TEP, VEP, and WEP, it can be concluded that the effects of TEP and VEP on energy loss are stronger than that of WEP. In other words, the energy loss caused by VEP and TEP are the dominant factors influencing the energy conversion in the novel self-priming pump.

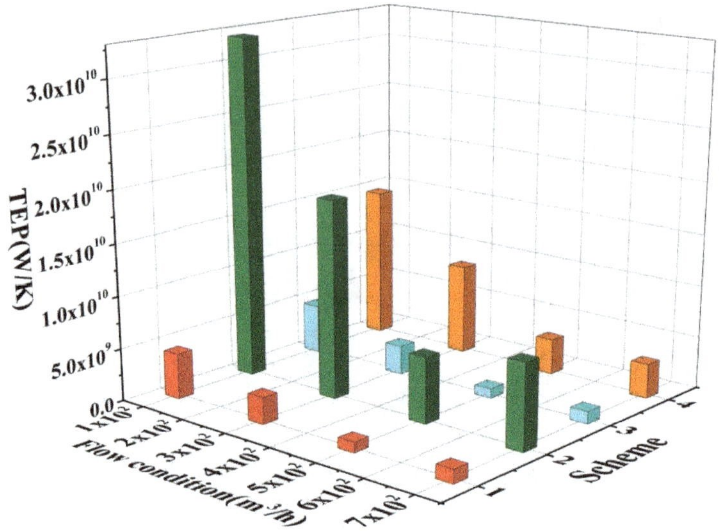

(**a**) The turbulence entropy production (TEP) of the impeller at various flow conditions

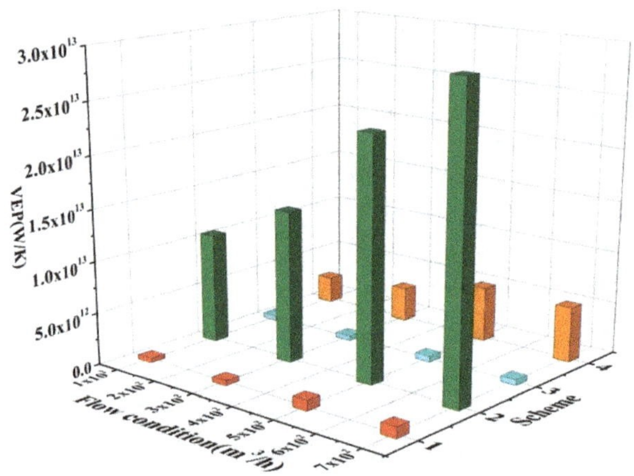

(**b**) The viscous entropy production (VEP) of the impeller at various flow conditions

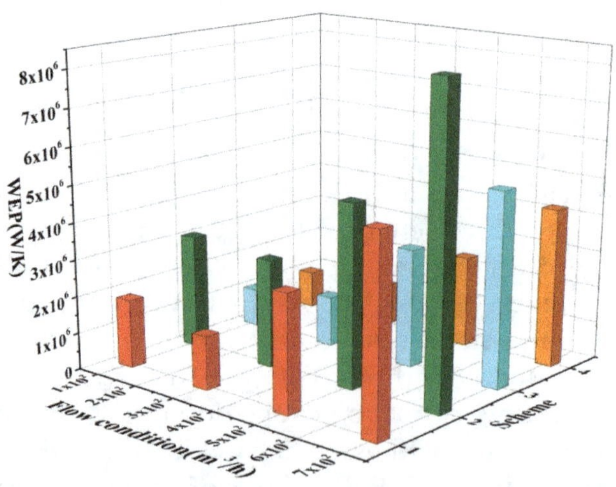

(**c**) The entropy production due to the dissipation of kinetic energy and potential energy on the wall (WEP) of the impeller at various flow conditions.

Figure 13. Impeller entropy production at various flow conditions.

As shown in Figure 14, the distribution of the VEP rate coincides with the VDV distribution rate. It can be noticed that the energy loss caused by the VEP rate is mainly concentrated near the trailing edge for a partial-load condition. There is a negative correlation between the VEP and flow conditions in the range $0.2Q_d$ to $1.0Q_d$. This means the magnitude of the VEP rate gradually decreases when the flow condition increases to the design flow condition. On the other hand, as the flow condition further increases beyond the design flow condition, the VEP rate shows an increasing trend and the location of energy loss changes from the trailing edge to the leading edge of the blade. These results can be attributed to the change in VEP due to viscous diffusion; greater viscous diffusion can be found at the leading edge under partial-load conditions, which leads to separation of the boundary layer. Moreover, the area and magnitude of viscous diffusion in Scheme 3 are obviously lower than those in Scheme 2; in other words, utilizing the blade with increasing thickness from the leading edge (Scheme 3) is a good solution to reduce the boundary layer separation and its effect.

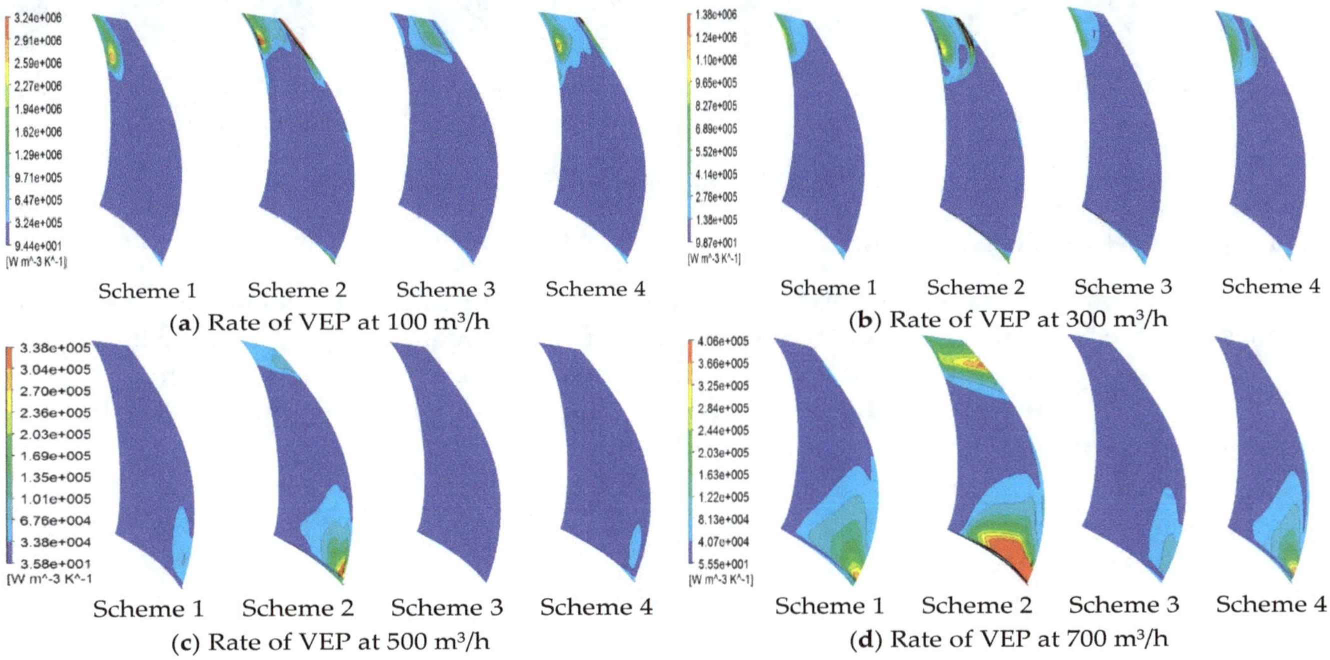

Figure 14. Rate of VEP at various flow conditions [24].

Figure 15 presents the distribution of the TEP rate at various flow conditions. It can be noticed that the TEP rate distribution has a consistent variation trend with the VEP rate, but its intensity is relatively smaller. Since the TEV is created under partial-load conditions, it can not only reduce the flow area of the impeller but also can induce turbulent dissipation at the trailing edge. In addition, by comparing the TEP rates of each schemes, it can be found that a larger TEP rate occurs at the trailing edge of the thick blade (Schemes 2 and 4), and that the TEP of Scheme 3 is closer to the ideal flow field.

Thus, based on the above observations, it can be concluded that the flow condition creates negative effect on the TEP and VEP rate in the range $0.2Q_d$ to $1.0Q_d$, and the lowest TEP and VEP rates are obtained for $1.0Q_d$. Furthermore, as the flow condition increases to the over-load condition, the TEP and VEP rates gradually increase and the energy loss caused by the VEP and TEP decrease to the minimum of the design condition. In addition, the entropy production from the thick blade is larger than that from the thin blade. Therefore, it can be concluded that the blade thickness distribution plays an essential role in the energy loss of the self-priming pump.

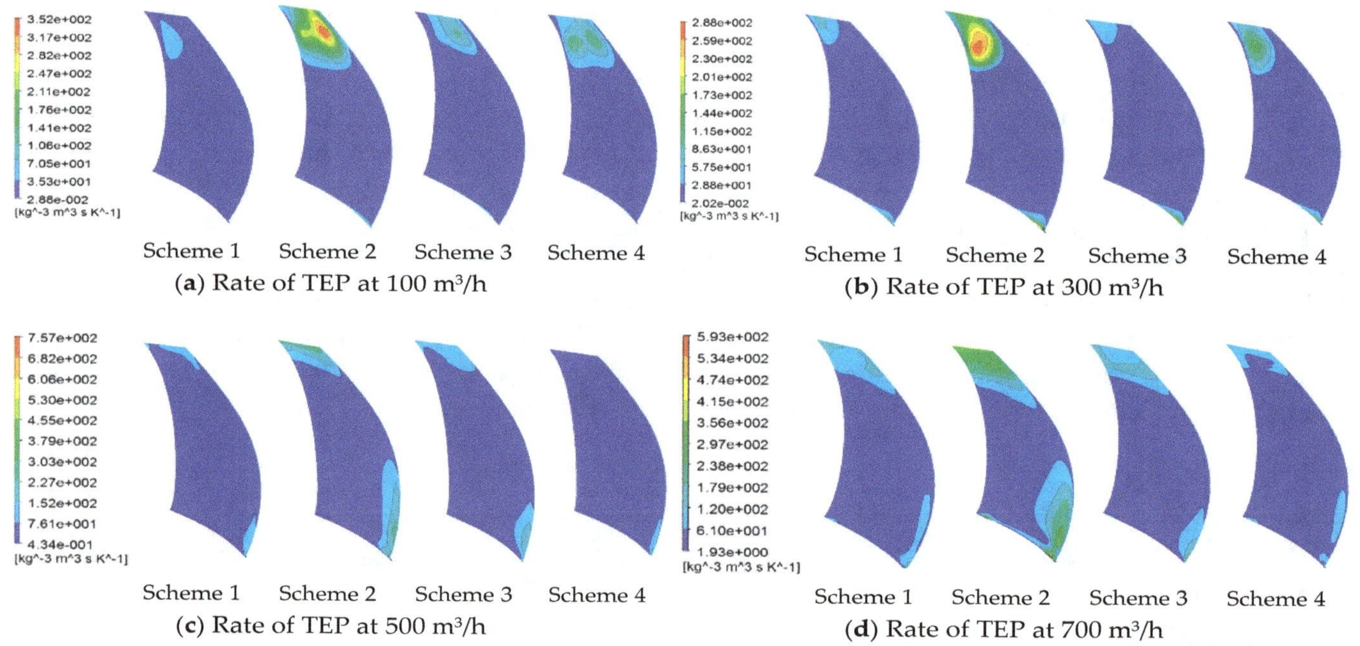

(**a**) Rate of TEP at 100 m³/h (**b**) Rate of TEP at 300 m³/h (**c**) Rate of TEP at 500 m³/h (**d**) Rate of TEP at 700 m³/h

Figure 15. Rate of TEP at various flow conditions [24].

5. Hydraulic Performance of the Pump

To compare the hydraulic performance of the pump with different thicknesses of the blade (Schemes 1–4), the head and efficiency are selected as the two key quantities for evaluating its hydraulic characteristics. The head indicates the hoisting height of water when it receives energy from the pump, while the efficiency expresses the energy conversion ability of the self-priming pump. The head and efficiency are defined as:

$$H = \frac{P_O - P_I}{\rho g} + \frac{V_O^2 - V_I^2}{2g} + (Z_O - Z_I) \tag{16}$$

and

$$\eta = \frac{60 g H Q_F}{\pi M n} \tag{17}$$

where $P_O - P_I$, $V_O^2 - V_I^2$ and $Z_O - Z_I$ denote the pressure difference, velocity difference, and height difference obtained when the water flow through the pump, M is the torque, and Q_F is the flow conditions of the self-priming pump. In the analysis and presentation of results, the head and flow conditions are converted into the dimensionless parameters given below [24]:

$$\Phi = \frac{60 Q_F}{\pi^2 D_2 b_2 n} \tag{18}$$

$$\Psi = \frac{7200 H g}{\pi^2 D_2^2 n^2}. \tag{19}$$

Figure 16 shows the differences in hydraulic performance of different schemes. It shows the formation of energy loss in the partial-load condition and over-load condition, which becomes minimal in the design condition. This trend in Figure 16 can be mainly attributed to the generation of TEV under a part-load condition and LEV under an over-load condition; however, the energy loss in the partial-load condition is more severe than in the over-load condition. Furthermore, it can be noted that the head of the Scheme 2 is lower than that of Scheme 1 overall except under the partial-load condition. Since a large TEV is generated in Scheme 2, the flow area reduces sharply, which results in large changes in the internal flow. Due to the effect of incoming flow, the head increases but results in large energy dissipation as shown in Figure 16b. Thus, the efficiency of Scheme 2 is obviously lower than that of

Scheme 1 for flow condition of $0.6Q_d$. Although the energy loss of the thinner blades (Scheme 1) is lower, its energy conversion ability is weaker than that of the blade with thicker trailing edge. More importantly, by comparing the efficiency of each scheme, it can be concluded that the energy conversion ability of Scheme 3 is superior to that of other schemes. Due to the advantages of the thin leading edge and thick trailing edge, Scheme 3 can not only significantly reduce the impact of the incoming flow in over-load condition, but also can promote the work capability of the blade on the fluid.

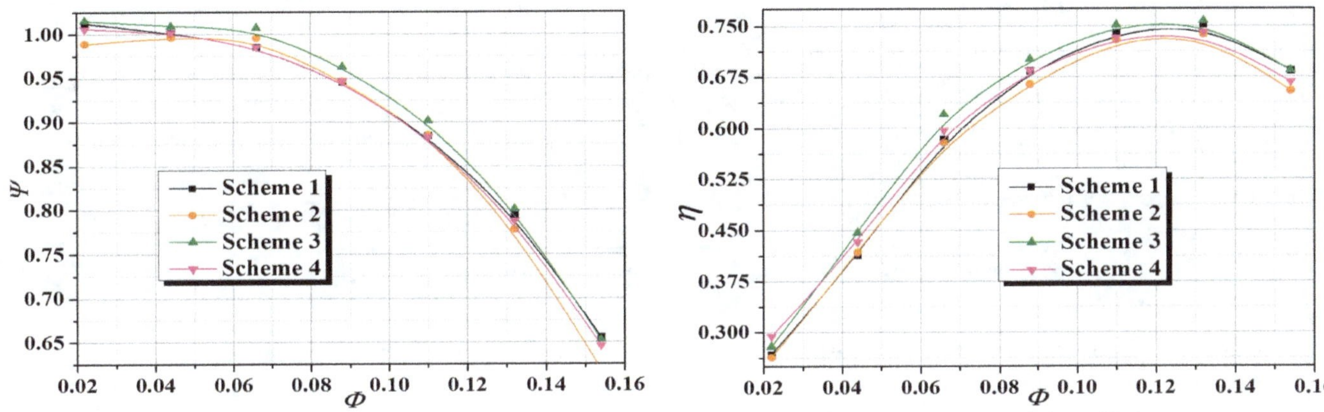

(**a**) Comparison of the head (**b**) Comparison of the efficiency

Figure 16. Head and efficiency under various flow conditions.

6. Experiment Results

To verify the accuracy of the calculations reported in Section 5, the hydraulic performance experiment for Scheme 3 (since it was found to be the best scheme in Section 5) was conducted at Jiangsu University in China. The measurement deviation of the pressure transducer (D_P) is ± 0.5 %. Considering that the monitoring points were set on the inlet and outlet tank and the measurement deviation of dynamic head can be ignored ($D_D = 0$), the measurement deviation of head (D_H) is obtained as follows:

$$D_H = \sqrt[2]{D_P^2 + D_D^2} \tag{20}$$

Furthermore, the deviation of the transducer turbine flowmeter (D_Q) is ± 1.0 %, the measurement deviations of torque (D_M) and speed transducer (D_N) are ± 0.5 % and ± 0.1 %, respectively. Therefore, the measurement deviation of shaft power (D_T) can be determined by:

$$D_T = \sqrt[2]{D_N^2 + D_M^2} \tag{21}$$

Finally, the measurement deviation of experimental system (D_S) can be expressed as follows:

$$D_S = \sqrt[2]{D_H^2 + D_Q^2 + D_T^2} \tag{22}$$

Therefore, the measurement deviation of the experiment is within ±1.22 % which quantifies the uncertainty in experimental measurements. The experimental results are shown in Figure 17. It can be seen from this figure that overall the experimental and numerical results for pump head and efficiency are in reasonable agreement with minor discrepancies. It should be noted that in the calculations, the internal structure of the pump was simplified. This meant the pump cavity, seal, and bearing were not considered in the numerical model. As a result, the pump performance results obtained in the calculations are slightly larger than those in the experiment. However, the maximum discrepancy between the numerical results and experimental results is within 3 %, which is in the acceptable range. It can therefore be concluded that the numerical model and computations reported in this paper are accurate and the results are reliable.

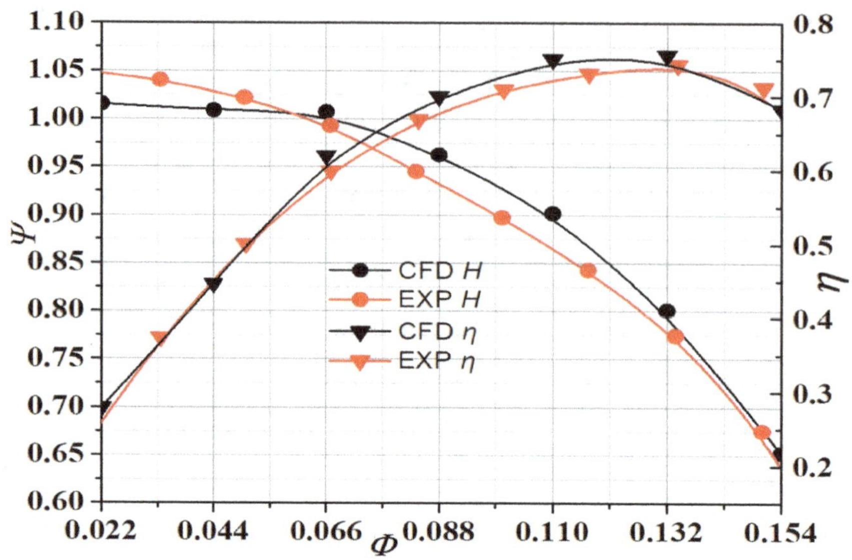

Figure 17. Comparison of experimental and numerical results for pump head and efficiency for Scheme 3.

In addition, to validate the capability of a novel self-priming pump, the experiments with waterless start were conducted at different altitudes at Jiangsu University in China.

Figure 18 shows the results of self-priming pump experiment at 5.3 m, 6 m, 7 m, and 8 m altitudes. It should be noted that the self-priming time reaches 62 s at 5.3 m altitude, which is remarkably shorter than the standard Chinese self-priming pump time of 120 s. Furthermore, it can be noted from Figure 18 that the increase rate in self-priming time R with the altitude can be separated into two stages, namely the altitude from 0 m to 6 m and from 6 m to 8 m. The first stage shows the non-linear variation while the second stage shows the linear variation in start-up time with the altitude. When the self-priming altitude is below 6 m, with increase in the altitude more resistance of water needs to be overcome, therefore negative pressure is generated in the pump which induces the air in the water to separate and leads to the air in the pump increasing. Thus, the line of the increase rate R in Figure 18 shows a nonlinear increase. On the other hand, as the altitude continues to increase, the air in the water is no longer discharged. Therefore, the relationship between air volume in the pump and altitude shows a linear increase, which is the reason for the linear growth in the second stages.

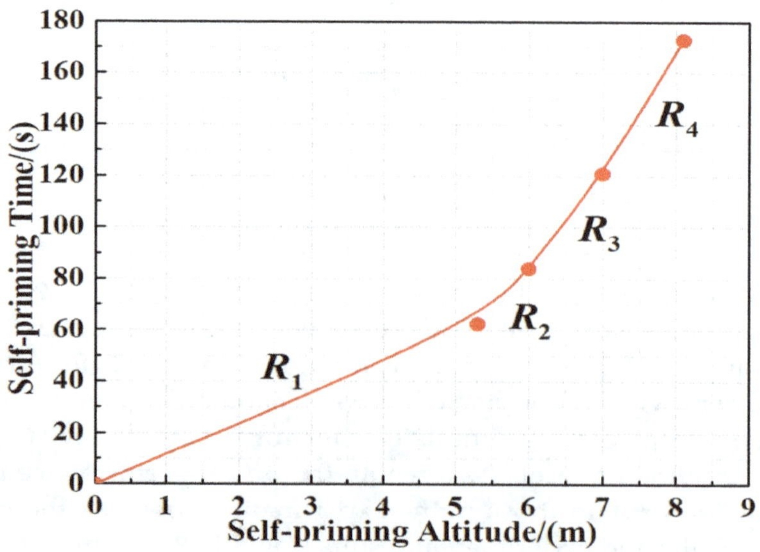

Figure 18. Self-priming time at different altitudes.

7. Conclusions

In this paper, by using the Q criterion with vorticity transport equation and analyzing the entropy production in the flow, the vortex structures and energy characteristics of a novel self-priming pump with different blade thicknesses were systematically investigated by numerical simulations which are validated for one of the most promising blade thickness distributions in terms of the pump head and efficiency. The following conclusions can be drawn.

A new self-priming pump has been designed which can achieve the pump start without water. Under the effect of the self-priming system, negative pressure is generated in the pump which results in rapid suction of the water in the pump. It is demonstrated that the self-priming capability of the pump is remarkably superior to the traditional pump.

The spatial structure of vortices in the pump is investigated by employing the Q criterion with the numerical solution of the vorticity transport equation. Based on the morphology, the vortices be separated into three categories: Trailing Edge Vortex (TEV), Leading Edge Vortex (LEV), and Gap Leakage Vortex (GLV). Generally, the morphology of the TEV is more disorderly than that of LEV and GLV, and the pressure and intensity of TEV are significantly higher than the other two vortex structures. TEV cannot only block the passage of the impeller but can also increase instability in the internal flow. Analysis of the formation mechanism of the vortices using the numerical solution of the vorticity transport equation shows that the Viscous Diffusion Variable (VDV) plays an essential role in the separation of the boundary layer which results in the generation of Tip Leakage Vortex (TLV). In addition, the thicker the blade is, the more severe the flow separation becomes.

To determine the magnitude and distribution of energy loss in the pump, entropy production analysis is applied to study the influence of blade thickness on energy characteristics. Based on the analysis, greater energy loss is generated in both the partial-load condition and the over-load condition relative to design condition; however, the energy loss in partial-load condition is larger than that in the over-load condition due to differences in the size and intensity of TLV. Furthermore, in other words, the energy loss caused by viscous entropy production (VEP) and turbulence entropy production (TEP) are the dominant factors influencing the energy loss in the novel self-priming pump. The energy loss of the thicker blade is larger than that of the thinner blade; however, employing the blade with a thin leading edge and thick trailing edge can not only significantly reduce the impact of incoming flow under over-load condition but can also increase the work capability of the blade on the fluid. Thus, increase in thickness of the blade from the leading edge to the trailing edge is beneficial for pump performance.

Finally, the hydraulic test confirmed the accuracy of the simulations. Therefore, the numerical results presented in this paper can be considered as accurate and reliable.

Author Contributions: This is a joint work and the authors contributed according to their expertise and capability: H.C. conducted the investigation and analysis, and writing and revisions of the paper; R.K.A. ontributed to methodology and revisions of the paper; W.L. Performed the validation and revision; L.Z. Performed the data analysis and W.S. was involved in revision of the manuscript.

Nomenclature

H	head m
Q_d	design flow condition, m^3/h
n	rotation speed, r/min
n_s	specific speed
D_1	impeller inlet diameter, mm
D_2	impeller outlet diameter, mm
Z	blade number

φ	blade wrap angles, °
b_1	volute inlet width, mm
D_4	volute outlet diameter, mm
SST	shear stress transition
t	rotation period, s
LEV	leading edge vortex
TEV	trailing edge vortex
GLV	gap leakage vortex
PSB	suction surface of the blade
VDV	viscidity diffusion variable
VSV	vortex stretching variable
WEP	wall entropy production
VEP	viscidity entropy production
TEP	turbulence entropy production
ρ	density of water, kg/m^3
ε	turbulence dissipation
M	torque, Nm
Q_F	flow condition, m^3/h
D_P	deviation of the pressure transducer
D_D	measurement deviation of dynamic head
D_H	deviation of transducer turbine flowmeter
D_M	measurement deviation of torque
D_N	deviation of speed transducer
D_T	measurement deviation of shaft power
D_S	measurement deviation of experiment system

References

1. Turkyilmazoglu, M. Performance of direct absorption solar collector with nanofluid mixture. *Energy Convers. Manag.* **2016**, *114*, 1–10. [CrossRef]
2. Wang, T.; Kong, F.; Xia, B.; Bai, Y.; Wang, C. The method for determining blade inlet angle of special impeller using in turbine mode of centrifugal pump as turbine. *Renew. Energy* **2017**, *109*, 518–528. [CrossRef]
3. Elyamin, G.R.H.A.; Bassily, M.A.; Khalil, K.Y.; Mohamed, S.G. Effect of impeller blades number on the performance of a centrifugal pump. *Alex. Eng. J.* **2019**, *58*, 39–48. [CrossRef]
4. Shao, C.; Zhou, J.; Cheng, W. Effect of viscosity on the external characteristics and flow field of a molten salt pump in the view of energy loss. *Heat Mass Transf.* **2019**, *55*, 711–722. [CrossRef]
5. Nejad, J.; Riasi, A.; Nourbakhsh, A. Efficiency improvement of regenerative pump using blade profile modification: Experimental study. *Proc. Inst. Mech. Eng. Part E J. Process Mech. Eng.* **2019**, *233*, 448–455. [CrossRef]
6. Han, X.; Kang, Y.; Li, D.; Zhao, W. Impeller optimized design of the centrifugal pump: A numerical and experimental investigation. *Energies* **2018**, *11*, 1444. [CrossRef]
7. Jeon, S.Y.; Yoon, J.Y.; Jang, C.M. Optimal design of a novel 'S-shape' impeller blade for a microbubble pump. *Energies* **2019**, *12*, 179. [CrossRef]
8. Nejad, J.; Riasi, A.; Nourbakhsh, A. Parametric study and performance improvement of regenerative flow pump considering the modification in blade and casing geometry. *Int. J. Numer. Methods Heat Fluid Flow* **2017**, *27*, 1887–1906. [CrossRef]
9. Wang, C.; Hu, B.; Zhu, Y.; Wang, X.; Luo, C.; Cheng, L. Numerical study on the gas-water two-phase flow in the self-priming process of self-priming centrifugal pump. *Processes* **2019**, *7*, 330. [CrossRef]
10. Wang, C.; He, X.; Zhang, D.; Hu, B.; Shi, W. Numerical and experimental study of the self-priming process of a multistage self-priming centrifugal pump. *Int. J. Energy Res.* **2019**, 1–19. [CrossRef]
11. Wang, C.; He, X.; Shi, W.; Wang, X.; Wang, X.; Qiu, N. Numerical study on pressure fluctuation of a multistage centrifugal pump based on whole flow field. *AIP Adv.* **2019**, *9*, 035118. [CrossRef]
12. Wang, C.; Zhang, Y.; Hou, H.; Zhang, J.; Xu, C. Entropy production diagnostic analysis of energy consumption for cavitation flow in a two-stage LNG cryogenic submerged pump. *Int. J. Heat Mass Transf.* **2019**, *129*, 342–356. [CrossRef]

13. Hou, H.; Zhang, Y.; Zhou, X.; Zuo, Z.; Chen, H. Optimal hydraulic design of an ultra-low specific speed centrifugal pump based on the local entropy production theory. *Proc. Inst. Mech. Eng. Part A J. Power Energy* **2019**, *233*, 715–726. [CrossRef]
14. Gu, Y.; Pei, J.; Yuan, S.; Wang, W.; Zhang, F.; Wang, P.; Appiah, D.; Liu, Y. Clocking effect of vaned diffuser on hydraulic performance of high-power pump by using the numerical flow loss visualization method. *Energy* **2019**, *170*, 986–997. [CrossRef]
15. Li, D.; Wang, H.; Qin, Y.; Han, L.; Wei, X.; Qin, D. Entropy production analysis of hysteresis characteristic of a pump-turbine model. *Energy Convers. Manag.* **2017**, *149*, 175–191. [CrossRef]
16. Chang, H.; Li, W.; Shi, W.; Liu, J. Effect of blade profile with different thickness distribution on the pressure characteristics of novel self-priming pump. *J. Braz. Soc. Mech. Sci. Eng.* **2018**, *40*, 518. [CrossRef]
17. Bai, L.; Zhou, L.; Han, C.; Zhu, Y.; Shi, W. Numerical study of pressure fluctuation and unsteady flow in a centrifugal pump. *Processes* **2019**, *7*, 354. [CrossRef]
18. Bai, L.; Zhou, L.; Jiang, X.; Pang, Q.; Ye, D. Vibration in a multistage centrifugal pump under varied conditions. *Shock Vib.* **2019**, 2057031. [CrossRef]
19. Alfonsi, G.; Primavera, L. Temporal evolution of vortical structures in the wall region of turbulent channel flow. *Flow Turbul. Combust.* **2009**, *83*, 61–79. [CrossRef]
20. Ji, B.; Luo, X.; Arndt, R.E.A.; Wu, Y. Numerical simulation of three-dimensional cavitation shedding dynamics with special emphasis on cavitation–vortex interaction. *Ocean Eng.* **2014**, *87*, 64–77. [CrossRef]
21. Liu, Y.; Tan, L. Tip clearance on pressure fluctuation intensity and vortex characteristic of a mixed flow pump as turbine at pump mode. *Renew. Energy* **2018**, *129*, 606–615. [CrossRef]
22. Wu, J.Z.; Ma, H.Y.; Zhou, M.D. *Vorticity and Vortex Dynamic*; Springer Science & Business Media: Berlin/Heidelberg, Germany, 2007.
23. Hucan, H.; Zhang, Y.; Li, Z. A numerical research on energy loss evaluation in a centrifugal pump system based on local entropy production method. *Therm. Sci.* **2017**, *21*, 1287–1299.
24. Chang, H.; Shi, W.; Li, W.; Liu, J. Energy loss analysis of novel self-priming pump based on the entropy production theory. *J. Therm. Sci.* **2019**, *28*, 306–318. [CrossRef]

4

Using PSO Algorithm to Compensate Power Loss due to the Aeroelastic Effect of the Wind Turbine Blade

Ying Zhao [1,2,3], Caicai Liao [1,4,5,*], Zhiwen Qin [1,3,4] and Ke Yang [1,3,4]

1. Institute of Engineering Thermophysics, Chinese Academy of Sciences, Beijing 100190, China; zhaoyingiet@163.com (Y.Z.); qinzhiwen@iet.cn (Z.Q.); yangke@iet.cn (K.Y.)
2. University of Chinese Academy of Sciences, Beijing 100190, China
3. Key Laboratory of Wind Energy Utilization, Chinese Academy of Sciences, Beijing 100190, China
4. National Laboratory of Wind Turbine Blade Research & Development Center, Beijing 100190, China
5. Dalian National Laboratory for Clean Energy, Chinese Academy of Sciences, Dalian 116023, China

* Correspondence: liaocaicai@iet.cn

Abstract: Power loss due to the aeroelastic effect of the blade is becoming an important problem of large-scale blade design. Prior work has already employed the pretwisting method to deal with this problem and obtained some good results at reference wind speed. The aim of this study was to compensate for the power loss for all of the wind speeds by using the pretwisting method. Therefore, we developed an aeroelastic coupling optimization model, which takes the pretwist angles along the blade as free variables, the maximum AEP (annual energy production) as the optimal object, and the smooth of the twist distribution as one of the constraint conditions. In this optimization model, a PSO (particle swarm optimization) algorithm is used and combined with the BEM-3DFEM (blade element momentum—three-dimensional finite element method) model. Then, the optimization model was compared with an iteration method, which was recently developed by another study and can well compensate the power loss at reference wind speed. By a design test, we found that the power loss can be reduced by pretwisting the origin blade, whether using the optimization model or the iteration method. Moreover, the optimization model has better ability than the iteration method to compensate the power loss with lower thrust coefficient while keeping the twist distribution smooth.

Keywords: aeroelastic effect; pretwisting method; power loss; optimization model; pretwist angle

---

1. Introduction

In modern times, in order to reduce environmental contamination and carbon emission, clean energy such as wind energy are quickly developing. To get more energy from wind energy, many researchers have done lots of works to improve the power output. These works include aerodynamic optimization [1,2], adding flow control devices like vortex generators (VGs) and gurney flaps (GFs) on the blade [3–5], control rules design [6,7] and so on. Besides, some researchers have even applied intelligent algorithms such as reinforcement learning (RL) [6] and evolutionary strategy algorithm [7] to increase aerodynamic performance. These studies are very useful for raising the ratio of wind energy utilization. However, power loss due to the aeroelastic effect is not dedicatedly studied.

Practically, with the increasing unit capacity of wind turbines, blades become longer and softer. Under the action of the different kinds of loads, especially aerodynamic forces, the flexible blade generates non-negligible deflections. These deflections, in turn, change the aerodynamic loads on the blade. Particularly, the induced twist due to the aeroelastic effect has a significant impact on the load and the power performance of the blade. Many researchers have studied the aeroelastic effect and found that an induced twist would not only reduce the blade load, but also decrease the power output [8–12]. This phenomenon is contrary to design objectives, one of which is to capture as much wind energy as possible. To deal with this problem, Lobitz et al. [9] suggested compensating the power loss by pretwisting the blade to adjust for coupling-induced twist. Lee et al. [11] also thought a pretwisting design of blade geometry could provide power performance superior to that of the origin blade after they found that the power loss was more than 13% in some wind speeds. Therefore, understanding how to obtain a suitable pretwist angle distribution along the blade is very important to output the maximum power. Recently, Stäblein et al. [8] presented an iteration procedure to get the pretwist angle distribution at specific wind speed. It can efficiently increase the power at the selected wind speed and make it close to the power without considering the aeroelastic effect. However, the power performances at many other wind speeds are not taken into account. In fact, the induced twist distribution changes with the operational state. This pretwist angle distribution calculated at one wind speed may not suitable for other wind speeds. Hence, it cannot get the maximum annual energy production (AEP), and it even reduces the AEP. Besides, the smooth of the blade profile, which makes the blade feasible in practical, has not been considered yet. Generally, the pretwist angle varies from the blade root to the blade tip. If the pretwist angle distribution is not restrained, after pretwisting the origin blade using these pretwist angles, the twist distribution of the blade will become out of control, and the blade may get an irregular aerodynamic profile.

Therefore, the aim of this study was to compensate for the power loss for all the wind speeds by using the pretwisting method. Firstly, an aeroelastic coupling optimization model, combining the PSO (particle swarm optimization) algorithm with the BEM-3DFEM (blade element momentum—three-dimensional finite element method) model, was built and verified. In this model, we considered the pretwist angle distribution as free variables, taking the maximum AEP (annual energy production) as the optimal object and the smooth of the twist distribution as one of the constraint conditions. Secondly, the optimization model was applied to a 100 kW wind turbine and compared with an iteration method [8], which was recently developed by another study and can well compensate the power loss at reference wind speed. Thirdly, we carried out some analysis and discussion about the results.

2. Aeroelastic Model

In this section, a steady-state aeroelastic coupling model, called the BEM-3DFEM (blade element momentum—three-dimensional finite element method) model, was developed to calculate the power output during the normal operating state.

2.1. Aerodynamic Model and Verification

There are many methods for calculating the aerodynamic performance of a wind turbine, such as the BEM (blade element momentum) method [13,14], the vortex model [14,15], and the CFD (computational fluid dynamics) model [15,16]. Among them, the BEM method is the most widely used, due to its superior comprehensive performance on the computation cost and simulation precision. In this paper, the BEM model, considering the tip loss and hub loss, was used to get the steady aerodynamic performance of the blade. This aerodynamic model was programmed by using the MATLAB language (Matlab R2012a, The MathWorks, Inc., Novi, MI, USA, 2012). To certificate this procedure, it was used to calculate the aerodynamic performance of the NREL (natioanal renewable energy laboratory) phase VI wind turbine (NREL, Golden, CO, USA) [17,18]. The results of this BEM procedure, compared with the experiment results and the results using GH Bladed software [19], are shown in Figure 1. As can be seen, the power coefficients and the powers from the above three methods are close. So, this BEM procedure can be used to evaluate the aerodynamic performance of the wind turbine blade.

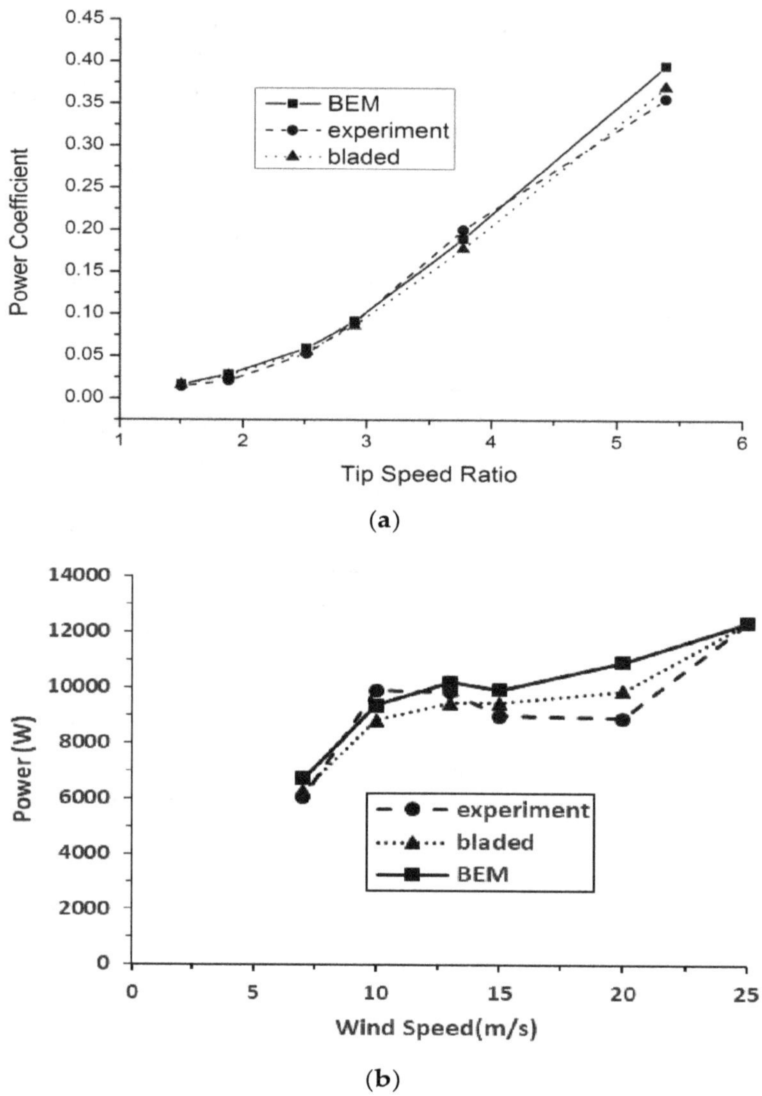

Figure 1. Aerodynamic performance of the NREL phase VI wind turbine from different methods: (**a**) Power coefficient changing with tip speed ratio; (**b**) power changing with wind speed. Note: "BEM" stands for the results obtained from the BEM procedure; "experiment" stands for the measurement results in ref [8]; "bladed" stands for the results by the use of GH Bladed software.

2.2. Structure Model and Verification

There are also lots of structure models for simulating the blade structure properties. These models can be roughly categorized into three groups. They are FEM (finite element method) model, the multi-body model, and the 1D (one-dimensional) equivalent beam model [11,15,20]. Compared with the other two models, the 3D (three-dimensional) FEM model has higher computational precision. This is because it can correctly describe the varieties of the layers in detail. Hence, the 3DFEM model was applied in this article to ensure the accuracy of structure deform analysis. The blade was built by shell elements, which are suitable to characterize composite laminates and sandwich structures (see Figure 2). To help the datum transfer between aerodynamics and the structure model, the structure model was programmed by APDL (ANSYS Parametric Design Language), which includes blade parametric model, deformation calculating, and results processing. To get an accurate result, we can see from Figure 2 that the flap-wise and edge-wise direction loads were distributed on the nodes of spar caps to avoid stress concentration. Because the root stiffnesses of the blade are strong enough to endure the blade loads and moments, the deformations at the root are very small. The six degrees of freedom at the blade root is fixed to simply the analysis.

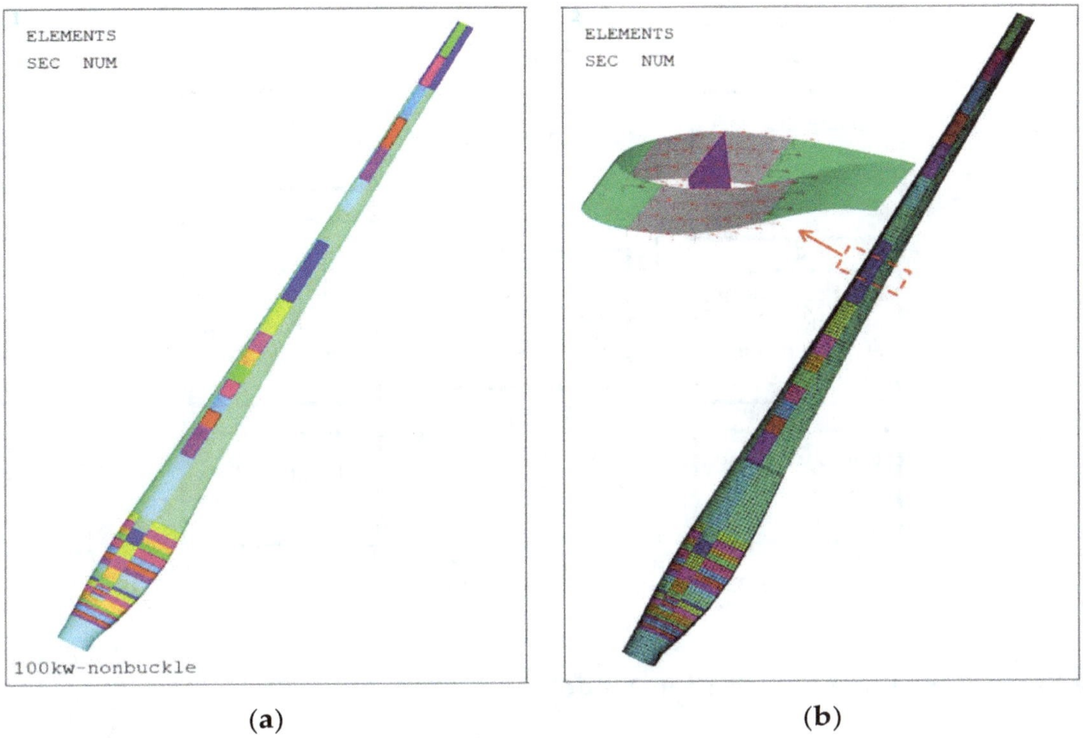

Figure 2. The 3DFEM model of a 100 kW blade: (**a**) Blade laminates in different colors; (**b**) meshes and loading of the blade model.

To verify this FEM procedure, it was also applied to the 100 kW wind turbine blade, which is subjected to a full-scale static test [21]. The numerical simulation results by using this FEM procedure, together with the test results, are shown in Table 1. We can see the tip deflections are very close. It indicates that this 3DFEM model has superior computational accuracy and can be applied to build the aeroelastic model.

Table 1. Tip deflections of a 100 kW blade.

Experiment (m)	FEM (m)	Relative Error (%)
1.369	1.330	2.8

2.3. Building the BEM-3DFEM Model

The aeroelastic process is very complicated. To analyze it, many coupling models have been developed [22–24]. The weak coupling method, due to its advantage of keeping the independence of the aerodynamic model and structure model and its expansibility, has been widely used in fluid-structure coupling analysis cases. In this paper, based on the above BEM method and 3DFEM model, an aeroelastic coupling model, named as BEM-3DFEM, was established by using a weak coupling method. The calculation process of the BEM-3DFEM is shown in Figure 3. During the calculating process, the loads were firstly calculated by the BEM method and loaded on the 3DFEM model of the blade to get the deflections of the blade. Then, the induced twist of every section used in BEM were gotten by interpolating the information of the nodes nearby. After that, the induced twist of every section were used in BEM computation to change the loads on the structure model. So, the deformations of the blade also change by the loads. If the difference of the deformation between two adjacent iterations, expressed as Δ, is less than the tolerance Δ_a, the analysis process ends. At the same time, the final pretwist angle distribution is obtained. At this point, a steady aeroelastic analysis has been done.

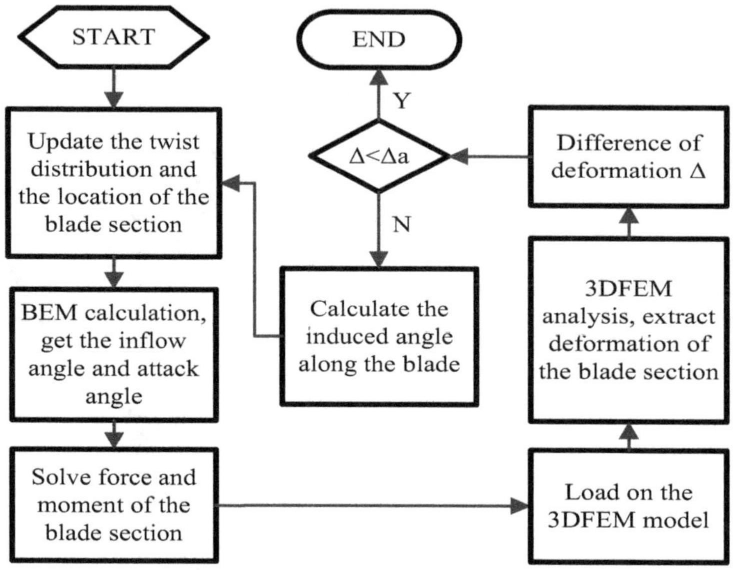

Figure 3. Flow chart of the BEM-3DFEM model.

3. An Aeroelastic Coupling Optimization Model

3.1. Objective Function

The power output is one of the most important properties of the wind turbine, so the maximum *AEP* considering the aeroelastic effect was selected as the objective function. It is shown as follows:

$$\max\{AEP\} = \max\{N_h \int_{v_{in}}^{v_{out}} P(v) \cdot f(v)dv\}$$

where v_{in} and v_{out} are cut-in and cut-out wind speed, respectively; N_h is the number of hours in a year; $P(v)$ is the power of the wind turbine at the wind speed of v, taking the aeroelastic effect into account; and $f(v)$ is the probability of the wind speed of v in a year.

In this article, the Rayleigh distribution with the average wind speed of 8 m/s in a year was used to calculate the $f(v)$.

3.2. Free Variables

The twist plays an important role in the aerodynamic design of a blade. Its change can reflect the aeroelastic effect. Hence, in this optimization model, pretwist angles at different cross-sections of the blade were selected as free variables, which are named as $x_i(i = 1, 2, 3 \cdots n)$, where x_i is the pretwist angle at section i and n is the number of the sections used to compute the power.

After getting the x_i, the origin twist at the section i is pretwisted by x_i.

3.3. Constraints

In this optimization model, to achieve the excellent aerodynamic performance of the blade, some constraints about the x_i are given.

Firstly, on the one hand, x_i should not be too big. This is because a big x_i makes the attack angle of the section i exceed the stall attack angle. On the other hand, x_i should not be too small. This leads to the aerodynamic twist becoming bigger than the allowed value, which is limited by the production and transportation. So, it is expressed as follows:

$$Lb_i - c \leq x_i \leq Ub_i + c$$

where Lb_i is the minimum induced twist from computing the twist change at the different wind speed, considering the aeroelastic effect for the origin blade; Ub_i is the maximum induced twist from computing the twist change at the different wind speed, considering the aeroelastic effect for the origin blade; and c is a constant value, which is used to expand the searching range of x_i. In this article, it was decided by the difference of the Ub_i to the stall attack angle at section i.

Secondly, to keep the smoothness of the aerodynamic shape of the blade, the twist distribution along the blade after pretwisting should meet a curve of nine times polynomial. It is written as:

$$x_{i0} - x_i = a_0 + a_1 r_i + a_2 {r_i}^2 + a_3 {r_i}^3 + a_4 {r_i}^4 + a_5 {r_i}^5 + a_6 {r_i}^6 + a_7 {r_i}^7 + a_8 {r_i}^8 + a_9 {r_i}^9$$

where x_{i0} is the twist at the section i of the origin blade; r_i corresponds to the location of the section i span-wise from root to tip; and $a_i (i = 0, 1, \cdots, 9)$ is the coefficient, which is gotten by fitting the twist at different sections after pretwisting.

3.4. Optimization Process

The flow chart of the aeroelastic coupling optimization model to get the maximum AEP, as well as the optimal pretwist angle-distribution, is given in Figure 4.

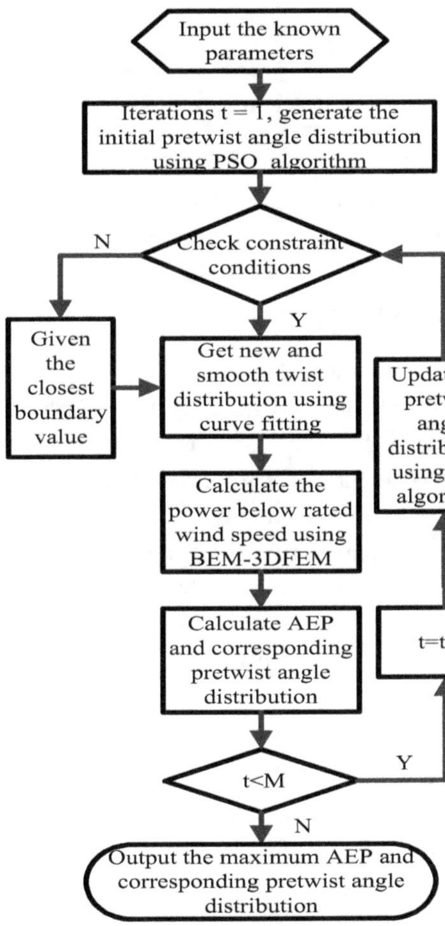

Figure 4. The flow chart of the aeroelastic coupling optimization model.

During the optimization, we generated the initial pretwist angle distribution using the PSO algorithm at first. After checking the constraints about the pretwist angle, we obtained a series of smooth twist distributions for different blades. Then, we used the BEM-3DFEM to evaluate the power performance at different wind speeds by considering the aeroelastic effect. The maximum AEP was gotten at the same probability distribution of annual wind speed. With the increase of iteration number, more suitable pretwist angle distributions were generated using the PSO algorithm [25–28]. Besides, the AEP was calculated again to find better results. While the iteration number was equal to the maximum iteration number M, the optimum AEP, as well as the corresponding pretwist angle distribution, were output.

Therefore, the pretwist angle distribution changed in every iteration by using a linear change inertia weight PSO algorithm. In this way, the attack angle of the section increased in most wind speeds. The power loss due to the aeroelastic influence was made up.

4. Results & Discussion

In this paper, a 100 kW wind turbine blade was analyzed using the above aeroelastic coupling optimization model to compensate the power loss due to the aeroelastic effect. It is a field test wind turbine operated by National Energy Wind Turbine Blade R&D Center and located in Zhangbei County, China. It is also a VSVP (variable speed variable pitch) wind turbine, like the large scale wind turbines. So, the power under the rated wind speed is mainly studied. Besides, the induced twists of the blade are small due to the relatively large stiffness of the blade. A high precision model, especially the structure model, is needed to simulate the deflection of the blade. Fortunately, the BEM-3DFEM model can efficiently reflect the aeroelastic influence of this blade because it simulates the blade layers in detail.

4.1. Pre-Assigned Variables

During the computing process, some of the parameters were fixed. They are as follows:

1. PSO algorithm, such as inertial weight, accelerating factors;
2. Aerodynamic profile, such as chord distribution, twist distribution, relative thickness distribution;
3. Structure layers, such as the number, size, location, materials; and
4. Some other parameters about the wind turbine, such as the hub height, hub radius, cut-in wind speed, cut-out wind speed, rated wind speed.

Some typical parameters among them are given in Table 2. In Table 2, the first six parameters were used in the PSO algorithm. The inertial weights and accelerating factors were selected according to the Ref. [26–28]. The maximum number of iterations and number of individuals were relatively small to reduce the calculating time due to the high computing cost of FEM analysis. The 7th to 10th parameters in this table were used to describe the 100 kW wind turbine. So, their values were fixed. The last parameter was used for evaluating the convergence of the steady aeroelastic analysis. Its value was small enough for 100 kW wind turbine to get the correct blade deflection. However, we should be concerned; this value cannot be too small as this would lead to the non-convergence of the FEM-3DFEM analysis. In addition, this value changes for different wind turbines.

Table 2. Some typical parameters used in this optimization model.

Parameter Names	Values
Maximum inertial weight	1
Minimum inertial weight	0
accelerating factors c_1	1
accelerating factors c_2	1
Maximum number of iterations M	25
Number of individuals	22
Number of the blade sections	16
Radius of the rotor (m)	10.292
Hub height (m)	26.2
Number of blades	3
Δ_a	1.00×10^{-6}

4.2. Results and Analysis

To analyze the benefits of the aeroelastic coupling optimization model, we applied the optimization model and the iteration method in Ref. [8] to the origin blade. Two new blades were gotten, and named as the PSO blade and the iteration blade. The AEP under four different cases was computed. The cases are as follows:

1. The origin blade without considering the aeroelastic coupling effect, which is represented as ucpl.;
2. The origin blade considering the aeroelastic coupling effect, which is represented as cpl.;
3. The PSO blade considering the aeroelastic effect, which is represented as PSO + cpl.; and
4. The iteration blade considering the aeroelastic effect, which is represented as pret. + cpl.

When the AEP was calculated, wind speeds from 4 to 25 m/s with an interval of 1 m/s, as well as the rated wind speed of 10.6 m/s, were used. However, only wind speeds from 4 to 10.6 m/s consider the aeroelastic effect, because the blade was used on a VSVP wind turbine. Besides, for the different reference wind speeds, different pretwisting angle distributions were gotten using the iteration method. This led to different AEP. In this article, lots of AEP using the iteration method were calculated by varying the reference wind speed from 4 to 10.6 m/s. Then, the maximum value of them, with reference wind speed of 10 m/s, was selected as the result of the iteration blade.

The twist distribution along the blade for the origin blade, the PSO blade, and the iteration blade are given in Figure 5. The relative difference in AEP for different cases compared to the ucpl. case is shown in Figure 6. We can see that the AEP of the origin blade decreases due to the aeroelastic effect. This feature is the same as the result of other researches [8–11]. In addition, the AEP of the PSO blade and the iteration blade are higher than that of the origin blade considering the aeroelastic effect. It indicates that the optimization model and the iteration method can both compensate the power loss and make the AEP close to that of the ucpl. case. Based on the varieties of the twist distribution in Figure 5, we can conclude that reducing the twists of the blade, especially that of the outboard part of the blade, leads to the increase of AEP to some extent. By comparing the twist distribution of the PSO blade with that of the iteration blade, it indicates that the AEP does not continuously rise while the twist keeps going down. This is because the ratio of the lift to drag, as well as the lift coefficient, cannot continue to increase while the attack angle gets bigger. So, there exists an optimal twist distribution to get the maximum AEP under the consideration of the aeroelastic influence.

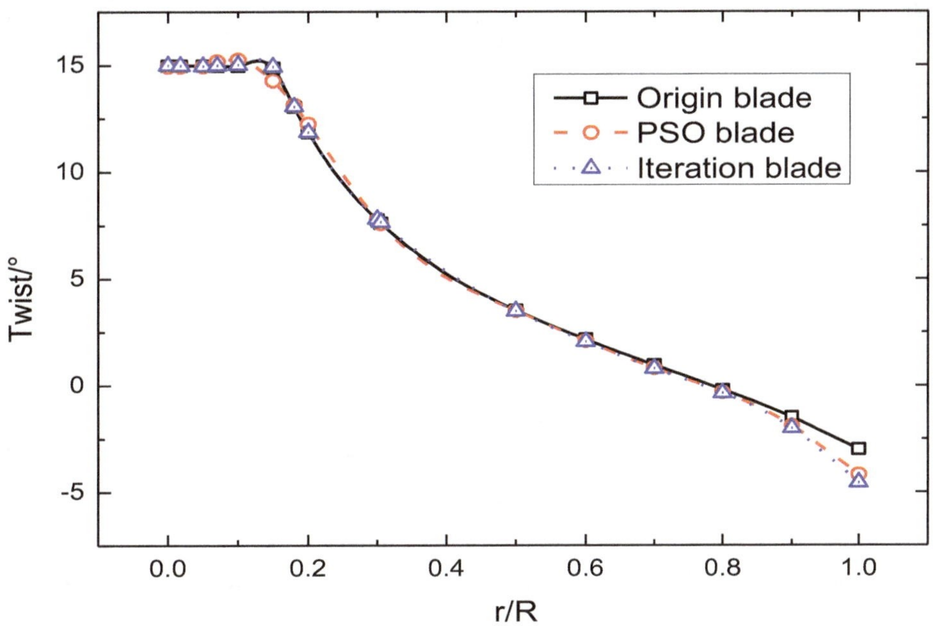

Figure 5. Twist distribution along the blade.

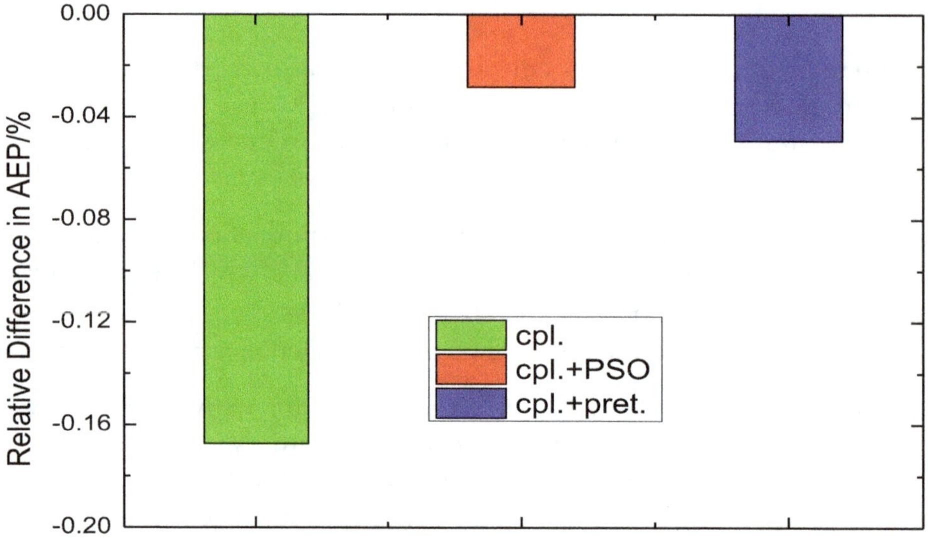

Figure 6. The relative difference in annual energy production (AEP) compared to the ucpl. Case.

From Figures 5 and 6, we can see that by using the optimization model, the PSO blade gets the maximum AEP and its twist distribution is very smooth. It shows that the optimization model is more efficient to compensate for power loss than the iteration method. Therefore, it gives us a good choice to improve the power output of the blade, while the aeroelastic effect cannot be negligible.

Figures 7 and 8 illustrate the relative difference in power and thrust coefficient compared to the cpl. Case, respectively. From Figure 7, we can see, in the case of considering the aeroelastic effect, the powers of the PSO blade and the iteration blade are higher than that of the origin blade at high wind speed. However, the comparison results about the power are converse at low wind speed. It can be concluded that the enhancements of the powers at high wind speed play the leading role in raising the AEP of the PSO blade and the iteration blade. This may be caused by the stronger aeroelastic coupling at high wind speed.

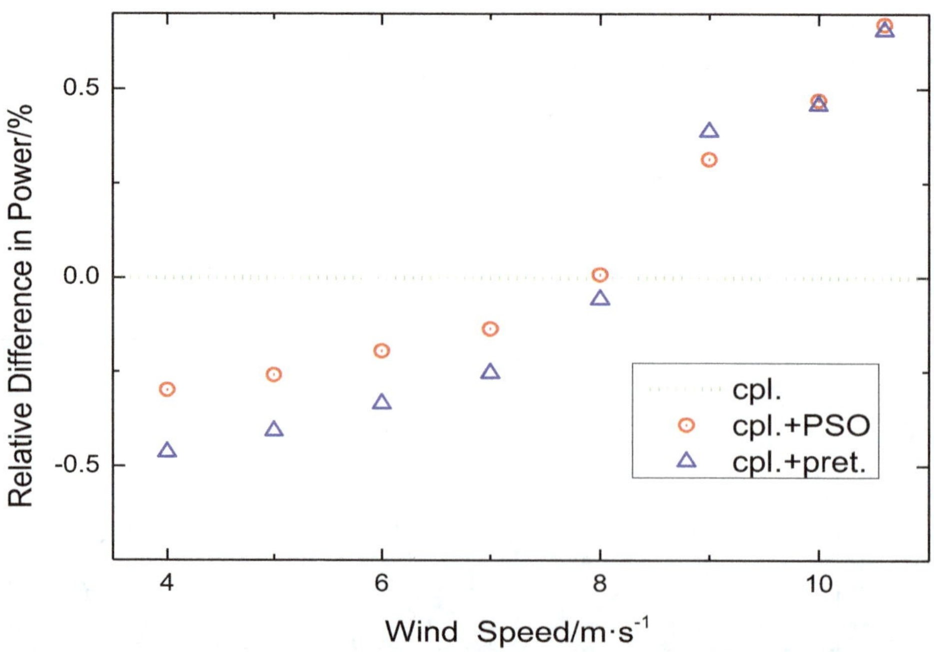

Figure 7. The relative difference in power compared to the cpl. Case.

Figure 8 shows the increase of the thrust coefficients of the PSO blade and the iteration blade at every wind speed. This is mainly due to the increase of the attack angle by pretwisting the origin blade.

In the future design, this should be a constraint condition for getting the maximum AEP. By comparing the PSO blade with the iteration blade, we can find that the thrust coefficients of the PSO blade are less than that of the iteration blade, while the powers of the PSO blade are larger than that of the iteration blade, at most of the wind speeds. This is mainly because the attack angles and the inflow angles of different sections change at the same time after pretwisting. However, the optimization model can help us to find out the more suitable values for them. This is a good advantage of the optimization model.

Certainly, the above result are based on the steady aeroelastic analysis. This is because we just applied the steady BEM model in the BEM-3DFEM. For the evaluation of AEP, this steady model is suitable. However, the wind turbine always operates in unsteady wind conditions. This needs an unsteady aerodynamic model to simulate the aerodynamic force. Hence, to calculate the aeroelastic loads in unsteady conditions, the steady BEM model needs to be corrected by using the dynamic stall model, or even replaced by the vortex method or unsteady CFD model. The BEM-3DFEM model will be improved in the future to compute the effect of the PSO blade under turbulence wind.

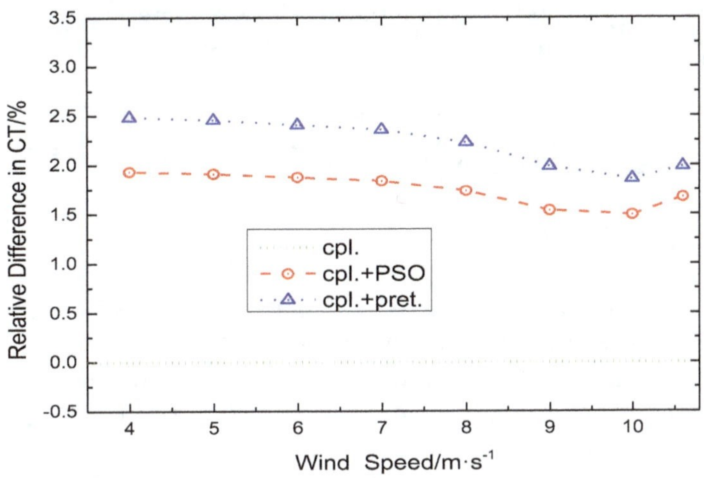

Figure 8. Relative difference in thrust coefficient compared to the cpl. Case.

5. Conclusions

In this article, to get the maximum AEP and the optimal pretwist angle distribution, an aeroelastic coupling optimization model, combining the PSO algorithm with the BEM-3DFEM model, was built to compensate power loss due to the aeroelastic effect.

Then, this aeroelastic coupling optimization model was applied to a 100 kW wind turbine blade to reduce the power loss. The iteration method developed to solve this problem from Ref. [8] was also used for the same blade. Two new blades with different aerodynamic profiles were gotten, respectively. After that, different aerodynamic properties for these two new blades, as well as the origin blade, were calculated and compared.

It is shown that the AEP of the origin blade reduces due to the aeroelastic effect. This feature is the same as the result of other researches [8–11]. In addition, the AEP of the origin blade with the aeroelastic effect can be improved by pretwisting the origin blade, whether using the aeroelastic coupling optimization model or the iteration method. Compared with the iteration method, the optimization model is more efficient to compensate for the power loss, with a lower thrust coefficient while keeping the twist distribution smooth. Therefore, it gives us a good choice to improve the power output of the origin blade, while the aeroelastic effect cannot be negligible.

According to the results, the enhancements of the powers at high wind speed play the leading role in raising the AEP for both the PSO blade and the iteration blade. This may be caused by the stronger aeroelastic process at high wind speed. So, we can conclude that the improvement of the AEP is more obvious when the wind turbine is used in a better wind resource area. However, the thrust coefficients increase after pretwisting the origin blade, since the attack angles enlarge. Hence, to avoid the thrust of the blade exceeding the allowable value, the thrust coefficient should be taken as a constraint condition while getting the maximum AEP in the future.

Besides, the results of compensating for the power loss on the 100 kW wind turbine blade are relatively small. This may be because the aeroelastic effect of the 100 kW wind turbine blade is weak. To further certify this optimization model, it will be used on some large-scale wind turbines such as the NREL 5MW reference wind turbine and the DTU (Danmarks Tekniske Universitet) 10MW reference wind turbine. Certainly, this article focuses on steady aeroelastic analysis. This is because we just applied the steady BEM model in the BEM-3DFEM. To simulate aeroelastic loads in unsteady conditions, the steady BEM model needs to be corrected by using the dynamic stall model, or replaced by the vortex method or even the unsteady CFD model. This BEM-3DFEM model will be improved in the future to compute the effect of the PSO blade under turbulence wind.

Author Contributions: Y.Z. developed most of the procedure, did the numerical computation, and wrote the article; C.L. developed the new idea, did the analysis of the results, and reviewed the article; Z.Q. developed the APDL procedure for FEM analysis and modified the article; K.Y. supervised the work and gave some useful comments for writing.

Acknowledgments: The authors would like to thank the China Scholarship Council (CSC) and the International Clean Energy Talent (iCET) Program. This work was also supported by the Strategic Priority Research Program of the Chinese Academy of Science, Grant No. XDA21050303.

References

1. Bavanish, B.; Thyagarajan, K. Optimization of power coefficient on a horizontal axis wind turbine using bem theory. *Renew. Sustain. Energy Rev.* **2013**, *26*, 169–182. [CrossRef]
2. Liu, X.W.; Wang, L.; Tang, X.Z. Optimized linearization of chord and twist angle profiles for fixed-pitch fixed-speed wind turbine blades. *Renew. Energy* **2013**, *57*, 111–119. [CrossRef]
3. Saenz-Aguirre, A.; Fernandez-Gamiz, U.; Zulueta, E.; Ulazia, A.; Martinez-Rico, J. Optimal Wind Turbine Operation By Artificial Neural Network-based Active Gurney Flap Flow Control. *Sustainability* **2019**, *11*, 2809. [CrossRef]
4. Fernandez-Gamiz, U.; Zulueta, E.; Boyano, A.; Ansoategui, I.; Uriarte, I. Five Megawatt Wind Turbine Power Output Improvements By Passive Flow Control Devices. *Energies* **2017**, *10*, 742. [CrossRef]
5. Fernandez-Gamiz, U.; Zulueta, E.; Boyano, A.; Ramos-Hernanz, J.A.; Lopez-Guede, J.M. Microtab Design And Implementation On A 5 Mw Wind Turbine. *Appl. Sci.* **2017**, *7*, 536. [CrossRef]
6. González-González, A.; Etxeberria-Agiriano, I.; Zuluet, E.; Oterino-Echavarri, F.; Lopez-Guede, J.M. Pitch Based Wind Turbine Intelligent Speed Setpoint Adjustment Algorithms. *Energies* **2014**, *7*, 3793–3809. [CrossRef]
7. Kusiak, A.; Zheng, H.Y.; Song, Z. Power optimization of wind turbines with data mining and evolutionary computation. *Renew. Energy* **2010**, *35*, 695–702. [CrossRef]
8. Stäblein, A.R.; Tibaldi, C.; Hansen, M.H. Using Pretwist to Reduce Power Loss of Bend-Twist Coupled Blades. In Proceedings of the 34th Wind Energy Symposium, San Diego, CA, USA, 4–8 January 2016; American Institute of Aeronautics & Astronautics: San Diego, CA, USA, 2016; AIAA 2016-1010.
9. Lobitz, D.W.; Veers, P.S. Load Mitigation with Bending/Twist-coupled Blades on Rotors using Modern Control Strategies. *Wind Energy* **2003**, *6*, 105–117. [CrossRef]
10. Verelst, D.R.; Larsen, T.J. *Load Consequences When Sweeping Blades—A Case Study of a 5 MW Pitch Controlled Wind Turbine*; Risø-R-Report; Risø National Laboratory for Sustainable Energy: Roskilde, Denmark, 2010; Risø-R-1724(EN).
11. Lee, Y.J.; Jhan, Y.T.; Chung, C.H. Fluid–structure interaction of FRP wind turbine blades under aerodynamic effect. *Compos. Part B Eng.* **2012**, *43*, 2180–2191. [CrossRef]
12. Bottasso, C.L.; Campagnolo, F.; Croce, A.; Tibaldi, C. Optimization-Based Study of Bend-Twist Coupled Rotor Blades for Passive and Integrated Passive/Active Load Alleviation. *Wind Energy* **2013**, *16*, 1149–1166. [CrossRef]

13. Hansen, M.O.L. *Aerodynamics of Wind Turbines*; Earthscan Publications Ltd.: London, UK, 2008.
14. Hansen, M.O.L.; Sørensen, J.N.; Voutsinas, S.; Sørensen, N.; Madsen, H.A. State of the art in wind turbine aerodynamics and aeroelasticity. *Prog. Aerosp. Sci.* **2006**, *42*, 285–330. [CrossRef]
15. Wang, L.; Liu, X.W.; Kolios, A. State of the art in the aeroelasticity of wind turbine blades: Aeroelastic modelling. *Renew. Sustain. Energy Rev.* **2016**, *64*, 195–210. [CrossRef]
16. Bai, C.J.; Wang, W.C. Review of computational and experimental approaches to analysis of aerodynamic performance in horizontal-axis wind turbines (HAWTs). *Renew. Sustain. Energy Rev.* **2016**, *63*, 506–519. [CrossRef]
17. Hand, M.M.; Simms, D.A.; Fingersh, L.J.; Jager, D.W.; Cotrell, J.R.; Schreck, S.; Larwood, S.M. *Unsteady Aerodynamics Experiment Phase VI: Wind Tunnel Test Configurations and Available Data Campaigns*; Techinical Report; National Renewable Energy Laboratory: Golden, CO, USA, 2001; NREL/TP-500-29955.
18. Simms, D.; Schreck, S.; Hand, M.; Fingersh, L.J. *NREL Unsteady Aerodynamics Experiment in the NASA-Ames Wind Tunnel: A Comparison of Predictions to Measurements*; Techinical Report; National Renewable Energy Laboratory: Golden, Colorado, USA, 2001; NREL/TP-500-29494.
19. Bossanyi, E.A. *GH Bladed Theory Manual*; Garrad Hassan and Partners Limited: Bristol, UK, 2008.
20. Hayat, K.; Ha, S.K. Load mitigation of wind turbine blade by aeroelastic tailoring via unbalanced laminates composites. *Compos. Struct.* **2015**, *128*, 122–133. [CrossRef]
21. Chen, X.; Qin, Z.W.; Yang, K.; Zhao, X.L.; Xu, J.Z. Numerical Analysis and Experimental Investigation of Wind Turbine Blades with Innovative Features: Structural Response and Characteristics. *Sci. China Tech. Sci.* **2014**, *58*, 1–8. [CrossRef]
22. Li, Y.; Castro, A.M.; Sinokrot, T.; Prescott, W.; Carrica, P.M. Coupled multi-body dynamics and CFD for wind turbine simulation including explicit wind turbulence. *Renew. Energy* **2015**, *76*, 338–361. [CrossRef]
23. Mo, W.W.; Li, D.Y.; Wang, X.N.; Zhong, C.T. Aeroelastic Coupling Analysis of the Flexible Blade of a Wind Turbine. *Energy* **2015**, *89*, 1001–1009. [CrossRef]
24. Lee, J.W.; Lee, J.S.; Han, J.H.; Shin, H.K. Aeroelastic analysis of wind turbine blades based on modified strip theory. *J. Wind Eng. Ind. Aerodyn.* **2012**, *110*, 62–69. [CrossRef]
25. Liao, C.C.; Zhao, X.L.; Xu, J.Z. Blade layers optimization of wind turbines using FAST and improved PSO algorithm. *Renew. Energy* **2012**, *42*, 227–233.
26. Shi, Y.; Eberhart, R.C. Empirical study of particle swarm optimization. Evolutionary Computation. In Proceedings of the 1999 Congress on Evolutionary Computation-CEC99, Washington, DC, USA, 6–9 July 1999; pp. 1945–1950.
27. Shi, Y.; Eberhart, R.C. A modified particle swarm optimizer. Evolutionary Computation. In Proceedings of the IEEE World Congress on Computational Intelligence, Anchorage, AK, USA, 4–9 May 1998; pp. 69–73.
28. Eberhart, R.; Kennedy, J. A New Optimizer Using Particle Swarm Theory. In Proceedings of the Sixth International Symposium on Micro Machine and Human Science, Nagoya, Japan, 4–6 October 1995; pp. 39–43.

5

Mathematical Modeling and Simulation on the Stimulation Interactions in Coalbed Methane Thermal Recovery

Teng Teng [1,2,*], Yingheng Wang [3], Xiang He [1] and Pengfei Chen [1]

1 School of Energy and Mining Engineering, China University of Mining and Technology, Beijing 100083, China
2 State Key Laboratory of Coal Resources and Safe Mining, China University of Mining and Technology, Beijing 100083, China
3 School of Electronics and Information Engineering, Tongji University, Shanghai 201804, China
* Correspondence: T.Teng@cumtb.edu.cn

Abstract: Heat stimulation of coalbed methane (CBM) reservoirs has remarkable promotion to gas desorption that enhances gas recovery. However, coalbed deformation, methane delivery and heat transport interplay each other during the stimulation process. This paper experimentally validated the evolutions of gas sorption and coal permeability under variable temperature. Then, a completely coupled heat-gas-coal model was theoretically developed and applied to a computational simulation of CBM thermal recovery based on a finite element approach of COMSOL with MATLAB. Modeling and simulation results show that: Although different heat-gas-coal interactions have different effects on CBM recovery, thermal stimulation of coalbed can promote methane production effectively. However, CBM thermal recovery needs a forerunner heating time before the apparent enhancement of production. The modeling and simulation results may improve the current cognitions of CBM thermal recovery.

Keywords: coalbed methane thermal recovery; thermal stimulation interaction; heat-gas-coal model; modeling and simulation

---

1. Introduction

Coalbed methane (CBM) is a cleaner and cheaper resource among the fossil fuels [1–3]. It is reported that the accumulated reserve of Chinese CBM is about 10 billion cubic meters, and the recoverable resource represents about 47% [4]. In earlier years, CBM is treated as a nerve-wracking hazard in coal mining. However, people's attitudes are changing that CBM becomes an efficiently and environmentally friendly fuel now. However, the production of CBM meets a great challenge of lower reservoir permeability and higher methane sorption capacity in China. A survey research shows that the permeability of Chinese coal ranges from 1×10^{-4} mD to 1×10^{-3} mD while it ranges from 0.1 mD to 1 mD of American or Australian coal [5]. As a result, the traditional direct-recovery method of CBM cannot satisfy the demands of effective production. Therefore, researchers and engineers in this field tried some unconventional methods by using a series of manual treatments to the methane reservoirs, such as the fracturing by water and gas, the displacement of adsorption by carbon dioxide and the stimulation of increasing temperature.

Thermal stimulation to deep CBM reservoirs is an effective method to promote gas recovery. Researchers have explored many different basic theories and mining technologies [6–10]. Before the production of methane, water with high temperature of 80 °C is assumed to be injected into a hypothetical CBM reservoir to causes helpful coal-gas interactions [11]. The simulation results indicate that the reservoir temperature increases 30 °C in twelve years to finally expand the methane production by 58% compared to the conventional method of direct recovery. Li et al. [12] established a mathematical model before demonstrating the complicated couplings among coalbed, methane and temperature. The model was then applied to a CBM thermal recovery of microwave heating. Research results show that thermal stimulation with microwave expands methane recovery more than 40% due to the deformation induced by gas sorption. Khoshnevis et al. [13] investigated the synergy type of gas production by injecting geothermal water. They established a three-dimension model to discuss the production potential of gas field and the influences of injection rate, reservoir permeability, saturation condition on the ultimate methane production. Lu [14] proposed a numerical simulation of heat injection into a three-dimensional temperature field by ANSYS to evidence the increasing production of CBM with temperature due to the easier desorption with higher temperature. Shahtalebi et al. [15] point out that the costs in coalbed methane thermal production limit the economic effectiveness. Only when the price of natural gas is comparatively higher and the demand for clean energy is stronger, the method of thermal production can become an economically attractive option.

CBM thermal recovery benefits a lot from the enhanced gas adsorption behavior [16–18]. Sakurovs et al. [19] presented three sets of experimental sorption evolutions at variable temperature. The results show that the methane adsorption capacity has great dependency with coal temperature. The sorption of coal to methane at different temperature ranges may indicate different behaviors, and a traditional Langmuir equation cannot describe the different trends. For example, Guan et al. [20,21] estimated the adsorption isotherms at temperatures from 283 K to 343 K. They find that the adsorption capacity of CBM decreases linearly with increasing temperature at the range of 283 K to 323 K, and keeps constant at the range of 323 K to 343 K. Similarly, Crosdale et al. [22] also observed the influence of temperature on gas desorption. Besides, at temperature lower that 30 °C, the adsorption capacity is averagely elevated by 10%, where it has no significant influences at the temperatures that higher than 60 °C. After a series of detailed theoretical and experimental research, Zhang et al. [23] concluded that the hysteresis degree was to be influenced by the microstructures of coal, especially the surface area. Further, Liu et al. [24] find that the influence of temperature on methane adsorption is more remarkable in coals that have smaller pores rather than larger pores. The adsorption capacity of methane decreases by approximately 19%, 32% and 45% for coals at micro-pore with sizes of 0.7–0.9 nm, 1.0–1.3 nm and size that larger than 1.4 nm respectively. Although many researchers have pointed out the effects of temperature on gas sorption, a theoretical presentation based on experiments that can concisely and usefully describe the effects is still lack.

Change of reservoir temperature causes a succession of interactions to CBM seepage behavior and the evolution of coal permeability [25–27]. Wang et al. [28] found that the evolution of coalbed permeability with increasing temperature may be divided into two stages: coal permeability decreases due to the internal swelling of coal matrix at the first stage and increases at the domination of rising gas pressure. Li et al. [12] performed a set of thermo-hydro-mechanical experiments to observe coal permeability by a servo-controlled equipment. The results indicate that evolutions of coal permeability with rising temperature are closely related with coal deformation. Yin et al. [29] goes further to observe the changing of coal permeability at different deformation stages. Coal permeability decreases with the rising temperature remarkably before the peak deformation, and it keeps almost constant after the peak deformation. That is because the growth of fractures in coal has more influences than the temperature on coal permeability. Focusing on the fractured bituminous coal, Perera et al. [30] investigated the permeability evolution under five different temperatures that ranges from 25 °C to 70 °C. The testing result shows that the CO_2 permeability increases linearly with temperature when coal temperature is higher than 90 °C due to the temperature-sensitive sorption behavior of CO_2. Besides, temperature change causes thermal expansion and fracturing [31]. Teng et al. explored the mechanism of these coal-gas interactions and tried to establish a permeability model for methane flow under thermo-hydro-mechanical situations. To conclude, one can find that, the above experimental results are mainly qualitative, it is necessary to propose a theoretically permeability model that can better connect the experimental rules with CBM thermal production.

CBM thermal recovery donates a multi-physical issue of coal, gas and heat [32,33]. Abed et al. [34] described a thermo-hydro-mechanical (THM) framework that suitable for modeling the behavior of unsaturated soils and rock. Due to the CBM thermal recovery method of microwave heating, Gao et al. [35] illustrated the interactions among temperature field, coalbed compaction, and methane transfer. According to their thermo-hydro-mechanical model developed in a finite element environment, higher stimulation temperature results in larger recovery radius. Fan et al. [36] discussed the competitive sorption of carbon dioxide with methane and its interactions with water under a thermo-hydro-mechanical-chemical condition. A two-phase-flow model was developed for CO_2 enhanced CBM recovery. Xia et al. [37] established a coupled hydro-thermo-mechanical model for the spontaneous combustion of underground coal and quantitatively predicted the spontaneous combustion locations of Dongtan coal mine. These precursor works have enlightening meanings for the study of multi-physical interactions in CBM thermal recovery. However, a fully coupled heat-gas-coal model is still necessary.

Following on our previous establishment of coal permeability model, this paper developed a completely coupled heat-gas-coal model for deep CBM thermal recovery by considering the interactions among coalbed deformation, methane delivery and heat transport. To evaluate these interactions among three physical fields and the production efficiency of CBM recovery, a numerical simulation using the finite element approach was validated. Finally, a series of analysis work based on the modeling and simulation are carried out.

2. Experimental Observation of Methane Sorption Under Variable Temperature

2.1. Experimental Program

In this section, the sorption characteristics of coal to pure CH_4 under variable temperature is observed in a self-developed gas adsorption apparatus, see Figure 1. Coal samples that acquired from the exploratory borehole of Jinjia lignite mine China are powdered into size of 150 μm. The experimental temperature varies from 25 °C to 85 °C, concretely 25 °C, 45 °C, 65 °C and 85 °C as representatives. At each temperature, the adsorption content was tested in a pressure vessel at the pressure that increases from the atmospheric pressure to a maximum of 10 MPa.

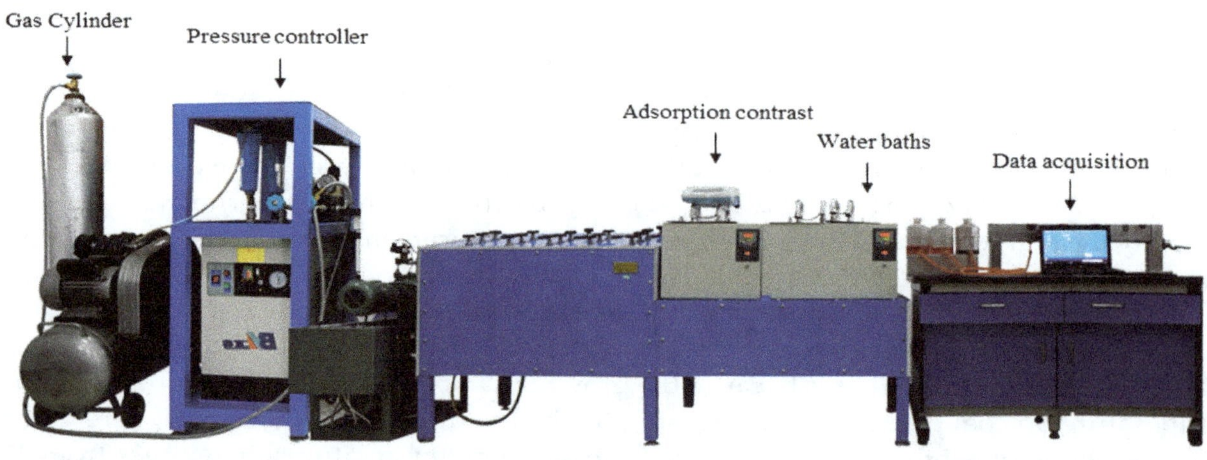

Figure 1. A self-developed isothermal gas adsorption apparatus.

2.2. *Effects of Temperature on Methane Adsorption Content*

The directly observed methane adsorption content under variable temperature is represented by the colorful spots in Figure 2. One can obtain that temperature has a significant influence on the adsorption capacity of Jinjia lignite. For example, when the temperature increases from 25 °C to

45 °C, 65 °C and 85 °C, the adsorption content at pressure of 10 MPa decreases 14.1%, 22.5%, 30.3%, respectively. It is because that the surface free energy of coal decreases when the temperature increases, as a result more CH_4 is released from the micro-pore.

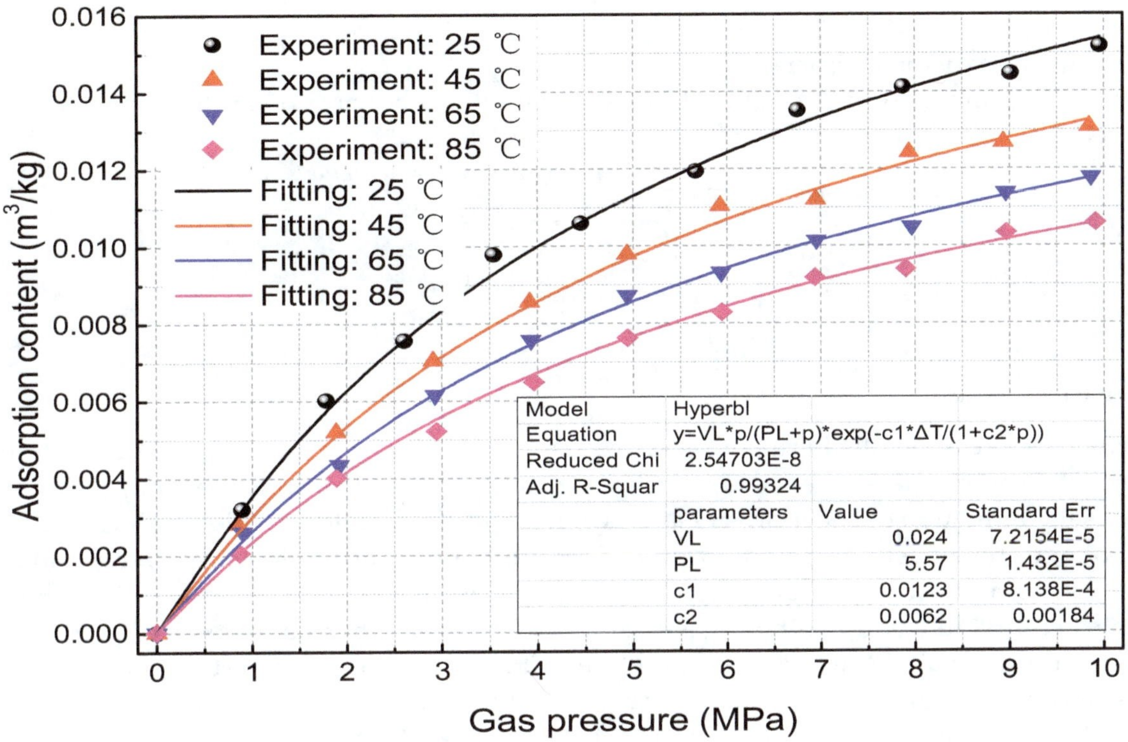

Figure 2. Validation of temperature modified Langmuir equation.

A Langmuir equation is often used to describe the methane adsorption content under constant temperature [38–40] as:

$$V_{sg} = V_L \frac{bp}{1+bp} \tag{1}$$

where, V_{sg} is the methane adsorption content, m^3/kg, V_L and b donate the Langmuir volume and pressure constants.

However, Equation (1) must be modified if the temperature changes. Here, an additional exponential term is introduced to revise the equation as:

$$V_{sg} = \frac{V_L p}{P_L + p} \exp\left[-\frac{c_2\left(T - T_{ref}\right)}{1 + c_1 p}\right] \tag{2}$$

where, P_L donates the Langmuir pressure constant, Pa. T_{ref} is the reference temperature for methane sorption, K. c_1 and c_2 are the coefficients for pressure and temperature, Pa^{-1} and K^{-1}, respectively. Figure 2 also shows the validation results of Equation (2) by the obtained experimental data. The well match of the fitting curves with the experimental results indicates that the modified Langmuir Equation (2) can effectively describe the evolution of methane sorption under variable temperature. This domino effect of gas sorption with changeable temperature indicates useful implication for CBM thermal recovery.

2.3. Volumetric Stain Induced by Methane Desorption

Coal matrix expands when it adsorbs gas. According to [41], the gas sorption-induced volumetric strain has a linear relationship with the gas adsorption content as:

$$\varepsilon_s = \alpha_{sg} V_{sg} \tag{3}$$

where α_{sg} is the expansion coefficient.

Thus, the volume deformation of coal caused by methane adsorption at different temperature can be expressed as:

$$\varepsilon_s = \frac{\alpha_{sg} V_L p}{P_L + p} \exp\left(-\frac{c_2 \Delta T}{1 + c_1 p}\right) - \frac{\alpha_{sg} V_L p_0}{P_L + p_0} \tag{4}$$

where $\Delta T = T - T_0$ shows the change of temperature.

3. Mathematical Model

Thermal recovery of CBM is a coupled process of coalbed deformation, methane delivery and heat transport. Among these three fields, one affects another. For example, the compaction of coalbed block causes the reduction of coal porosity and permeability for gas flow while the decrease of methane pressure changes the effective stress to promote coalbed deformation. Moreover, thermal expansion that caused by temperature change affects both coal deformation and gas flow. Following section is to establish a completely coupled heat-gas-coal model for CBM thermal recovery by considering the interactions among coalbed deformation, methane flow and heat transport. Before any derivation, coalbed is assumed as one kind of homogeneous, isotropic continuum while methane is treated as ideal gas. Methane transport in porous coalbed is treated as a Darcy's flow.

3.1. Coalbed Deformation Equation

Based on rock elasticity theory, the deformation of coal reservoir can be governed by a Navier-type equilibrium equation. In this paper, the equilibrium equation is optimized by the variable temperature [42] as:

$$G u_{i,kk} + \frac{G}{1 - 2\nu} u_{k,ki} - \alpha \cdot p_{,i} - K \varepsilon_{s,i} - K \alpha_T T_{,i} + F_i = 0 \tag{5}$$

in which F_i and p represent the body force of coal and the pressure of methane, MPa. $G = 0.5E/(1+\nu)$ and $K = E/(3(1-2\nu))$ are the shear and bulk modulus, respectively. E and ν represent the elasticity modulus and Poisson's ratio, respectively. $\alpha = 1 - K/K_s$, $(\alpha < 1)$ means the Biot's coefficients, where K_s donates the modulus of coal grains.

The volumetric stain that induced by temperature change is defined [43,44] as:

$$\varepsilon_T = \alpha_T \Delta T \tag{6}$$

where, α_T is the thermal expansion coefficient, K^{-1}.

3.2. Methane Flow Equation

The CBM flow in reservoirs obeys a mass conservation equation [37]:

$$\frac{\partial m}{\partial t} + \nabla \cdot (\rho_g v_g) = Q_s \tag{7}$$

in which Q_{sK} represents the methane source. $\boldsymbol{v}_g$ is the velocity vector of flow, m/s; m is the content of methane in coalbed that can be expressed as:

$$m = \rho_g \phi + \rho_c \rho_{ga} \frac{V_L p}{P_L + p} \exp\left[-\frac{c_2 \left(T - T_{ref}\right)}{1 + c_1 p} \right] \tag{8}$$

where, the first and second terms represent the free and adsorbed components, respectively. ϕ is the porosity of coal, ρ_c represents coal density while ρ_{ga} represents the methane density at standard conditions. The real gas density can be expressed as:

$$\rho_g = \frac{M_g p}{RT} \tag{9}$$

in which, R donates the universal gas constant and M_g means the molar mass.

Darcy's velocity vector $\boldsymbol{v}_g$ in Equation (7) is proportional to the pressure gradient ∇p and permeability k:

$$\boldsymbol{v}_g = \frac{-k}{\mu} \cdot \nabla p \tag{10}$$

where μ represents the viscosity coefficient of methane.

Following on our previous work [31], the permeability of coalbed can be expressed as:

$$\frac{k}{k_0} = \left(\frac{\phi}{\phi_0}\right)^3 = \left\{1 - \frac{3K}{\phi_0 K + 3K_f}\left[(\varepsilon_s + \varepsilon_T - \varepsilon_v) - (\varepsilon_{s0} + \varepsilon_{T0})\right]\right\}^3 \tag{11}$$

in which, K_f donates the equivalent modulus of coal fracture. Volume strain ε_v that induced by effective stress is expressed as:

$$\varepsilon_v = \frac{1}{K}(\overline{\sigma} + \alpha p) + \varepsilon_s + \varepsilon_T \tag{12}$$

in which, $\overline{\sigma}$ is the mean stress of coal.

Substituting Equations (4)–(6) into Equation (12), the permeability model reads as

$$\frac{k}{k_0} = \left(\frac{\phi}{\phi_0}\right)^3 = \left\{1 - \frac{3K}{\phi_0 K + 3K_f}\left[\left(\frac{\alpha_{sg} V_L p}{P_L + p} e^{-\frac{c_2 \Delta T}{1 + c_1 p}} + \alpha_T \Delta T - \varepsilon_v\right) - \frac{\alpha_{sg} V_L p_0}{P_L + p_0}\right]\right\}^3 \tag{13}$$

In the experimental testing of coal permeability, Equation (13) can be modified as:

$$\frac{k}{k_0} = \left(\frac{\phi}{\phi_0}\right)^3 = \left\{A - Be^{-C\Delta T} - D\Delta T\right\}^3 \tag{14}$$

where,

$$\begin{aligned} A &= 1 - \tfrac{3K\varepsilon_v}{\phi_0 K + 3K_f} - \tfrac{\alpha_{sg} V_L p_0}{P_L + p_0} \\ B &= \tfrac{3K}{\phi_0 K + 3K_f} \tfrac{\alpha_{sg} V_L p}{P_L + p} \\ C &= \tfrac{c_2}{1 + c_1 p} \\ D &= \tfrac{3K\alpha_T}{\phi_0 K + 3K_f} \end{aligned} \tag{15}$$

Figure 3 shows the fitting results of the experimental data [30] by the proposed permeability model. Table 1 lists the fitting parameters of Equation (14). The matching result shows that the proposed coal permeability model can be well used to describe the evolution of coal permeability under variable temperature.

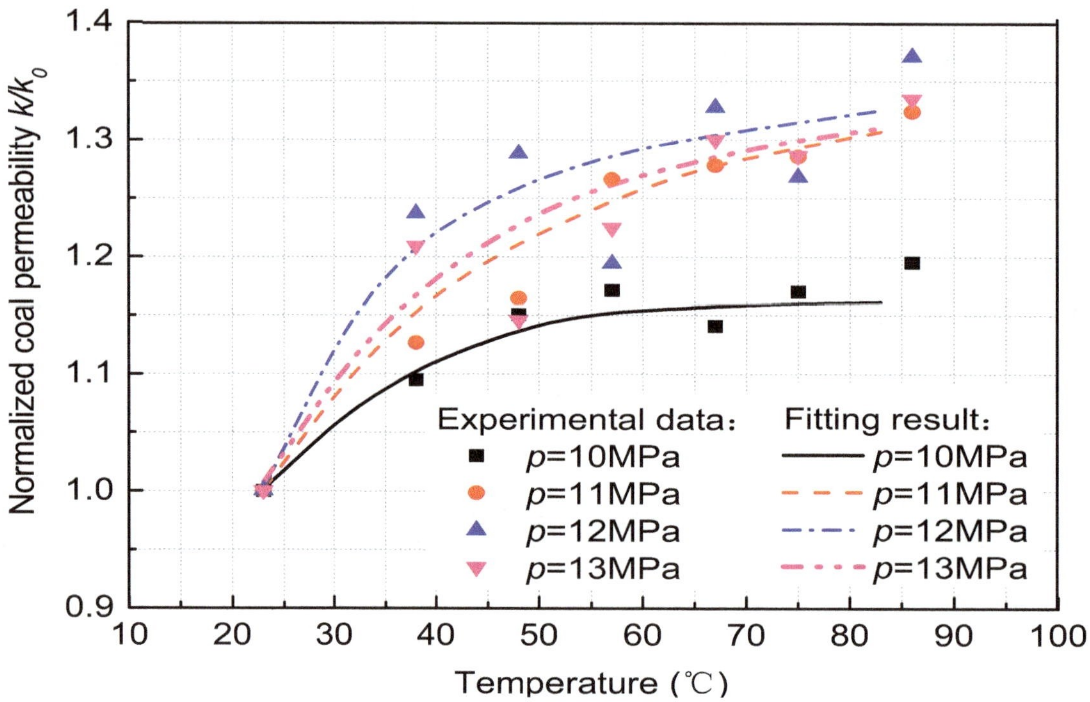

Figure 3. Validation of coal permeability model with experimental data.

Table 1. Fitting parameters for coal permeability evolution.

Coal Samples		**Fitting Parameters**				
		A	**B**	**C**	**D**	**R^2**
Gas pressure (MPa)	10	5.00×10^{-2}	0.09	6.14×10^{-4}	1.05×10^{-5}	0.915
	11	4.92×10^{-2}	0.11	8.23×10^{-4}	1.03×10^{-5}	0.854
	12	4.83×10^{-2}	0.08	7.01×10^{-4}	1.16×10^{-5}	0.975
	13	4.76×10^{-2}	0.10	1.50×10^{-3}	1.50×10^{-5}	0.962

Substituting Equations (8)–(13) into Equation (7), we can obtain the methane flow equation:

$$\begin{aligned}&\left\{\frac{\phi}{T}-\frac{N\varepsilon_L p}{T}\cdot\exp\left(-\frac{c_2\left(T-T_{ref}\right)}{1+c_1 p}\right)\cdot\left[\frac{P_L}{\left(P_L+p\right)^2}+\frac{p}{P_L+p}\cdot\frac{c_1 c_2\left(T-T_{ref}\right)}{\left(1+c_1 p\right)^2}\right]\right\}\frac{\partial p}{\partial t}\\&+\left\{\frac{N\varepsilon_L p}{T}\cdot\exp\left(-\frac{c_2\left(T-T_{ref}\right)}{1+c_1 p}\right)\cdot\left[\frac{p}{P_L+p}\cdot\frac{c_2}{\left(1+c_1 p\right)^2}\right]-\frac{\phi p}{T^2}-\frac{Np}{T}\left[\alpha_T+\frac{\vartheta\phi_o}{\left(1-\vartheta\left(T-T_0\right)\right)^2}\right]\right\}\frac{\partial T}{\partial t}\\&+\nabla\cdot\left(-\frac{kp}{\mu T}\nabla p\right)+\nabla\cdot\left[-D\nabla\left(\frac{p}{T}\phi\right)\right]+\frac{Np}{T}\frac{\partial\varepsilon_v}{\partial t}=0\end{aligned} \tag{16}$$

in which, $N = 3\phi_0 K/\left(\phi_0 K + 3K_f\right)$.

3.3. Heat Transport Equation

The conservation of energy for CBM thermal recovery obeys an equilibrium [45] as:

$$\frac{\partial\left(C_{eq}T\right)}{\partial t}+\nabla\left(-K_{eq}\nabla T\right)+p\nabla\cdot v_g+K\alpha_T T\frac{\partial\varepsilon_v}{\partial t}=Q_T \tag{17}$$

in which, the specific heat capacity of coalbed reads:

$$C_{eq}=\phi\rho_g C_g+(1-\phi)\rho_c C_c \tag{18}$$

K_{eq} is the effective thermal conductivity. C_g and C_c represent the specific heat coefficients for methane and coal. Q_T is the heat source.

Combining Equation (18) with Equation (17), one gets the modified conservation equation of energy as:

$$C_{eq}\frac{\partial T}{\partial t} + p\nabla\cdot\left(-\frac{k}{\mu}\nabla p\right) + K\alpha_T T\frac{\partial \varepsilon_v}{\partial t} = K_{eq}\nabla^2 T + \rho_g C_g \frac{k}{\mu}\nabla p \nabla T \quad (19)$$

The Equations (5), (16) and (19) make up a completely coupled heat-gas-coal model. The mathematical model is used in simulating CBM thermal production.

4. Modeling on CBM thermal Recovery

Figure 4 chooses a modeling and simulation domain to represent a half area of CBM thermal recovery from the research of Shahtalebi et al. [15]. Based on a partial-differential-equation solver of COMSOL with MATLAB that runs on Windows 10 environment with Intel Core i7-6500U and RAM 16.0 GB hardware, the simulator is implemented in 1049 s. In Figure 4, the simulation domain consists of 4208 elements. The computation time is 10^9 s that are derived into 100 steps.

The length and width of the simulation domain are one hundred meters and forty meters, respectively. The depth of methane recovery well into coalbed is seventy meters. A thermal stimulation well in size of thirty meters locates in the center. For the CBM reservoir, the initial methane pressure and temperature are 3.5 MPa and 298 K, respectively. For coal deformation, the boundary AB and BC are confined to normal force of 15 and 8 MPa, while the boundary AD and CD are constrained by normal displacement. For gas flow, the boundary AB, BC, CD and DE are symmetric boundary with no flow, while AE is common pressure boundary of 0.1 MPa. For heat transfer, the boundary AB, BC, CD and DA are thermal insulation, while FG is common temperature boundary of 373 K. To contrast simulation results, comparison sites, no. 1, 2, 3 and 4, are pre-set. Other simulation parameters are listed in Table 2.

Table 2. Parameters used in simulation of coalbed methane thermal recovery.

Variable	Parameter	Value Used	Sources
E	Young's modulus of coal, (MPa)	2713	[45]
E_s	Young's modulus of coal grains, (MPa)	4070	[45]
v	Poisson's ratio of coal	0.339	[45]
ρ_c	Density of coal, (kg/m^3)	1.25×10^3	[41]
ρ_{ga}	Density of gas at standard condition, (kg/m^3)	0.717	-
ε_L	Sorption coefficient for volumetric strain, (kg/m^3)	0.0156	[41]
k_0	Initial permeability of coal, (mD)	0.001	[31]
ϕ_0	Initial porosity of coal	0.02	Given
μ	Dynamic viscosity coefficient of gas, (Pa·s)	1.84×10^{-5}	[41]
V_L	Langmuir volume constant, (m^3/kg)	0.048	[46]
P_L	Langmuir pressure constant, (MPa)	1.57	[46]
c_1	Pressure coefficient, (MPa^{-1})	0.07	[44]
c_2	Temperature coefficient, (K^{-1})	0.02	[44]
T_a	Temperature at standard condition, (K)	273	-
T_0	Temperature of coal seam, (K)	298	Given
p_0	Initial value of gas pressure, (MPa)	3.5	Given
p_a	Pressure at standard condition, (MPa)	0.103	-
α_T	Thermal expansion coefficient, (K^{-1})	2.4×10^{-5}	[45]
C_c	Specific heat capacity of coal, (kJ/(kg·K))	1.25	[44]
C_g	Specific heat capacity of gas, (kJ/(kg·K))	1.62	[44]
K_{eq}	Effective thermal conductivity of coal, (J/(m·s·K))	0.2	[44]

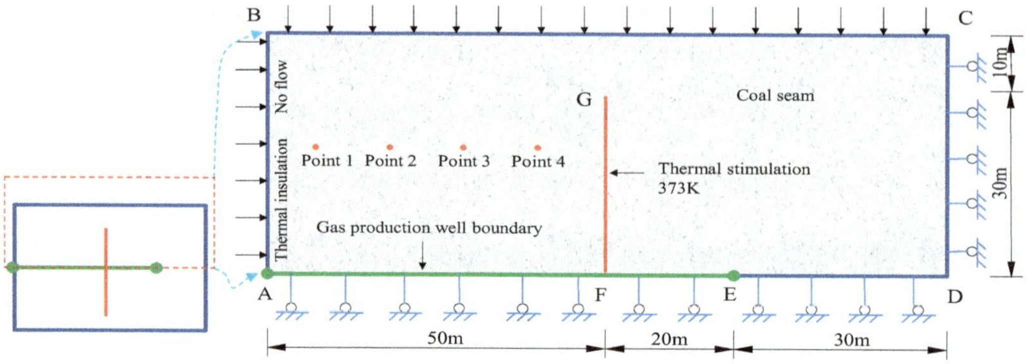

Figure 4. Numerical simulation model.

5. Modeling Results

5.1. Distribution of Coalbed Temperature

Figures 5 and 6 show the contour distribution of temperature in reservoirs during CBM thermal recovery, especially the temperature at four observation points.

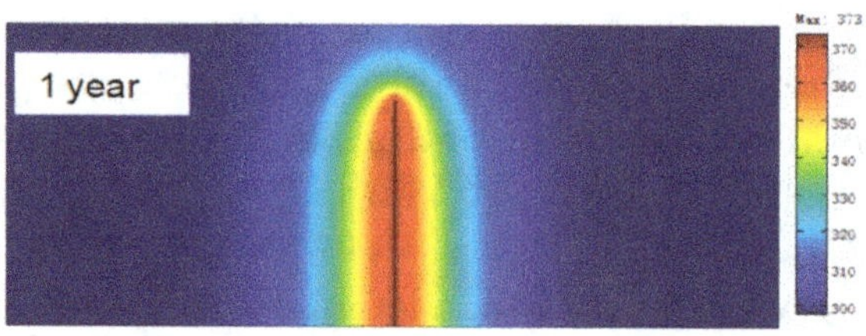

(**a**) temperature distribution after 1 year

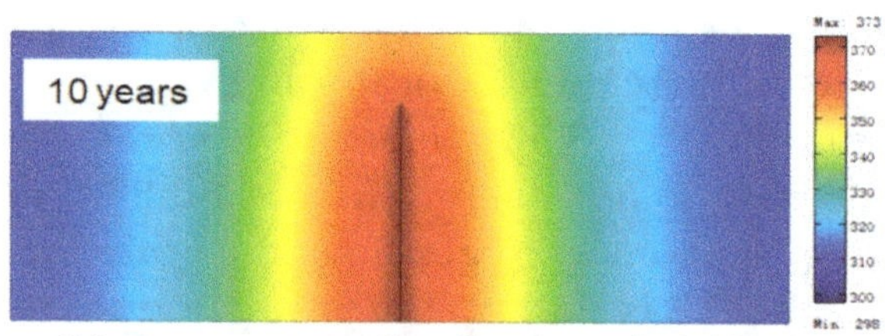

(**b**) temperature distribution after 10 years

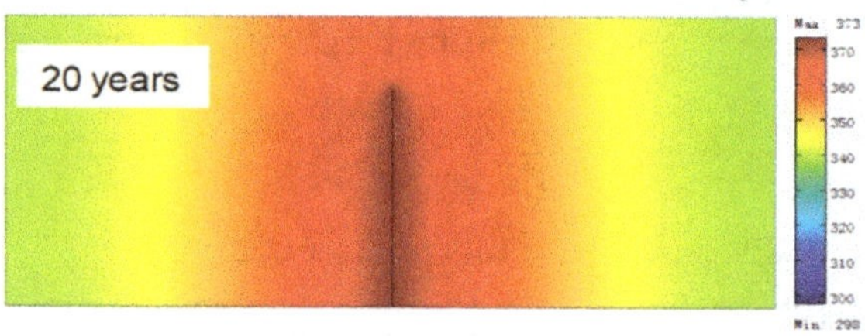

(**c**) temperature distribution after 20 years

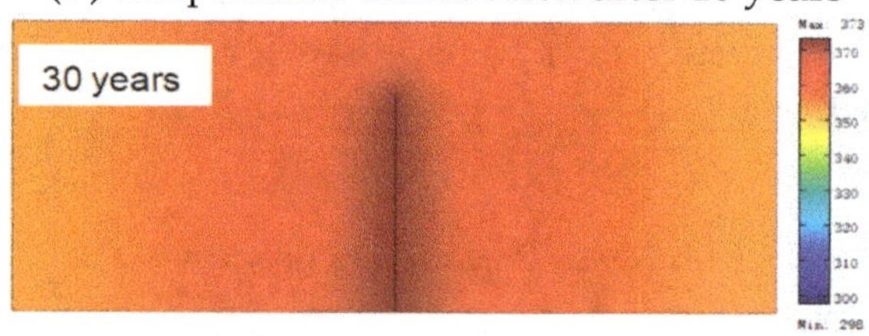

(**d**) temperature distribution after 30 years

Figure 5. Distribution of coal temperature (K) after different production years.

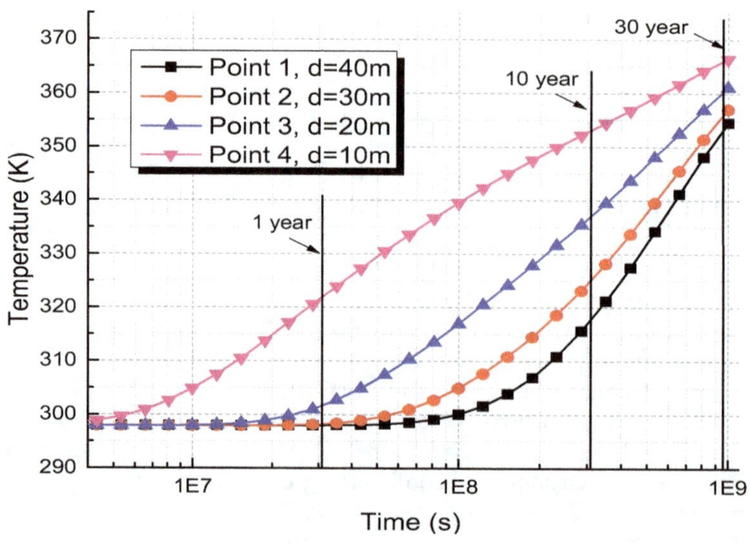

Figure 6. Evolution of reservoir temperature (K) at different observation sites.

From the figures, one can obviously find that the larger distances from thermal stimulation well correspond to lower increment of reservoir temperature. For earlier years, temperature change is only confined to the vicinity of thermal well. It indicates that CBM thermal recovery needs a long period of heating time to get beneficial output. For example, after one year (about 3×10^7 s), the reservoir temperature keeps almost it initial temperature, the temperature at point 4 is 320 K while the temperatures at point 3, 2 and 1 are lower than 300 K. However, when the heating time is longer than 10 years, obvious increment of temperature is observed in each observation point. After a heating time of 30 years, the reservoir temperature is higher than 350 K.

5.2. Evolution of Methane Pressure

The distribution of methane pressure in reservoirs changes a lot during CBM thermal recovery. It can be seen from the contour distribution of methane pressure after 1, 10, 20 and 30 years in Figure 7 that the methane pressure decreases in the first 10 years. From the figure, we can easily find that the mean gas pressure is about 2.1 MPa after 10 years recovery, whereas 0.65, 0.33 and 0.24 MPa after 20, 30 and 40 years.

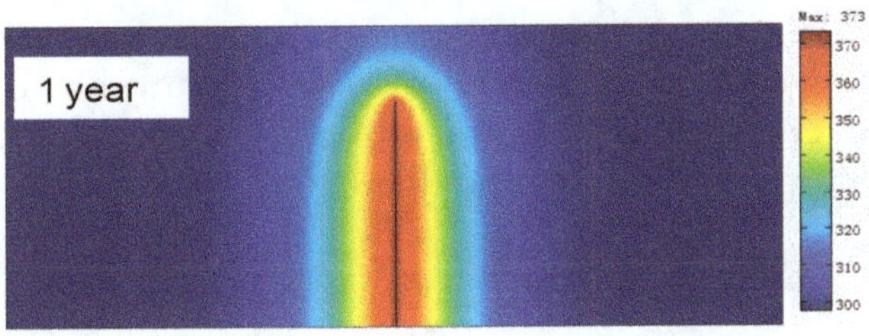

(**a**) distribution of gas pressure afterl year

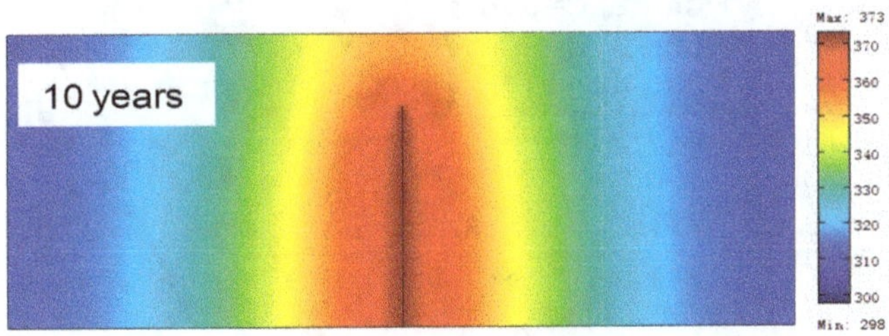

(**b**) distribution of gas pressure after10 years

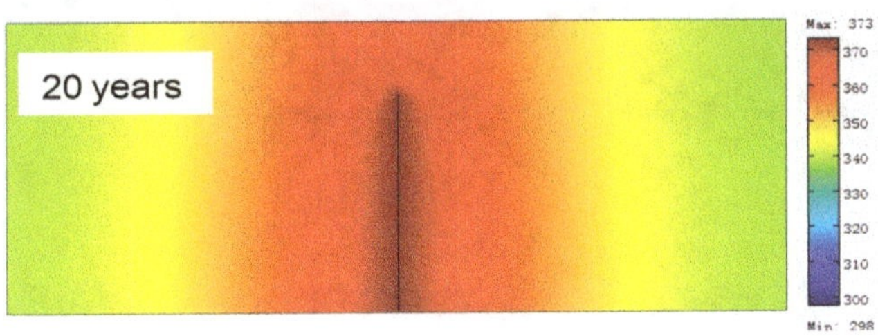

(**c**) distribution of gas pressure after 20 years

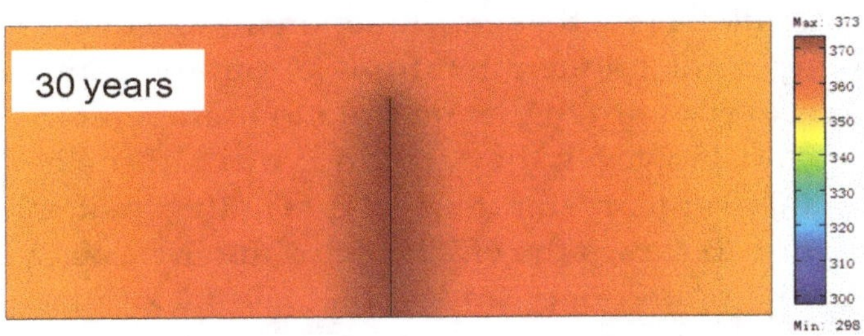

(**d**) distribution of gas pressure after 30 years

Figure 7. Distribution of gas pressure (MPa) after different production times.

According to the state equation of ideal gas, higher temperature leads to a higher gas pressure in a confined space. Figure 7 shows that higher coalbed temperature doesn't have prominent domination to methane pressure in reservoirs, even in the area near the thermal stimulation source. This is because the CBM is not confined in coal blocks seriously, but strongly transported out from the narrow fracture network.

5.3. Evolution of Coalbed Permeability

Figure 8 shows the evolution of coalbed permeability at different comparison sites. From the figure, one can draw two conclusions. The first is that coal permeability enlargers firstly and then decreases a little tiny bit. The second is that the permeability ratio increases with the decreasing distance from thermal stimulation well. Further, the normalized permeability ratio enlargers to the maximal value of about 1.53 firstly and then decreases. Although the evolution trends of normalized permeability ratio at 4 different observation points are similar, the corresponding production time node for the maximum permeability ratio delays with the increasing distance from the heat injection well, circumstantiate 8×10^8, 7×10^8, 5×10^8 and 2×10^8 s, respectively. From Equation (8), and the numerical results of Figures 6 and 8, one can explain the permeability evolution trend as that the accelerative effect of thermal desorption due to increased temperature plays a dominant role in enlarging coal permeability in the increasing stage, however the effective stress induced coal compaction gradually takes over the dominant role to reduce coal permeability in the decreasing stage where the reservoir temperature is stabilized.

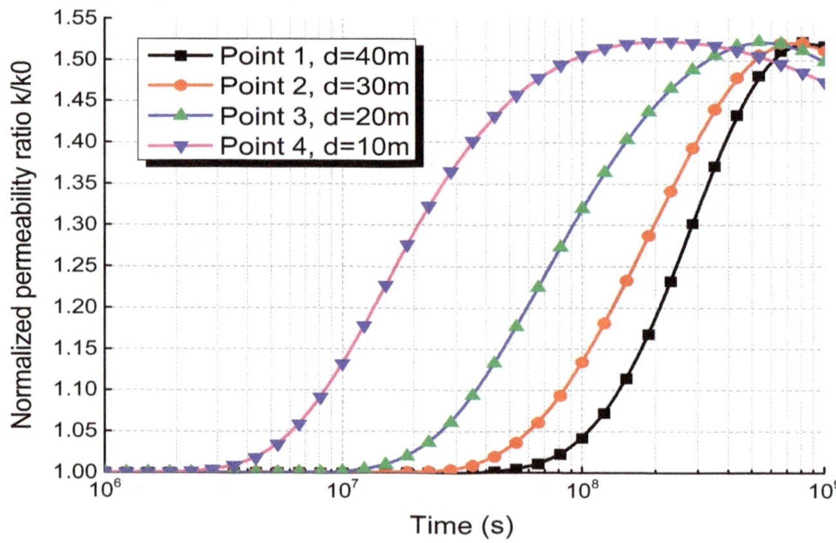

Figure 8. Evolution of coal permeability at different observation sites.

5.4. Methane Production with Different Thermal Stimulation Temperature

Figure 9a,b show the cumulative methane production and recovery efficiency in CBM thermal recovery cases with different stimulation temperatures. From the cumulative production, one can conclude that the thermal recovery with higher stimulation temperature has greater promotion to the final production of methane. However, the equal increment of coalbed temperature generates a larger promoting efficiency at lower temperature level due to the great enhancement of temperature to methane desorption. Figure 9b shows the efficiency of the increasing yield of coalbed methane thermal production. From the figure, one can see that the maximum production efficiencies of methane with stimulation temperatures of 50 °C, 75 °C and 100 °C are enlarged by 20%, 30% and 33% respectively compared with the conversional production method. However, Figure 9b also indicates that thermal recovery of coalbed methane needs a forerunner heating time to obtain apparent enhancement of production.

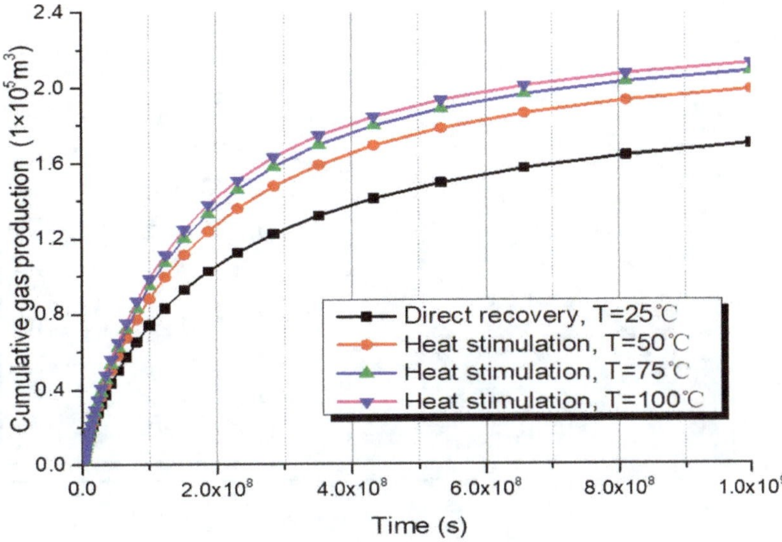

(a) Cumulative production.

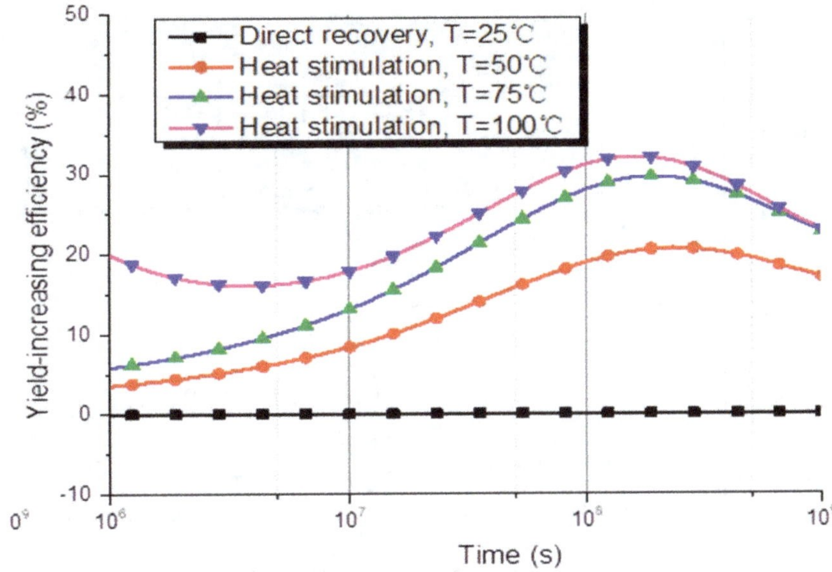

(b) production efficiency.

Figure 9. Gas production with different stimulation temperatures.

5.5. Methane Production with Different Initial Permeability

To evaluate the effects of initial reservoirs permeability on CBM thermal recovery, five numerical cases with different permeability and consistent stimulation temperature of 100 °C are carried out. Evolutions of total methane production are shown in Figure 10a. We can conclude that the methane production enlargers exponentially with production time. To measure the production efficiency, we define an index of output proportion as a percentage of cumulative production to total content of reservoir. Figure 10b shows the evolution of production time with the initial permeability when the output proportion percentage takes value of 25%, 50%, 75% and 90%, respectively. From Figure 10, one can obtain that larger value of initial permeability represents higher methane production in earlier time. For example, it takes 17 and 0.5 years respectively to recover 75% content of the total reserve when the reservoir permeability is 5×10^{-19} m^2 and 5×10^{-17} m^2, respectively.

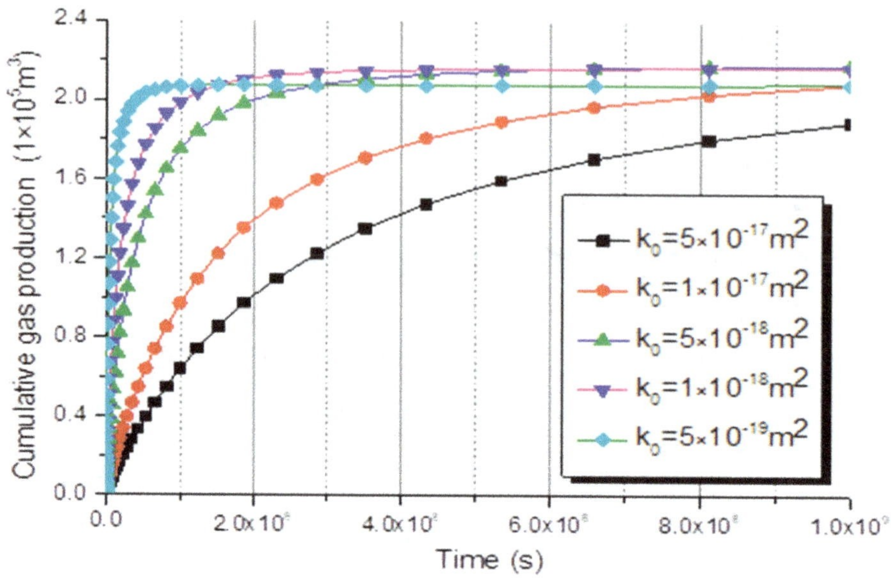

(**a**) Cumulative production.

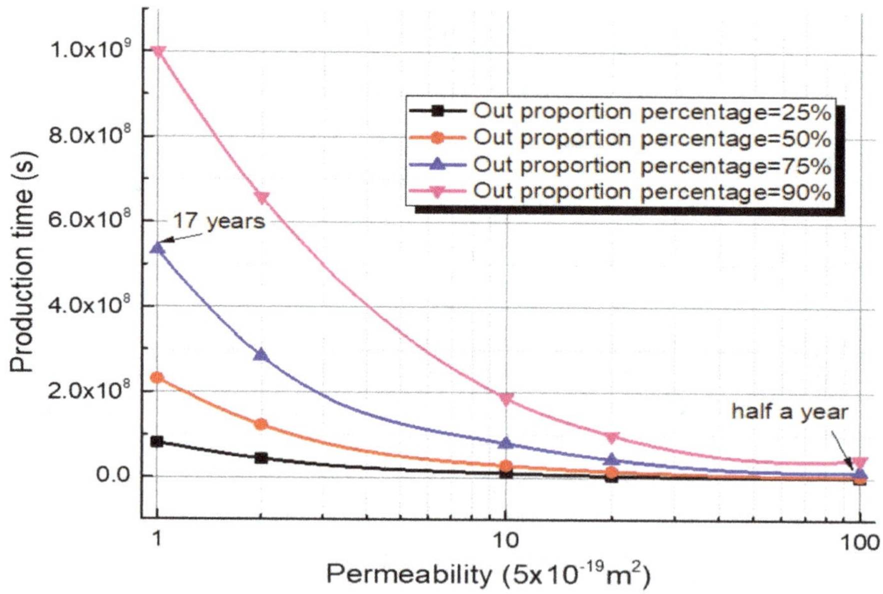

(**b**) production time for different output proportion

Figure 10. Gas production with different initial reservoir permeability.

6. Conclusions

Methane adsorption and coal permeability decrease with increasing temperature. This paper experimentally validated the evolutions of gas sorption and coal permeability under variable temperature. It also established a completely coupled heat-gas-coal model including multi-physics of coalbed deformation, methane delivery and heat transport. The mathematical model was applied to a computational simulation of CBM thermal recovery and solved by COMSOL with MATLAB in a finite element approach environment. To evaluate the thermal recovery process and the production efficiency, a series of analysis work were carried out. The results show that: (1) Thermal stimulation with higher temperature contributes to greater promotion to CBM recovery production. However, it needs a forerunner heating time before an apparent enhancement of production. (2) The normalized coal permeability ratio increases to the maximal value of about 1.53 firstly, and then decreases slightly.

The corresponding production time node for the maximum permeability ratio delays with the distance from thermal stimulation well. This completely coupled heat-gas-coal model can improve the current understandings of coalbed methane thermal recovery.

Author Contributions: T.T. established the model and performed the numerical simulation; X.H. wrote the original draft preparation; Y.W. and P.C. revised the paper.

References

1. Al-Jubori, A.; Johnston, S.; Boyer, C.; Lambert, S.W.; Bustos, O.A.; Pashin, J.C.; Wray, A. Coalbed methane: Clean energy for the world. *Oilfield Rev.* **2009**, *21*, 4–13.
2. Yuan, L. Theory and practice of integrated coal production and gas extraction. *Int. J. Coal Sci. Technol.* **2015**, *2*, 3–11. [CrossRef]
3. Towler, B.; Firouzi, M.; Underschultz, J.; Rifkin, W.; Garnett, A.; Schultz, H.; Witt, K. An overview of the coal seam gas developments in Queensland. *J. Nat. Gas Sci. Eng.* **2016**, *31*, 249–271. [CrossRef]
4. Zhou, F.; Xia, T.; Wang, X.; Zhang, Y.; Sun, Y.; Liu, J. Recent developments of coal mine methane extraction and utilization in china: A review. *J. Nat. Gas Sci. Eng.* **2016**, *31*, 437–458. [CrossRef]
5. Noack, K. Control of gas emissions in underground coal mines. *Int. J. Coal Geolo.* **1998**, *35*, 57–82. [CrossRef]
6. Slastunov, S.V.; Prezent, G.M.; Kolikov, K.S. Experimental work on methane recovery at the Lenin Mine of the Karaganda Basin. *J. Min. Sci.* **1999**, *35*, 531–535. [CrossRef]
7. Mutyala, S.; Fairbridge, C.; Paré, J.J.; Bélanger, J.M.; Ng, S.; Hawkins, R. Microwave applications to oil sands and petroleum: A review. *Fuel Process. Technol.* **2010**, *91*, 127–135. [CrossRef]
8. Singh, A.K.; Baumann, G.; Henninges, J.; Görke, U.J.; Kolditz, O. Numerical analysis of thermal effects during carbon dioxide injection with enhanced gas recovery: A theoretical case study for the Altmark gas field. *Environ. Earth Sci.* **2012**, *67*, 497–509. [CrossRef]
9. Wang, H.; Ajao, O.; Economides, M.J. Conceptual study of thermal stimulation in shale gas formations. *J. Nat. Gas Sci. Eng.* **2014**, *21*, 874–885. [CrossRef]
10. Xue, L.; Dai, C.; Wang, L.; Chen, X. Analysis of Thermal Stimulation to Enhance Shale Gas Recovery through a Novel Conceptual Model. *Geofluids* **2019**, *2019*. [CrossRef]
11. Salmachi, A.; Haghighi, M. Feasibility study of thermally enhanced gas recovery of coal seam gas reservoirs using geothermal resources. *Energy Fuels* **2012**, *26*, 5048–5059. [CrossRef]
12. Li, B.; Yang, K.; Xu, P.; Xu, J.; Yuan, M.; Zhang, M. An experimental study on permeability characteristics of coal with slippage and temperature effects. *J. Pet. Sci. Eng.* **2019**, *175*, 294–302. [CrossRef]
13. Khoshnevis, N.; Khosrokhavar, R.; Nick, H.; Bruhn, D.F.; Bruining, H. Injection of disposal water from a geothermal reservoir into a gas reservoir. *J. Pet. Sci. Eng.* **2019**, *178*, 616–628. [CrossRef]
14. Lu, B.X. Numerical simulations of temperature field of coalbed methane with heat injection based on ANSYS. In Proceedings of the 2012 International Conference on Computer Science and Electronics Engineering, Hangzhou, China, 23–25 March 2012; Volume 2.
15. Shahtalebi, A.; Khan, C.; Dmyterko, A.; Shukla, P.; Rudolph, V. Investigation of thermal stimulation of coal seam gas fields for accelerated gas recovery. *Fuel* **2016**, *180*, 301–313. [CrossRef]
16. Li, H.; Kang, J.; Zhou, F.; Qiang, Z.; Li, G. Adsorption heat features of coalbed methane based on microcalorimeter. *J. Loss Prev. Process Ind.* **2018**, *55*, 437–449. [CrossRef]
17. Miklová, B.; Staf, M.; Kyselová, V. Influence of ash composition on high temperature CO2 sorption. *J. Environ. Chem. Eng.* **2019**, *7*, 103017. [CrossRef]
18. Okolo, G.N.; Everson, R.C.; Neomagus, H.W.; Sakurovs, R.; Grigore, M.; Bunt, J.R. The carbon dioxide, methane and nitrogen high-pressure sorption properties of South African bituminous coals. *Int. J. Coal Geol.* **2019**, *209*, 40–53. [CrossRef]
19. Sakurovs, R.; Day, S.; Weir, S.; Duffy, G. Temperature dependence of sorption of gases by coals and charcoals. *Int. J. Coal Geol.* **2008**, *73*, 250–258. [CrossRef]
20. Charrière, D.; Pokryszka, Z.; Behra, P. Effect of pressure and temperature on diffusion of CO2 and CH4 into coal from the Lorraine basin (France). *Int. J. Coal Geol.* **2010**, *81*, 373–380. [CrossRef]
21. Guan, C.; Liu, S.; Li, C.; Wang, Y.; Zhao, Y. The temperature effect on the methane and CO2 adsorption capacities of Illinois coal. *Fuel* **2018**, *211*, 241–250. [CrossRef]

22. Crosdale, P.J.; Moore, T.A.; Mares, T.E. Influence of moisture content and temperature on methane adsorption isotherm analysis for coals from a low-rank, biogenically-sourced gas reservoir. *Int. J. Coal Geol.* **2008**, *76*, 166–174. [CrossRef]
23. Zhang, L.; Aziz, N.; Ren, T.X.; Wang, Z. Influence of temperature on coal sorption characteristics and the theory of coal surface free energy. *Procedia Eng.* **2011**, *26*, 1430–1439. [CrossRef]
24. Liu, Y.; Zhu, Y.; Liu, S.; Li, W.; Tang, X. Temperature effect on gas adsorption capacity in different sized pores of coal: Experiment and numerical modeling. *J. Pet. Sci. Eng.* **2018**, *165*, 821–830. [CrossRef]
25. Akbarzadeh, H.; Chalaturnyk, R.J. Structural changes in coal at elevated temperature pertinent to underground coal gasification: A review. *Int. J. Coal Geol.* **2014**, *131*, 126–146. [CrossRef]
26. Niu, S.; Zhao, Y.; Hu, Y. Experimental ivestigation of the temperature and pore pressure effect on permeability of lignite under the in situ condition. *Trans. Porous Media* **2014**, *101*, 137–148. [CrossRef]
27. Zhang, X.G.; Ranjith, P.G.; Perera, M.S.A.; Ranathunga, A.S.; Haque, A. Gas transportation and enhanced coalbed methane recovery processes in deep coal seams: A review. *Energy Fuels* **2016**, *30*, 8832–8849. [CrossRef]
28. Wang, K.; Du, F.; Wang, G. Investigation of gas pressure and temperature effects on the permeability and steady-state time of Chinese anthracite coal: An experimental study. *J. Nat. Gas Sci. Eng.* **2017**, *40*, 179–188. [CrossRef]
29. Yin, G.; Jiang, C.; Wang, J.G.; Xu, J. Combined effect of stress, pore pressure and temperature on methane permeability in anthracite coal: An experimental study. *Trans. Porous Media* **2013**, *100*, 1–16. [CrossRef]
30. Perera, M.S.A.; Ranjith, P.G.; Choi, S.K.; Airey, D. Investigation of temperature effect on permeability of naturally fractured black coal for carbon dioxide movement: An experimental and numerical study. *Fuel* **2012**, *94*, 596–605. [CrossRef]
31. Teng, T.; Wang, J.G.; Gao, F.; Ju, Y.; Jiang, C. A thermally sensitive permeability model for coal-gas interactions including thermal fracturing and volatilization. *J. Nat. Gas Sci. Eng.* **2016**, *32*, 319–333. [CrossRef]
32. Pandey, S.N.; Chaudhuri, A.; Kelkar, S. A coupled thermo-hydro-mechanical modeling of fracture aperture alteration and reservoir deformation during heat extraction from a geothermal reservoir. *Geothermics* **2017**, *65*, 17–31. [CrossRef]
33. Tong, F.; Jing, L.; Zimmerman, R.W. A fully coupled thermo-hydro-mechanical model for simulating multiphase flow, deformation and heat transfer in buffer material and rock masses. *Int. J. Rock Mech. Min. Sci.* **2010**, *47*, 205–217. [CrossRef]
34. Abed, A.A.; Sołowski, W.T. A study on how to couple thermo-hydro-mechanical behaviour of unsaturated soils: Physical equations, numerical implementation and examples. *Comput. Geotech.* **2017**, *92*, 132–155. [CrossRef]
35. Gao, F.; Xue, Y.; Gao, Y.; Zhang, Z.; Teng, T.; Liang, X. Fully coupled thermo-hydro-mechanical model for extraction of coal seam gas with slotted boreholes. *J. Nat. Gas Sci. Eng.* **2016**, *31*, 226–235. [CrossRef]
36. Fan, C.; Elsworth, D.; Li, S.; Zhou, L.; Yang, Z.; Song, Y. Thermo-hydro-mechanical-chemical couplings controlling CH4 production and CO2 sequestration in enhanced coalbed methane recovery. *Energy* **2019**, *173*, 1054–1077. [CrossRef]
37. Xia, T.; Zhou, F.; Liu, J.; Kang, J.; Gao, F. A fully coupled hydro-thermo-mechanical model for the spontaneous combustion of underground coal seams. *Fuel* **2014**, *125*, 106–115. [CrossRef]
38. Zhang, R.; Liu, S. Experimental and theoretical characterization of methane and CO2 sorption hysteresis in coals based on Langmuir desorption. *Int. J. Coal Geol.* **2017**, *171*, 49–60. [CrossRef]
39. Rani, S.; Padmanabhan, E.; Prusty, B.K. Review of gas adsorption in shales for enhanced methane recovery and CO2 storage. *J. Pet. Sci. Eng.* **2018**, *175*, 634–643. [CrossRef]
40. Kim, J.; Jang, Y.; Seomoon, H.; Lee, H.; Sung, W. A new sorption-corrected deconvolution method for production data analysis in a shale gas reservoir containing adsorbed gas. *Int. J. Oil Gas Coal Technol.* **2019**, *20*, 55–68. [CrossRef]
41. Zhang, H.; Liu, J.; Elsworth, D. How sorption-induced matrix deformation affects gas flow in coal seams: A new FE model. *Int. J. Rock Mech. Min. Sci.* **2008**, *45*, 1226–1236. [CrossRef]
42. Wu, Y.; Liu, J.; Chen, Z.; Elsworth, D.; Pone, D. A dual poroelastic model for CO_2-enhanced coalbed methane recovery. *Int. J. Coal Geol.* **2011**, *86*, 177–189. [CrossRef]
43. Liang, B. Study on temperature effects on the gas absorption performance. *J. Heilongjiang Min. Inst.* **2000**, *10*, 20–22.

44. Zhu, W.C.; Wei, C.H.; Liu, J.; Qu, H.Y.; Elsworth, D. A model of coal-gas interaction under variable temperatures. *Int. J. Coal Geol.* **2011**, *86*, 213–221. [CrossRef]
45. Qu, H.; Liu, J.; Chen, Z.; Wang, J.; Pan, Z.; Connell, L.; Elsworth, D. Complex evolution of coal permeability during CO2 injection under variable temperatures. *Int. J. Greenh. Gas Control* **2012**, *9*, 281–293. [CrossRef]
46. Ibrahim, A.F.; Nasr-El-Din, H.A. A comprehensive model to history match and predict gas/water production from coal seams. *Int. J. Coal Geol.* **2015**, *146*, 79–90. [CrossRef]

6

Optimized Energy Management Strategies for Campus Hybrid PV–Diesel Systems during Utility Load Shedding Events

Jacques Maritz

Department of Engineering Sciences, University of the Free State, P.O. Box 339, Bloemfontein 9300, South Africa; maritzjm@ufs.ac.za

Abstract: The unique situation of utility power curtailment unveils opportunities in the fields of energy management and digital resource management. During utility load shedding events, campuses are typically driven as Photo Voltaic (PV)–diesel generator hybrid systems, of which the main fossil resource driver is diesel. With the appropriate Supervisory Control and Data Acquisition (SCADA) systems, discrete departmental energy policies along with control, forecasting and Internet of Things (IoT) infrastructure, the campus hybrid system could be optimized on a short timescale during the shedding event. In this paper the optimization methodology, required technology infrastructure, possible forecasting algorithms and potential implementation will be discussed.

Keywords: hybrid PV–diesel generator systems; digital resource management; energy management

---

1. Introduction

The scenario of power loss from the grid, whether for short or long periods, places a unique operational constraint on the electric network of a campus [1]. Typically, this mode of operation is governed by a hybrid PV–diesel generator system, that allows ample room for optimization of its main resource metric, namely fossil fuel [2]. Campus curtailment and demand side management programs could benefit from this scheme, especially when energy managers engage in campus disconnect from utility. The choice of hybrid Photo Voltaic (PV)–diesel generator system for optimization in this paper is based on local utilization trends and that diesel generators are considered an underutilized asset. The optimization framework of modern hybrid systems is fed by several elements that culminate in the field of digital resource management, such as energy Internet (EI) [3], energy Internet of Things (EIoT), energy forecasting techniques (EFT) [4], intelligent electrical control systems [5], the ability to digitally manage the consumer's energy purchasing strategy [5] and coherent energy data formats generated by both loads and generators. However, few high-resolution digital energy management strategies exist that make use of energy forecasting (with forecasting uncertainties) to optimize fuel usage, while simultaneously increasing PV penetration of the electrified campus grid on a small time-step basis.

Hybrid generation systems on campuses mimic the setup of micro-grids, in islanded operational mode (loss of utility power), where loads and generators interact in an intelligent mode to generate savings and possibly other avenues of revenue for campuses [6], without feeding back into the powerless grid (for situations where feedback is not allowed). As part of such a campus micro grid of the future, it is assumed that the hypothetical campus presented in this paper, is interwoven with IoT sensors (multi-faceted) and equally capable communication networks [7]. The EIoT infrastructure on campus injects unique data streams into the energy management (EM) local/cloud server that could be leveraged for the purposes of enhancement of decision-making structures regarding energy usage, dispatch and forecasting on campus. For example, the EIoT network could feed departmental energy managers with information about campus dynamics that would, in turn, influence their scheduling of lecture halls (or partial use) and the utilization of research lab equipment. The same argument holds for Heating, Ventilation and Air-conditioning (HVAC) scheduling of lecture halls based on the local weather. The campus EIoT network therefore would be able to supply rich layers of information that could feed decision support systems used by departmental energy managers (DEMs) and lead campus energy managers (LCEMs). An EIoT network on campus also feeds the concept of the energy Internet, whereby all consumers and generators are connected by some efficient communication network (such as Zigbee) and allows for full control, high resolution data acquisition, energy data analysis, energy management and departmental energy policy monitoring (all while communicating in an energy efficient manner) [7].

Real-time and optimized energy management of smart campuses relies on the full control of the campus energy system, real-time energy data acquisition and demand/generation horizon forecasting. This type of energy generation/demand forecasting is driven by several state-of-the-art machine learning techniques that leverage the EI [8]. The forecasting techniques rely on historical data with different training set lengths (depending on the forecasting method) in order to forecast energy usage/generation with uncertainty based on a plethora of different training metrics and unique data streams. A typical example of such a forecasting strategy is that of solar irradiance based on the underlying metrics of humidity, temperature and cloud cover as can be found in Reference [9]. Complex connections exist between different metrics that can only be recovered using machine learning techniques that learn from experience, given a typical large energy dataset. As a state-of-the-art example, consider the scenario where LCEMs strategize a particular renewable dispatch strategy based on historic data and half-hourly newly acquired data. The LCEMs train models on historic data and uses the newly acquired data to firm up on the energy dispatch system. It remains inefficient to retrain the entire model on the historic data and newly added data, thus the LCEMs use online forecasting methods (such as online Gaussian Process (GP) forecasting methods that are based on multi in–multi out systems) to retrain only on the new acquired data. In this manner, LCEMs can firm up their energy dispatch system, whilst remaining computationally efficient with quick response strategies [10].

The nervous system of a smart campus will typically by governed by an electrical control system (Building Automated/Management System), to ensure that a high resolution of control is achieved when imposing energy management strategies and critical control. This nervous system interconnects the largest consumers such as HVAC, hot water supply (part of HVAC) and research labs [11]. High-resolution low-level departmental control becomes more expensive due to either non-existing control circuits or mandatory control retrofits, thus in many cases a control retrofit is needed [12]. However, building automated systems (BASs) will be able to perform macro control of the largest energy consumers.

A critical concept in digital resource management of smart campuses is that of coherent energy data structures and exchange formats. As an example, the campus building control systems typically generate data that aids decision support systems in shifting or eliminating certain electrical loads during critical periods. Separate from the building control systems, the PV generation data is shared on a separate platform and so is the distributed diesel generators (both control and electrical output). However, when dealing with such large energy ecosystems such as campuses, the energy management strategies become largely data driven and depend on coherent multi-faceted data streams culminated on a single platform. The energy data analysis backend is hosted on the same platform, allowing for easy debugging and amendment [13].

Smart trading of resources feature in several markets and will be equally beneficial as part of the smart campus scenario (one smart campus selling to a neighboring smart campus). A smart campus with the necessary digital infrastructure (EIoT, EI, EM etc.) will easily overcome the requirements to manage its own resources and interact with other receiving clients [5]. In order to trade resources in real-time, a campus needs to be able to control and forecast resource usage and generation with certainty. Added to this requirement, the LCEMs need to utilize the EIoT data influx in a real-time manner that adapts to market responses and predictions [14].

This paper proposes an umbrella scheme/concept that utilizes smart campus elements in order to optimize fossil fuel usage associated with PV–diesel generator hybrid systems, from the perspective of energy forecasting techniques, energy campus IoT and the necessary control systems. Special focus was given towards the critical role-players and data flow within the proposed scheme. The emphasis of this paper is therefore on the formulation and discussion of the umbrella optimization scheme (not to determine the equation of energy utilization) and attempt to include all the possible elements that are considered role-players within the final optimization. Future work will include the process of finding the function of energy utilization based on real campus data that will be conditioned to feed the hypothetical scheme. The mathematical framework (or data driven methodology) will follow future testing of the scheme on real campus data.

2. Holistic Optimization Framework

The opportunity generated by emergency utility curtailment for a scheduled period realizes in the interplay of the two main non-utility contributing sources of energy on campus, namely that of PV generation and fossil fuel generators, or PV–diesel hybrid systems. Battery storage is omitted from this proposed framework, due to non-favorable economic feasibility when considering large PV plants. Assuming that the diesel generators are of high enough quality to allow synchronicity with that of the renewable generator, and that the fossil fuel generators are distributed in such a manner to allow connection to any load collection on campus; typically, a characteristic of future smart campuses from the perspective of resources. Campus electrical networks can be classified as centralized PV with decentralized diesel generators islands or centralized PV with centralized diesel generators, however, the technical influence of the configuration on the proposed optimization framework is not part of the scope of the paper. The following scenario is proposed for the optimization of fuel usage, the reader is referred to Figures 1–3 for a visual summary of each step of the global optimization scheme.

The entire network (generation and load) is optimized in discrete time-steps (that could become real-time in future) based on an influx of data from the PV plant, diesel generators, load banks and the campus EIoT in which the elements are embedded. The main aims of the optimization scheme will be to determine the fixed critical load bank and fixed serving diesel generator bank, after which the remaining variable banks (both load and generator) could be optimized based on a list of critical metrics harvested from the PV plant and campus IoT network. The main energy savings metric in this proposed scenario is driven by the savings of fuel (diesel) and possibly the energy saved by comparing to "normal operations" if the utility curtailment event did not occur. The optimization scheme of the load/generation banks is driven from the viewpoint of the diesel generators with the added benefit of critical heat/electric generation from hydrogen plants on campus.

The optimization of the load banks is done on an energy manager server that utilizes building automated systems (BAS), building control systems (BCS), EIoT data influx (departmental and campus wide), renewable power generation and departmental energy policies, all of which are embedded on a local or cloud framework that holds enough resources and memory capacity to perform the optimization process (typically a server located on campus or cloud driven services). The local EM server (if this is the case) will also be restricted to a campus EM policy to perform computations during the appropriate tariff times (as not to be power intrusive itself). It is assumed that the PV plant is connected to the direct current (DC) bus and both the loads and diesel generators are connected to the alternating current (AC) bus. Converters reside between the AC and DC bus. Furthermore, it is assumed that generator-load pairs are not islanded and that every component in the network is linked.

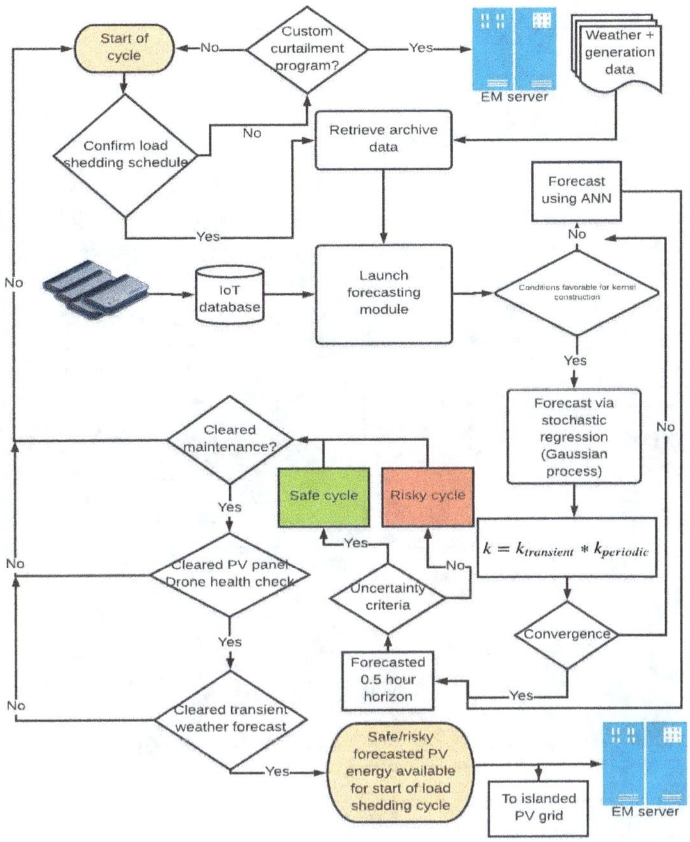

Figure 1. Optimization scheme for the determination of the available PV power output for a 0.5 h horizon. The PV scheme feeds both load and diesel generator schemes.

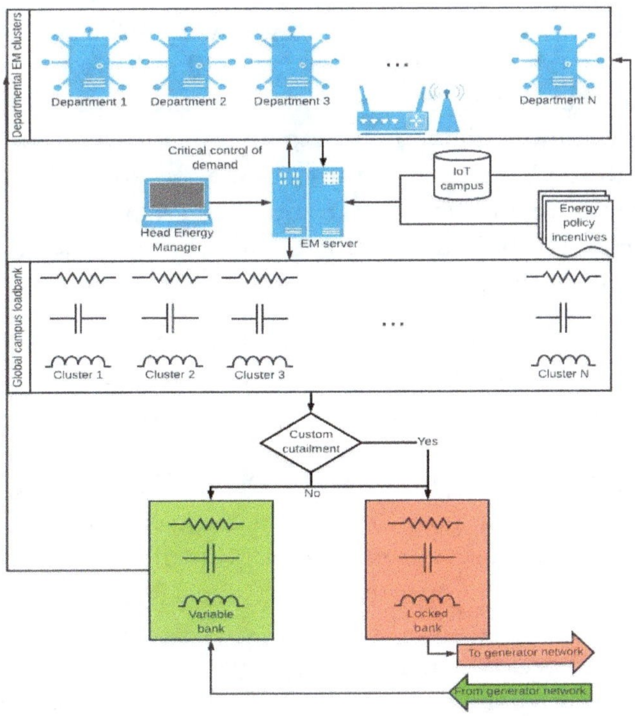

Figure 2. Optimization scheme to determine the variable/locked load banks to be fed to the diesel generator optimization scheme. This scheme depends on low-level departmental control and energy management policies.

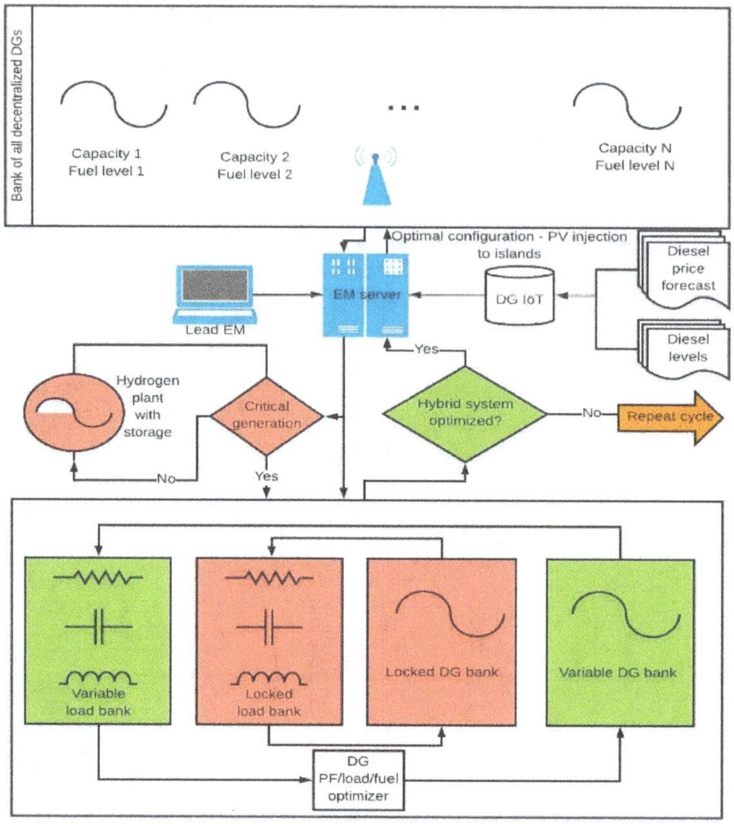

Figure 3. Optimization scheme for diesel generator—load interactions based two classes, namely variable and locked banks.

The PV optimization scheme (see Figure 1) uses actual environmental data together with generation forecasting to produce an estimated available PV energy over a 0.5 h horizon. The estimated available energy is complimented with a known uncertainty and will classify the possible PV output as either a risky cycle or a safe cycle (depending on the chosen critical uncertainty). The PV optimization scheme uses several streams of unique data, including historic weather and generation data, IoT sensor network data associated with the PV plant, facilities management software (maintenance programs) and weather forecasts spanning the prediction horizon. The PV optimization scheme must be the first step to be completed as part of the holistic optimization scheme, since the chosen dispatch strategy primarily manages the remaining demand.

Prior to the utility curtailment event, all curtailment schedules and priorities are checked and the necessary data is gathered in order to initialize the forecasting module. The forecasting module is based on either a hybrid scheme that initializes neural nets (if no kernel can be constructed) or a Gaussian process when data conditions are favorable for kernel construction. The forecasting module is trained to achieve a 0.5 h horizon with forecasting uncertainty. Depending on the length of the training data set and regression resolution, forecasting algorithms can become computationally expensive and should be included as part of a dedicated EM server with enough computational capacity for quick response times on forecasting requirements. The output of the forecasted module will be the total renewable energy that can be dispatched to the campus electrical network with associated uncertainty (based on the classification of safe or risky cycle). The PV module (see Figure 1) acts as the first step to initialize the energy dispatch strategy within the horizon of 0.5 h. The optimization scheme will critically act on maintenance conditions that are flagged and if the utility curtailment process seized. Transient weather events will increase the uncertainty of the cycle and could lead to the determination of the optimization cycle.

The load bank optimizer (see Figure 2) initiates at departmental level by checking the current state of each department's digital EM strategy. Departmental EM strategies are typically schedule driven (lecture halls, research labs and offices), thus each departmental EM strategy could deliver a load bank comprising of variable and locked banks as preconditioned by the departmental EM champion or framework. A typical response from the departmental EM framework can be seen in Tables 1 and 2 (color coding corresponds to safe or risky cycle in optimization schemes).

Table 1. Simplistic response of a departmental energy policy framework over a 0.5 h time-step (without campus IoT intervention).

Load Type	Locked/Variable	Energy Estimation of Load (kWh)
Lecture Hall A	Locked	1.8 kWh
Offices 1–5	Locked	1.25 kWh
Offices 5–10	Variable (HVAC)	1.25 kWh
Research Lab A	Locked	10 kWh
Research Lab B	Variable (fume hoods)	10 kWh

Table 2. Simplistic response of a departmental energy policy framework over a 0.5 h time-step (with campus IoT intervention).

	Locked/Variable	Energy Estimation of Load (kWh)
Lecture Hall A	Locked	1.8 kWh
Offices 1–5	Variable (Lecturers not yet entered main gate of campus)	1.25 kWh
Offices 5–10	Variable (HVAC)	1.25 kWh
Research Lab A	Variable (Flagged due to power quality issue)	10 kWh
Research Lab B	Variable (fume hoods)	10 kWh

The lead EM framework accounts for all departmental locked/variable load banks, as well as the relative sizes of the loads. It is assumed that the departmental EM framework can account for the size, type of load, power factor and associated schedules of each load in the department (elements that future smart campuses will have). The final load bank that will partake in the 0.5 h cycle of energy optimization is finalized into a compatible data structure that awaits input from the fossil fuel generator optimization scheme (see Figure 3).

The fossil fuel optimization scheme is based on the assumption that all fossil fuel generators are equipped with two-way communication and control. Furthermore, any load could be connected to any generator on campus (dismissing any practical considerations). The fossil fuel IoT network of sensors captures data associated with the generators such as fuel levels, operating conditions and location. The lead EM server supplies additional information such as maintenance and refueling schedules to the global optimizer based on fuel price forecasting and diesel levels.

At this point global energy accounting can be performed (matching supply with load) and critical energy generation via hydrogen plants could be initiated to supply critical loads. A bank of variable/locked fossil fuel generators can now be generated and passed on to the final optimization phase. The optimizer then proceeds to match the locked/variable loads with the fossil fuel generators based on power factor and load size. Once the locked loads are matched with the locked fossil fuel generators, the algorithm proceeds to match the variable banks over the horizon of 0.5 h. This process repeats every 0.5 h with the aim of reducing fuel consumption for the span of the utility power loss.

3. Extremities of Optimization Framework

The following two scenarios are presented: (1) the remaining load (after PV quenching) is distributed among generators irrespectively of fuel levels, load type or fuel forecasts, (2) optimizer algorithm (see Figures 2 and 3) suggests two types of generators linked with two types of load collections (from either the locked or variable banks), calculated every 0.5 h over a 3 h horizon.

Two extreme poles/classes of option (2) are a generator with a low fuel level that is scheduled for refueling and a generator with high fuel levels. If the optimizer algorithm predicts favorable fuel, it will suggest that the low fuel level generator to be utilized at 50% and the high fuel level generator at the difference between maximum capacity and the spinning reserve. The aim here is to maximize fuel savings at the end of the three-hour utility curtailment cycle. Due to the efficiency curves of both typical generators described as extremities, fixed parameter loads are required to be served by both generators (having different fuel levels). The 50% utilized generator can serve loads from the variable bank (non-stringent constraints on power factor and load size requirements) and the fully utilized generator can serve loads from the locked bank (due to stringent power factor and load type requirements). The locked load bank can be assigned with certainty and all load sizes, types and Power Factor (PFs) are known.

Total savings generated by the optimization scheme is decomposed into the savings generated by refueling at a better tariff rate and actual fuel savings due to more efficient generator utilizations. Over the three-hour utility power loss event, the optimizer performs six optimization steps (0.5 h apart). The PV plant energy output could change over this horizon, thus changing the dynamics of the optimization. The spinning reserve of each generator is dependent on the uncertainty of the PV power output (risky or safe cycles, see Figure 1) divided among all the available generators. As an example, if the available PV power output contains large uncertainty, the spinning reserve for each generator will increase and vice versa. The spinning reserve of each generator is re-calculated after every cycle and determines the maximum utilization of the generators.

It is further assumed that the collection of generators are of variable type (constant frequency, but variable kW input). The typical variable generator is specified by a rating (e.g., 250 kVA, 0.8 PF, 200 kW and 150 kVAR) and will be connected to a collection of loads specified by type, size and PF. Specified by the generator's efficiency/PF curve, as the utilization of the generator increases, both the efficiency and PF requirements increases. Generators with a high forced utilization factor (determined by the optimization algorithm, see Figures 1–3), must be connected to cleaner collection of loads (from the perspective of PF) to safeguard possible breaching of the generator's kVAR rating. Thus, low fuel level generators could be utilized at e.g., 50% and will serve load collections with low power factors safely (vice versa for high fuel generators).

4. Conclusions

The holistic optimization framework exhibits potential of typical campus hybrid PV–fossil fuel generator systems generating fuel savings under powerless utility scenarios when optimized in discrete time steps using campus IoT, EM and generation/consumption forecasting techniques, e.g., forecasting of renewable power output (PV, wind and hydrogen) and campus load characteristics. This is an improved savings scheme as opposed to running the generators at fixed speed without any type of data/optimization feedback. The novelty of the scheme lies in the PV power output estimation with uncertainty. The known uncertainty associated with the available PV energy could classify the output as risky or safe. This will allow improved load distribution to generators that are sensitive to fuel price forecasts and refueling maintenance, as well as improved predictive maintenance schedules associated with the generators. The main aim of the optimization scheme is to maximize fuel savings during the utility power loss. Technical recommendations for a full-scale scenario, consisting of an EM server platform hosting all the forecasting algorithms, campus IoT and policy data will be discussed and commissioned in future.

The full-scale scenario will initially be commissioned using a campus department that utilizes decentralized renewable resources, together with fossil fuel generation and energy policies. The EM server that hosts all the tiers of the optimization scheme will be located in the department itself. Following the initial commissioning test, the platform will be rolled out to the entire campus network. This full-scale scenario will require full integration of controls, metering installation/retrofit, IoT campus integration, SCADA system implementation, energy policy reviews, energy policy integration and decentralization/centralization of required generators. All of the last-mentioned elements will have to be interwoven on an EM server that contains the optimization scheme in algorithmic form, as well as the necessary maintenance and energy policy framework. Since all the discrete optimization steps will have associated uncertainty, the final savings will be able to be verified with ease. The optimization scheme presented in this paper is an example of generating savings in the absence of grid power.

Future work will be based on data produced by actual smart campus elements, sensors, generators and loads. From there a study and framework will be formulated regarding the system time dependencies. However, emphasis must be placed on the practicality in the hypothetical scheme, especially since the sampling time, signal changes and signal delays found in the forecasting techniques, campus IoT network and internal data transport play a critical role in the success of this proposed global scheme. The process of illustrating the improvement of the campus power supply will be based on a PV penetration metric, that needs to be evaluated in conjunction with the scheme.

References

1. Van Merode, D.; Tabunshchyk, G.; Patrakhalko, K.; Yuriy, G. Flexible technologies for smart campus. In Proceedings of the 2016 13th International Conference on Remote Engineering and Virtual Instrumentation (REV 2016), Madrid, Spain, 24–26 February 2016.
2. Bajpai, P.; Dash, V. Hybrid renewable energy systems for power generation in stand-alone applications: A review. *Renew. Sustain. Energy Rev.* **2012**, *16*, 2926–2939. [CrossRef]
3. Appelrath, H.-J.; Terzidis, O.; Weinhardt, C. Internet of Energy. *Bus. Inf. Syst. Eng.* **2012**, *4*, 1–2. [CrossRef]
4. Suganthi, L.; Samuel, A.A. Energy models for demand forecasting-A review. *Renew. Sustain. Energy Rev.* **2012**, *16*, 1223–1240. [CrossRef]
5. Zhang, C.; Wu, J.; Long, C.; Cheng, M. Review of Existing Peer-to-Peer Energy Trading Projects. *Energy Procedia* **2017**, *105*, 2563–2568. [CrossRef]
6. Sikorski, J.J.; Haughton, J.; Kraft, M. Blockchain technology in the chemical industry: Machine-to-machine electricity market. *Appl. Energy* **2017**, *195*, 234–246. [CrossRef]
7. Xu, L.D.; He, W.; Li, S. Internet of things in industries: A survey. *IEEE Trans. Ind. Inform.* **2014**, *10*, 2233–2243. [CrossRef]
8. Deb, C.; Zhang, F.; Yang, J.; Lee, S.E.; Shah, K.W. A review on time series forecasting techniques for building energy consumption. *Renew. Sustain. Energy Rev.* **2017**, *74*, 902–924. [CrossRef]
9. Voyant, C.; Notton, G.; Kalogirou, S.; Nivet, M.L.; Paoli, C.; Motte, F.; Fouilloy, A. Machine learning methods for solar radiation forecasting: A review. *Renew. Energy* **2017**, *105*, 569–582. [CrossRef]
10. Csató, L.; Opper, M. Sparse on-line gaussian processes. *Neural Comput.* **2002**, *14*, 641–668. [CrossRef] [PubMed]
11. Laughman, C.; Lee, K.; Cox, R.; Shaw, S.; Leeb, S.; Norford, L.; Armstrong, P. Power signature analysis. *IEEE Power Energy Mag.* **2003**, *1*, 56–63. [CrossRef]
12. Agdas, D.; Srinivasan, R.S.; Frost, K.; Masters, F.J. Energy use assessment of educational buildings: Toward a campus-wide sustainable energy policy. *Sustain. Cities Soc.* **2015**, *17*, 15–21. [CrossRef]
13. Diamantoulakis, P.D.; Kapinas, V.M.; Karagiannidis, G.K. Big Data Analytics for Dynamic Energy Management in Smart Grids. *Big Data Res.* **2015**, *2*, 94–101. [CrossRef]
14. Siano, P. Demand response and smart grids-A survey. *Renew. Sustain. Energy Rev.* **2014**, *30*, 461–478. [CrossRef]

7

Productivity Models of Infill Complex Structural Wells in Mixed Well Patterns

Liang Sun *, Baozhu Li and Yong Li

Research Institute of Petroleum Exploration & Development, PetroChina, Beijing 100083, China; lbz@petrochina.com.cn (B.L.); liyongph@petrochina.com.cn (Y.L.)
* Correspondence: sunliang328@petrochina.com.cn

Abstract: The mathematical models of productivity calculation for complex structural wells mainly focus on the single well or the regular well pattern. Previous research on the seepage theory of complex structural wells and vertical wells in mixed well pattern is greatly insufficient. Accordingly, this article presents a methodology of evaluating the productivity of infill complex structural wells in mixed well patterns. On the basis of the mirror-image method and source–sink theory, two semi-analytical models are established. These models are applied to the productivity prediction of an infill horizontal well inhorizontal-vertical well pattern and an infill multilateral well inmultilateral-vertical well pattern, respectively, in which the interference of other wells, the randomicity of well patterns, and the pressure drawdown along the horizontal laterals are taken into account. The semi-analytical models' results are consistent with those calculated by the Eclipse reservoir simulator with the relative error of less than 15%. Results indicate that the bottom hole flowing pressure decreases logarithmically while the wellbore flow rate increases monotonically from the toe to the heel of the horizontal well. Due to the pseudo-hemispherical flow at each endpoint and the pseudo-linear flow at the center of the horizontal well, the drainage area at each endpoint is relatively larger than that at the center. The radial inflow at each endpoint of the horizontal segment is considerably greater than that at the center, which presents the U-shape distribution. The proposed methodology enhances and promotes the theory of productivity evaluation for complex structural wells in mixed well patterns.

Keywords: complex structural well; mixed well pattern; productivity evaluation; semi-analytical model; well location optimization

1. Introduction

Complex structural wells including horizontal wells and multilateral wells have become a popular alternative for the development of oil and gas fields around the world because of their high flow efficiency due to larger contact area made with the reservoir and lower pressure drawdown at the same liquid volume [1]. As the process of oilfield development enters into the intermediary and later phase, the mixed well patterns of complex structural wells and conventional vertical wells have been widely applied to the implementation of adjustment plans [2]. The infilling horizontal wells or multilateral wells in mixed well patterns are of great significance to optimization of development strategies. Due to the complexity of seepage mechanism near the horizontal wellbore, the coupling between reservoir flow and wellbore conduit flow, and the interference of other wells in mixed well patterns [3], the original mathematical models based on time invariant flow are no longer applicable as a result of the change of flow regimes. Therefore, it is necessary to establish new productivity models of infill complex structural wells in mixed well patterns.

The productivity evaluation of complex structural wells under different reservoir conditions is an important topic in the field of complex structural wells. At present, studies have been conducted on the methods of production calculation for horizontal wells [4–18] and multilateral wells [19–23], which are mainly divided into analytical methods and semi-analytical methods. The analytical model aims to directly build a calculation formula based on ideal assumptions. In the semi-analytical model, each branch of the complex structural well is divided into several infinitesimal sections, thus the productivity can be obtained by solving the system of linear equations combined with fluid flow rate and pressure drawdown for each infinitesimal in the wellbore.

However, these methods mainly focus on building models of single well or regular well patterns, and little research has been conducted on the seepage theory of mixed well patterns of vertical wells and complex structural wells [24–30]. In view of the field problems involving the productivity prediction of infill wells in irregular well patterns, there are limitations in current mathematical models [31–35]. On the one hand, they are only suitable for the productivity calculation of the entire well pattern not for single infill wells; on the other hand, they are merely applied to the regular five-spot pattern, seven-spot pattern, and nine-spot pattern not for the irregularly mixed well pattern. More importantly, the pressure drop caused by wall friction and fluid acceleration along the horizontal lateral is not comprehensively taken into account during the coupling of reservoir seepage and wellbore conduit flow.

In order to overcome the deficiencies of the existing models, Ye et al. [36] presented a productivity evaluation model for infill horizontal wells considering the interference of other wells and the wellbore friction, which is suitable for mixed horizontal injection and production patterns. Although this method did not take into consideration more complicated conditions, such as the mixed well pattern including multilateral-vertical wells, it provides us with an effective approach to solve these problems.

The objective of this article here is to present two semi-analytical models for the productivity evaluation of infill complex structural wells in mixed well patterns on the basis of the mirror-image method and source–sink theory. The first model is suitable for the infill horizontal well in horizontal-vertical well pattern, and the other for the infill multilateral well in multilateral-vertical well pattern. Then, the two models are applied to the study the seepage mechanism in terms of the bottom hole flowing pressure and the distribution of wellbore flow and radial flow along the horizontal segment. The main feature of this methodology is that the models take into account the interference of other wells, the randomicity of well patterns, and the pressure drawdown along the horizontal laterals. The application in the optimization of infill well location also indicates the significant practical value of the proposed models.

2. Productivity Model

2.1. Productivity Model of Mixed Horizontal-Vertical Well Pattern

2.1.1. Reservoir Flow Model

As is shown in Figure 1, in the deployment of mixed horizontal-vertical well pattern, we assume that the number of vertical producers including $P_1^z, P_2^z, \cdots, P_m^z$ is m^z, the number of horizontal producer including $P_1^s, P_2^s, \cdots, P_m^s$ is m^s, the number of vertical injectors including $I_1^z, I_2^z, \cdots, I_n^z$ is n^z, the number of horizontal injectors including $I_1^s, I_2^s, \cdots, I_n^s$ is n^s, and P_{new} is the infill horizontal producer. The top and bottom boundaries of the reservoir are both closed, and the surrounding area is infinite. In order to analyze the well performance, the vertical interval and horizontal interval are both divided into N segments (Figure 2) [2,36].

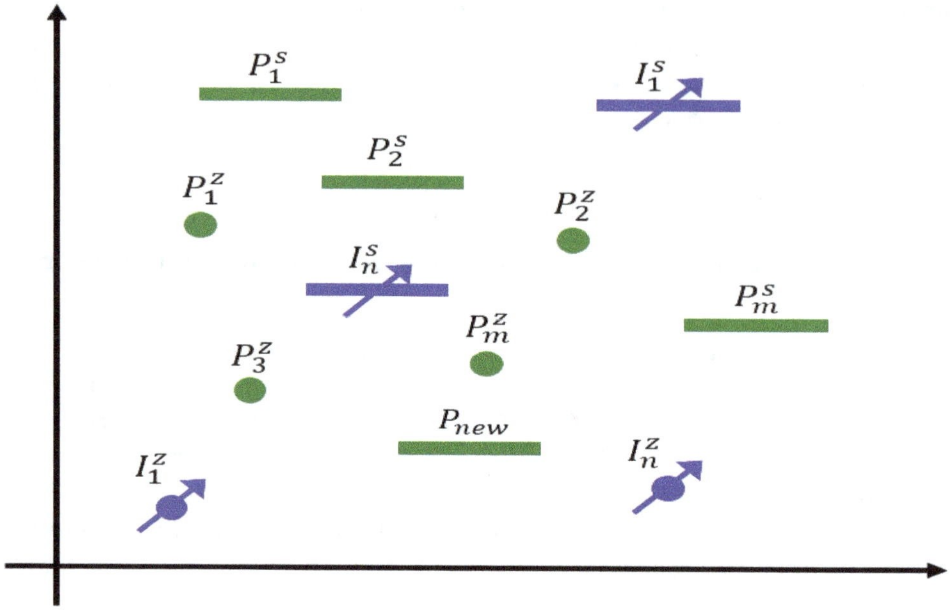

Figure 1. The deployment of mixed horizontal-vertical well pattern.

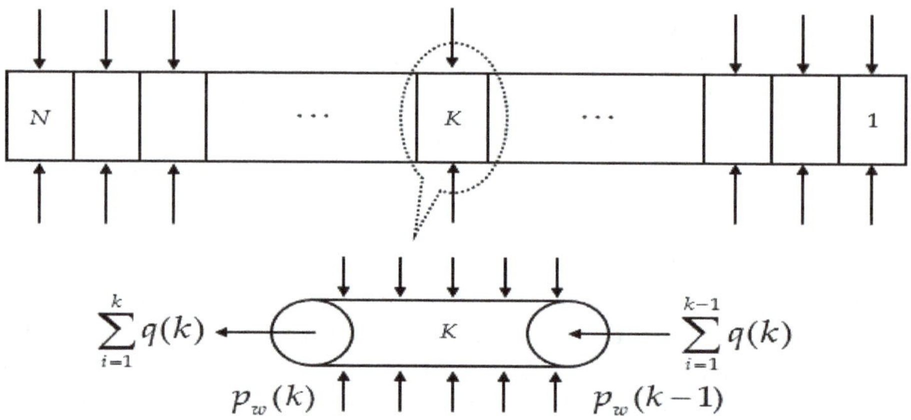

Figure 2. Schematic of the segmented horizontal well.

The coordinates of an arbitrary point M in the vertical segment k ($1 \le k \le N$) of the uth vertical producer can be expressed as follows:

$$\begin{cases} x_z^p(u,k,t) = x_z^p, \\ y_z^p(u,k,t) = y_z^p, \\ \vdots \\ z_z^p(u,k,t) = z_z^p + (k+t-1)L/N, \qquad 0 \le t \le 1 \end{cases} . \tag{1}$$

The coordinates of an arbitrary point M in the horizontal segment k ($1 \le k \le N$) of the uth horizontal producer can be expressed as follows:

$$\begin{cases} x_s^p(u,k,t) = x_s^p + (k+t-1)L/N, \\ y_s^p(u,k,t) = y_s^p, \\ \vdots \\ z_s^p(u,k,t) = z_s^p, \qquad 0 \le t \le 1 \end{cases} . \tag{2}$$

The coordinates of an arbitrary point M in the vertical segment k $(1 \leq k \leq N)$ of the vth vertical injector can be expressed as follows:

$$\begin{cases} x_z^i(v,k,t) = x_z^i, \\ y_z^i(v,k,t) = y_z^i, \\ \vdots \\ z_z^i(v,k,t) = z_z^i + (k+t-1)L/N, & 0 \leq t \leq 1 \end{cases} . \tag{3}$$

The coordinates of an arbitrary point M in the horizontal segment k $(1 \leq k \leq N)$ of the vth horizontal injector can be expressed as follows:

$$\begin{cases} x_s^i(v,k,t) = x_s^i + (k+t-1)L/N, \\ y_s^i(v,k,t) = y_s^i, \\ \vdots \\ z_s^i(v,k,t) = z_s^i, & 0 \leq t \leq 1 \end{cases} . \tag{4}$$

The coordinates of an arbitrary point M in the horizontal segment k $(1 \leq k \leq N)$ of the infill horizontal producer can be expressed as follows [37]:

$$\begin{cases} x(new,k,t) = x_{new}^p + (k+t-1)L/N, \\ y(new,k,t) = y_{new}^p, \\ \vdots \\ z(new,k,t) = z_{new}^p, & 0 \leq t \leq 1 \end{cases} , \tag{5}$$

where $x_z^p(u,k,t), y_z^p(u,k,t), z_z^p(u,k,t)$ are the coordinates of an arbitrary point in the segment k of the uth vertical producer; x_z^p, y_z^p, z_z^p are the coordinates of the left end (well heel) of the uth vertical producer; $x_s^p(u,k,t), y_s^p(u,k,t), z_s^p(u,k,t)$ are the coordinates of an arbitrary point in the segment k of the uth horizontal producer; x_s^p, y_s^p, z_s^p are the coordinates of the left end (well heel) of the uth horizontal producer; $x_z^i(v,k,t), y_z^i(v,k,t), z_z^i(v,k,t)$ are the coordinates of an arbitrary point in the segment k of the vth vertical injector; x_z^i, y_z^i, z_z^i are the coordinates of the left end (well heel) of the vth vertical injector; $x_s^i(v,k,t), y_s^i(v,k,t), z_s^i(v,k,t)$ are the coordinates of an arbitrary point in the segment k of the vth horizontal injector; x_s^i, y_s^i, z_s^i are the coordinates of the left end (well heel) of the vth horizontal injector; $x(new,k,t), y(new,k,t), z(new,k,t)$ are the coordinates of an arbitrary point in the segment k of the infill horizontal producer; $x_{new}^p, y_{new}^p, z_{new}^p$ are the coordinates of the left end (well heel) of the infill horizontal producer. L is the length of the entire horizontal section, m; N is the total number of the segments, dimensionless.

The mirror-image method and source–sink theory have been widely applied to the boundary effect reduction. As is shown in Figure 3a, assuming the constant pressure boundary is a mirror, we can project the producer (we call sink in the method) into an image injector (we call source in the method) at the symmetric coordinate to counteract the constant pressure boundary effect. As is shown in Figure 3b, in the same way, we can project the producer into an image producer at the symmetric coordinate to counteract the closed boundary effect [2].

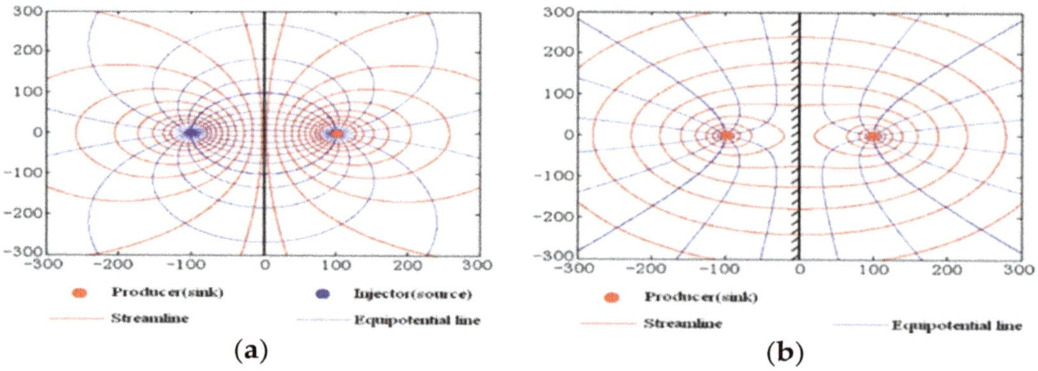

Figure 3. The mirror-image method and source–sink theory applied in different boundaries: (**a**) constant pressure boundary; (**b**) closed boundary.

On the basis of the method of images and the principle of superposition, we take the closed boundaries of the top and the bottom of the reservoir as mirrors. Thus, it is transformed into the problem of infinite well rows for horizontal producers, which is easy to solve. The potential of the producing segments of the infill horizontal well at any point M (x, y, z) in the infinite formation is:

$$\Phi(x,y,z) = \sum_{k=1}^{N} \Phi_k(x,y,z) = \frac{1}{4\pi\Delta L_k}\sum_{k=1}^{N}[q_r(k)\varphi_k(x,y,z)] + C. \tag{6}$$

With:

$$\begin{aligned}
\varphi_i(x,y,z) = \; & \sum_{n=-\infty}^{+\infty}\{\xi_k[x(new,k,t),y(new,k,t),2nh+z(new,k,t)]+ \\
& \xi_k[x(new,k,t),y(new,k,t),2nh-z(new,k,t)]+ \\
& \sum_{u=1}^{m^z}\frac{q^p_{u,z}(k)}{q_r(k)}\xi_k[x^p_z(u,k,t),y^p_z(u,k,t),2nh+z^p_z(u,k,t)]+ \\
& \sum_{u=1}^{m^z}\frac{q^p_{u,z}(k)}{q_r(k)}\xi_k[x^p_z(u,k,t),y^p_z(u,k,t),2nh-z^p_z(u,k,t)]+ \\
& \sum_{u=1}^{m^s}\frac{q^p_{u,s}(k)}{q_r(k)}\xi_k[x^p_s(u,k,t),y^p_s(u,k,t),2nh+z^p_s(u,k,t)]+ \\
& \sum_{u=1}^{m^s}\frac{q^p_{u,s}(k)}{q_r(k)}\xi_k[x^p_s(u,k,t),y^p_s(u,k,t),2nh-z^p_s(u,k,t)]- \\
& \sum_{v=1}^{n^z}\frac{q^i_{v,z}(k)}{q_r(k)}\xi_k[x^i_z(v,k,t),y^i_z(v,k,t),2nh+z^i_z(v,k,t)]- \\
& \sum_{v=1}^{n^z}\frac{q^i_{v,z}(k)}{q_r(k)}\xi_k[x^i_z(v,k,t),y^i_z(v,k,t),2nh-z^i_z(v,k,t)]- \\
& \sum_{v=1}^{n^s}\frac{q^i_{v,s}(k)}{q_r(k)}\xi_k[x^i_s(v,k,t),y^i_s(v,k,t),2nh+z^i_s(v,k,t)]- \\
& \sum_{v=1}^{n^s}\frac{q^i_{v,s}(k)}{q_r(k)}\xi_k[x^i_s(v,k,t),y^i_s(v,k,t),2nh-z^i_s(v,k,t)]\Big\}, \quad n=0,\pm1,\pm2,\cdots
\end{aligned} \tag{7}$$

where $q^p_{u,z}(k) \approx \frac{Q^p_{u,z}}{N}$, $q^i_{v,z}(k) \approx \frac{Q^i_{v,z}}{N}$, $q^p_{u,s}(k) \approx \frac{Q^p_{u,s}}{N}$, $q^i_{v,s}(k) \approx \frac{Q^i_{v,s}}{N}$; $Q^p_{u,z}$ is the production of the uth vertical producer, m^3/d; $Q^i_{v,z}$ is the injection rate of the vth vertical injector; $Q^p_{u,s}$ is the production of the uth horizontal producer, m^3/d; $Q^i_{v,s}$ is the injection rate of the vth horizontal injector, m^3/d.

For:

$$\xi_k[x(new,k,t),y(new,k,t),z(new,k,t)] = \ln\frac{r_1(new,k)+r_2(new,k)+\frac{L}{N}}{r_1(new,k)+r_2(new,k)-\frac{L}{N}},$$

$$r_1(new,k) = \sqrt{[x(new,k,t=0)-x]^2+[y(new,k,t=0)-y]^2+[z(new,k,t=0)-z]^2},$$
$$r_2(new,k) = \sqrt{[x(new,k,t=1)-x]^2+[y(new,k,t=1)-y]^2+[z(new,k,t=1)-z]^2}.$$

In the same way, we can obtain the expressions of $\xi_k[x(u,k,t),y(u,k,t),z(u,k,t)]$ and $\xi_k[x(v,k,t),y(v,k,t),z(v,k,t)]$.

Once the distributions of the flow rates of $q_r(k)$, $q^p_{u,z}(k)$, $q^i_{v,z}(k)$, $q^p_{u,s}(k)$, $q^i_{v,s}(k)$ are known, all terms of flow pressure $p_{wf}(k)$ of the infill horizontal producer can be calculated. Hence Equation (6) is simplified to a linear equation system, and can be arranged in the form as follows:

$$\begin{bmatrix} \varphi_{11} & \varphi_{12} & \varphi_{13} & \cdots & \varphi_{1N} \\ \varphi_{21} & \varphi_{22} & \varphi_{23} & \cdots & \varphi_{2N} \\ \varphi_{31} & \varphi_{32} & \varphi_{33} & \cdots & \varphi_{3N} \\ \vdots & \vdots & \vdots & \ddots & \vdots \\ \varphi_{N1} & \varphi_{N2} & \varphi_{N3} & \cdots & \varphi_{NN} \end{bmatrix} \begin{bmatrix} q_r(1) \\ q_r(2) \\ q_r(3) \\ \vdots \\ q_r(N) \end{bmatrix} = \frac{4\pi K\Delta L}{\mu_o} \begin{bmatrix} p_e - p_{wf}(1) \\ p_e - p_{wf}(2) \\ p_e - p_{wf}(3) \\ \vdots \\ p_e - p_{wf}(N) \end{bmatrix}, \tag{8}$$

where φ_{ij} is the value of φ_j at the midpoint of the segment i of the infill horizontal producer, dimensionless; $q_r(i)$ is the radial flow rate of the segment i of the infill horizontal producer, m^3/d; p_e is the reservoir pressure, MPa; $p_{wf}(i)$ is the flowing pressure of the segment i of the infill horizontal producer, MPa; K is the reservoir permeability, 10^{-3} μm^2; μ_o is the oil viscosity, mPa·s; h is the net pay thickness, m.

The wellbore flow rate along the horizontal well can be expressed as follows:

$$q_l(k) = \sum_{j=k}^{N} q_r(j), \tag{9}$$

where $q_l(k)$ is the wellbore flow rate of the segment k of the infill horizontal producer, m^3/d.

2.1.2. Wellbore Flow Model

The wellbore pressures of each segment along the horizontal segment are not independent of each other, instead, they are related to each other via wellbore hydraulics. More specifically, the pressure difference between two adjacent segment midpoints is dependent on the radial flow into the two segments, the local pressure, and the fluid property [2,36].

The pressure drop along the horizontal well is caused by the wall friction and the fluid acceleration [3,11,12,38], and can be calculated by the correlation of pressure drop in the horizontal interval under the condition of open-hole completion proposed by Liu et al. [12]:

$$\begin{aligned} \Delta p_{wf}(i) &= \tfrac{L}{N}\left\{\tfrac{2f\rho}{\pi^2 D^5}\left[2q_l(i) - q_r(i)\tfrac{L}{N}\right]^2 + \tfrac{16\rho q_r(i)}{\pi^2 D^4}\left[2q_l(i) - q_r(i)\tfrac{L}{N}\right]\right\}, \\ p_{wf}(i) &= p_{wf}(i-1) + 0.5\big(\Delta p_{wf}(i-1) + \Delta p_{wf}(i)\big), \quad (1 \le i \le N+1) \end{aligned} \tag{10}$$

where p_{wf} is the flowing pressure of the well heel, $p_{wf}(0) = p_{wf}$, $\Delta p_{wf}(0) = \Delta p_{wf}(N+1) = 0$, MPa; ρ is the fluid density, g/cm^3; f is the wall friction factor, and can be calculated as follows:

$$\begin{aligned} f &= \tfrac{64}{N_{\mathrm{Re}}}, & N_{\mathrm{Re}} \le 2100 \\ \tfrac{1}{\sqrt{f}} &= 1.14 - 2\lg\left(\tfrac{\tau}{D} + \tfrac{21.25}{N_{\mathrm{Re}}^{0.9}}\right), & N_{\mathrm{Re}} \ge 2100 \end{aligned}$$

By combing Equations (9) and (10), the following system of equations can be obtained:

$$f[q_r(i), p_{wf}(i)] = 0. \tag{11}$$

As expected, the total number of Equations (8) and (11) is equal to $2N$, which is the same as that of unknowns. Thus, the model has a unique solution.

2.1.3. Solution Procedure

The iterative algorithm is an efficient way to solve this problem, and the solution procedure is shown in Figure 4.

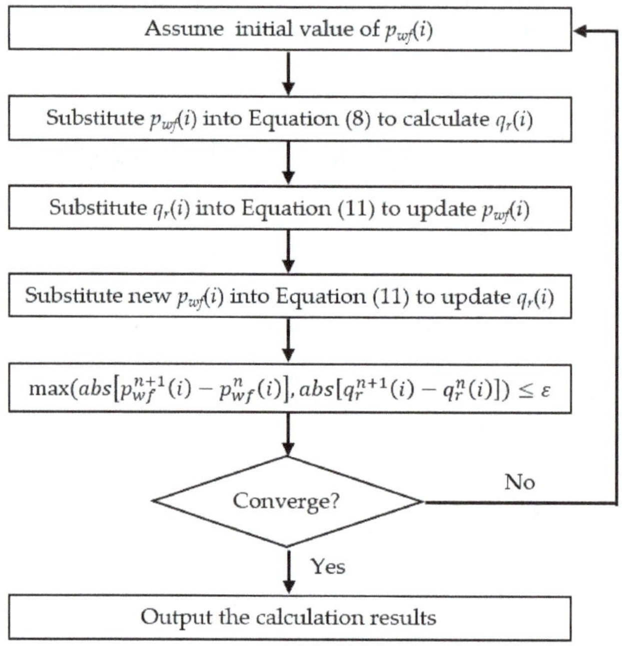

Figure 4. The flow chart of solution procedure.

2.2. Productivity Model of Mixed Multilateral-Vertical Well Pattern

2.2.1. Reservoir Flow Model

As is shown in Figure 5, in the deployment of mixed multilateral-vertical well pattern, we assume that the number of vertical producers including $P_1^z, P_2^z, \cdots, P_m^z$ is m^z, the number of vertical injectors including $I_1^z, I_2^z, \cdots, I_n^z$ is n^z, and P_{new} is the infill multilateral well. The top and bottom boundaries of the reservoir are both closed, and the surrounding area is infinite. In order to analyze the well performance, the vertical interval is divided into N segments.

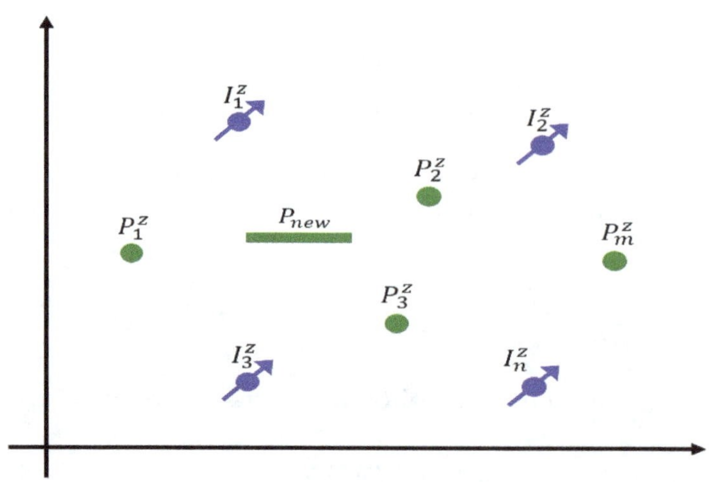

Figure 5. The mixed well pattern of multilateral-vertical wells.

The coordinates of an arbitrary point M in the vertical segment k $(1 \le k \le N)$ of the uth vertical producer can be expressed as follows:

$$\begin{cases} x_z^p(u,k,t) = x_z^p, \\ y_z^p(u,k,t) = y_z^p, \\ \vdots \\ z_z^p(u,k,t) = z_z^p + (k+t-1)L/N, \quad 0 \le t \le 1 \end{cases} . \tag{12}$$

The coordinates of an arbitrary point M in the vertical segment k $(1 \le k \le N)$ of the vth vertical injector can be expressed as follows:

$$\begin{cases} x_z^i(v,k,t) = x_z^i, \\ y_z^i(v,k,t) = y_z^i, \\ \vdots \\ z_z^i(v,k,t) = z_z^i + (k+t-1)L/N, \quad 0 \le t \le 1 \end{cases} \tag{13}$$

where $x_z^p(u,k,t), y_z^p(u,k,t), z_z^p(u,k,t)$ are the coordinates of an arbitrary point in the segment k of the uth vertical producer; x_z^p, y_z^p, z_z^p are the coordinates of the left end (well heel) of the uth vertical producer; $x_z^i(v,k,t), y_z^i(v,k,t), z_z^i(v,k,t)$ are the coordinates of an arbitrary point in the segment k of the vth vertical injector; x_z^i, y_z^i, z_z^i are the coordinates of the left end (well heel) of the vth vertical injector.

Figure 6 is the three-dimensional (3D) schematic of the infill multilateral well, we assume that each branch is symmetrically distributed on the same horizontal plane, the length of which is identical. The xoy coordinate system is established by taking the subpoint of the main hole in the xoy plane as the origin of coordinates, and the direction of the main hole as the z axis. Each branch is distributed counterclockwise around the z axis with the x axis as the starting direction.

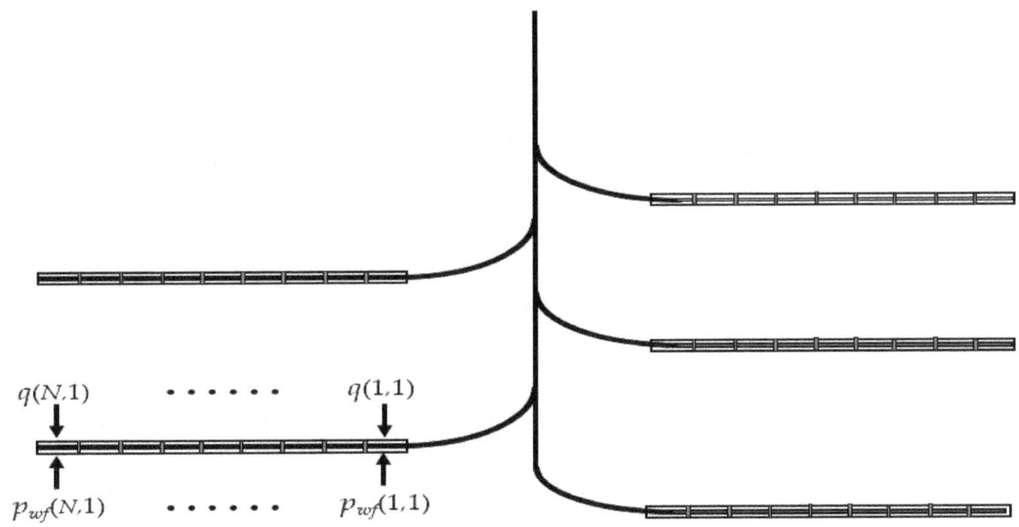

Figure 6. 3D schematic of infill multilateral well.

$M_0(x, y, z_0)$ is the bottom-hole coordinate of the main wellbore, and $M_i(x_{i,0}, y_{i.0}, z_{i,0})$ is the starting coordinate of the ith branch with the length of L_i $(1 \le i \le M)$. M is the number of branches. The ith branch of the infill multilateral well is divided into N_i $(1 \le i \le M)$ segments, thus the infinitesimal length of the ith branch is $\Delta L_i = L_i/N_i$.

Accordingly, the coordinates of an arbitrary point M in the horizontal segment i $(1 \le i \le M)$ of the infill multilateral well can be expressed as follows:

$$\begin{cases} x(i,j) = x_{i,0} + \Delta L_i[\sum_{k=1}^{j-1} (\sin\theta_{i,k} \cdot \cos\alpha_{i,k}) + t\sin\theta_{i,j} \cdot \cos\alpha_{i,k}], \\ y(i,j) = y_{i,0} + \Delta L_i[\sum_{k=1}^{j-1} (\sin\theta_{i,k} \cdot \sin\alpha_{i,k}) + t\sin\theta_{i,j} \cdot \sin\alpha_{i,k}], \\ z(i,j) = z_{i,0} + \Delta L_i[\sum_{k=1}^{j-1} \cos\theta_{i,k} + t\cos\theta_{i,j}], \quad (1 \le i \le M, 0 \le j \le N_i, 0 \le t \le 1) \end{cases} \tag{14}$$

where $x(i,j), y(i,j), z(i,j)$ are the coordinates of an arbitrary point in the ith branch of the infill multilateral producer; $x_{i,0}$, $y_{i,0}$, $z_{i,0}$ are the coordinates of the left end (well heel) in the ith branch of the infill multilateral producer; θ is the deviation angle of the ith segment of the infill multilateral producer; α is the azimuth angle of the ith segment of the infill multilateral producer.

As mentioned above, we can also transform this problem into the one of infinite well rows by using the method of images and the principle of superposition. Thus the potential of the producing segments of the infill multilateral well at any point $M(x,y,z)$ in the infinite formation is:

$$f(x,y,z) = \sum_{i=1}^{M} f_i(x,y,z) = \frac{1}{4\pi}\sum_{i=1}^{M}\sum_{j=1}^{N_j}\left[q_r(i,j)\varphi_{i,j}(x,y,z)\right] + C. \tag{15}$$

With:

$$\begin{aligned} \varphi_{i,j}(x,y,z) = & \sum_{n=-\infty}^{+\infty} [\tfrac{1}{\Delta L_i}\xi_{i,j}(x_{i,j}, y_{i,j}, 2nh + z_{i,j}) + \tfrac{1}{\Delta L_i}\xi_{i,j}(x_{i,j}, y_{i,j}, 2nh - z_{i,j}) + \\ & \tfrac{1}{\Delta L}\sum_{u=1}^{m^z} \tfrac{q^p_{u,z}(k)}{q_r(i,j)}\xi_k[x^p_z(u,k,t), y^p_z(u,k,t), 2nh + z^p_z(u,k,t)] + \\ & \tfrac{1}{\Delta L}\sum_{u=1}^{m^z} \tfrac{q^p_{u,z}(k)}{q_r(i,j)}\xi_k[x^p_z(u,k,t), y^p_z(u,k,t), 2nh - z^p_z(u,k,t)] - \\ & \tfrac{1}{\Delta L}\sum_{v=1}^{n^z} \tfrac{q^i_{v,z}(k)}{q_r(i,j)}\xi_k[x^i_z(v,k,t), y^i_z(v,k,t), 2nh + z^i_z(v,k,t)] - \\ & \tfrac{1}{\Delta L}\sum_{v=1}^{n^z} \tfrac{q^i_{v,z}(k)}{q_r(i,j)}\xi_k[x^i_z(v,k,t), y^i_z(v,k,t), 2nh - z^i_z(v,k,t)] \end{aligned} \tag{16}$$

where $q^p_{u,z}(k) \approx \frac{Q^p_{u,z}}{N}$, $q^i_{v,z}(k) \approx \frac{Q^i_{v,z}}{N}$; $Q^p_{u,z}$ is the production of the uth vertical producer, m^3/d; $Q^i_{v,z}$ is the injection rate of the vth vertical injector, m^3/d.

Once the distributions of the flow rates of $q_r(i,j)$, $q^p_{u,z}(k)$, $q^i_{v,z}(k)$ are known, all terms of flow pressure $p_{wf}(i,j)$ of the infill multilateral producer can be calculated. Hence Equation (15) is simplified to a linear equation system, and can be arranged in the form as follows:

$$A\begin{bmatrix} q_r(1,1) \\ q_r(1,2) \\ \vdots \\ q_r(1,N_1) \\ \vdots \\ q_r(M,N_M) \end{bmatrix} = \frac{4\pi\sqrt{k_h k_v}}{\mu_o}\begin{bmatrix} P_e - P_{wf}(1,1) \\ P_e - P_{wf}(1,2) \\ \vdots \\ P_e - P_{wf}(1,N_1) \\ \vdots \\ P_e - P_{wf}(1,N_M) \end{bmatrix}. \tag{17}$$

With:

$$A = \begin{bmatrix} \phi_{(1,1)(1,1)} - \phi_{e(1,1)} & \cdots & \phi_{(1,1)(1,N_1)} - \phi_{e(1,N)} & \phi_{(1,1)(2,1)} - \phi_{e(2,1)} & \cdots & \phi_{(1,1)(M,N_M)} - \phi_{e(M,N_M)} \\ \vdots & \ddots & \vdots & \vdots & \ddots & \vdots \\ \phi_{(1,N_1)(1,1)} - \phi_{e(1,1)} & \cdots & \phi_{(1,N_1)(1,N_1)} - \phi_{e(1,N)} & \phi_{(1,N_1)(2,1)} - \phi_{e(2,1)} & \cdots & \phi_{(1,N_1)(M,N_M)} - \phi_{e(M,N_M)} \\ \phi_{(2,1)(1,1)} - \phi_{e(1,1)} & \cdots & \phi_{(2,1)(1,N_1)} - \phi_{e(1,N)} & \phi_{(2,1)(2,1)} - \phi_{e(2,1)} & \cdots & \phi_{(2,1)(M,N_M)} - \phi_{e(M,N_M)} \\ \vdots & \ddots & \vdots & \vdots & \ddots & \vdots \\ \phi_{(M,N_M)(1,1)} - \phi_{e(1,1)} & \cdots & \phi_{(M,N_M)(1,N_1)} - \phi_{e(1,N)} & \phi_{(M,N_M)(2,1)} - \phi_{e(2,1)} & \cdots & \phi_{(M,N_M)(M,N_M)} - \phi_{e(M,N_M)} \end{bmatrix}$$

where $q_r(i, j)$ is the radial flow rate of the ith branch of the infill multilateral producer, $1 \le i \le M$, $1 \le j \le N_i$, m^3/d; P_e is the reservoir pressure, MPa; $P_{wf}(i, j)$ is the flow pressure of the ith branch of the infill multilateral producer, $1 \le i \le M$, $1 \le j \le N_i$, MPa; k_h is the reservoir horizontal permeability, 10^{-3} μm^2; k_v is the reservoir vertical permeability, 10^{-3} μm^2; μ_o is the oil viscosity, mPa·s.

2.2.2. Wellbore Flow Model

1. Pressure drop model of the horizontal lateral

The pressure drop along the horizontal interval is caused by the wall friction and the fluid acceleration. According to Equation (11), the pressure drop of the ith branch of the multilateral well along the horizontal interval can be calculated by:

$$F'_{i,j}\left[q_r(i,j), p_{wf}(i,j)\right] = 0. \tag{18}$$

2. Pressure drop model of the deviated segment

The branch holes enter the reservoir through the deviated segment. Although each deviated segment does not contribute to the production, the pressure drop in the deviated segments cannot be ignored. According to fluid seepage theory in the elbow, the pressure drop can be calculated by the following formula:

$$\Delta p_s = C_w \frac{4 \times 10^{-6} \rho \lambda_c R_c Q_W^2}{\pi d_w^5} + \frac{\rho g R_c}{10^6}, \tag{19}$$

where ΔP_s the pressure drop along the deviated segment, MPa; R_c is curvature radius of the deviated segment, m; ρ is the fluid density, kg/m^3; d_w is the wellbore diameter of the deviated segment, m; C_w is the similarity factor of the wellbore and deviated segment, dimensionless; λ_c is the friction factor of the deviated segment, dimensionless.

3. Pressure drop model of the vertical segment

According to the law of conservation of energy, the equation of pressure gradient in the pipe with inclination can be expressed as follows:

$$\frac{dP}{dz} = \rho g \sin\theta + \rho v \frac{dv}{dz} + f\frac{\rho v^2}{2d}, \tag{20}$$

where dP/dz is the pressure loss per unit length, MPa/m; $\rho g \sin\theta$ is the pressure drop caused by the fluid gravity, MPa/m; $\rho v \times dv/dz$ is the pressure drop caused by the fluid acceleration, MPa/m; $f \times \rho v^2/2d$ is the pressure drop caused by the friction, MPa/m; ρ is the fluid density, kg/m^3; d is the pipe diameter, m. The friction factor f can be calculated by:

$$f = \tfrac{64}{N_{\mathrm{Re}}}, N_{\mathrm{Re}} \le 2100$$
$$\frac{1}{\sqrt{f}} = 1.14 - 2\lg\left(\tfrac{\tau}{d} + \tfrac{21.25}{N_{\mathrm{Re}}^{0.9}}\right), N_{\mathrm{Re}} \ge 2100$$

where N_{Re} is the Reynolds number, dimensionless.

4. Internal pressure drop model of the branch hole

The flowing pressure at the heel of the ith branch can be calculated by simultaneous Equations (19) and (20):

$$P_{wf}(i) = P_{mb} - \Delta P_{vi} - \Delta P_{wi}(1 \le i \le M), \tag{21}$$

where P_{mb} is the flowing pressure of the main wellbore, MPa; ΔP_{vi} is the pressure drop of the vertical segment of the ith branch, MPa; ΔP_{wi} is the pressure drop of the deviated segment of the ith branch, MPa.

By combining Equations (20) and (24), the following system of equations can be obtained:

$$F_{i,j}\left[q_r(i,j), p_{wf}(i,j)\right] = 0, \left(1 \le i \le M, 1 \le j \le N_i\right). \tag{22}$$

As expected, the total number of equations of Equations (17) and (22) is equal to $2(M \times N_i)$, which is the same as that of unknowns. Therefore, the solution of the coupling model is unique.

2.2.3. Solution Procedure

The coupling model can also be solved by using the iterative algorithm, and the solution procedure is shown in Figure 7.

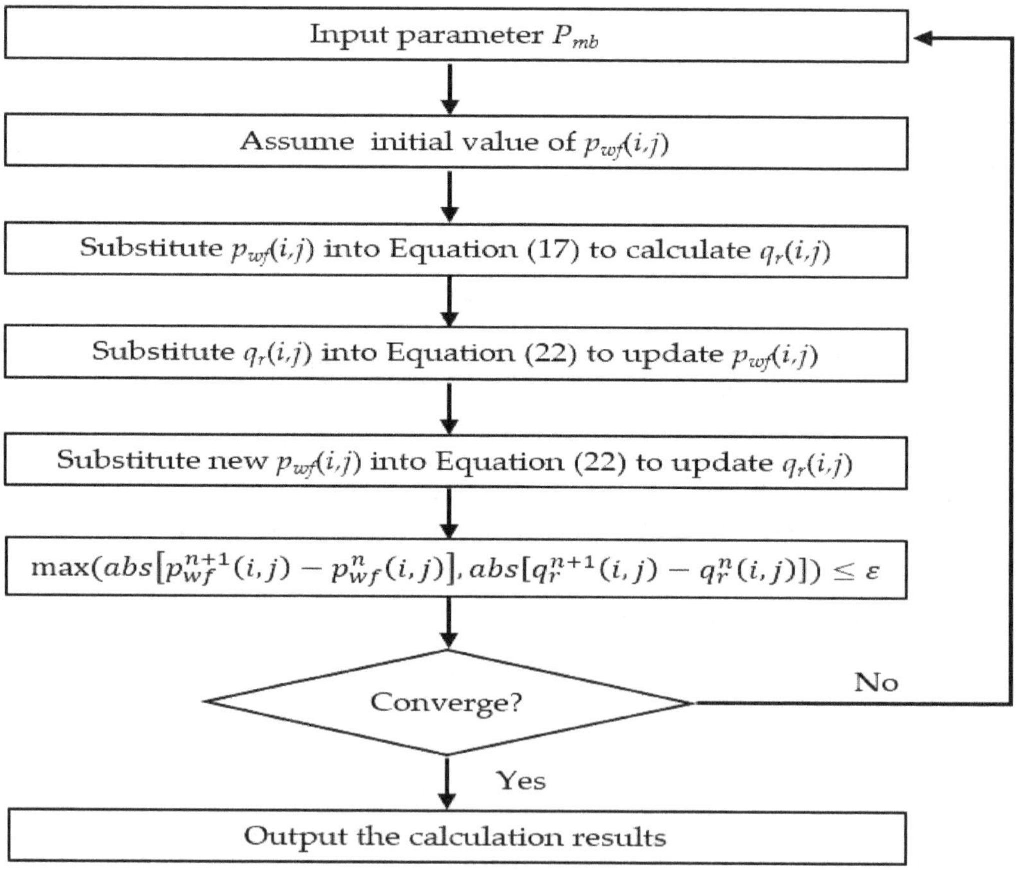

Figure 7. The flow chart of solution procedure.

3. Results and Discussion

3.1. Model Validation

The semi-analytical models can be applied to the productivity prediction of the infill horizontal well and the multilateral well in mixed well patterns. In order to verify the reliability of the models, two cases are considered, and the results calculated by the semi-analytical models are compared with those calculated by the Eclipse software (Schlumberger). In the first case, we selected a typical mixed well pattern of horizontal-vertical wells from a thin carbonate reservoir in the Middle East. There are three vertical producers (P_{11}, P_{12}, and P_{13}), one horizontal producer (P_{14}), one vertical injector (I_{11}), one horizontal injector (I_{12}), and P_{1new} is the infill horizontal producer. In the other case, the reservoir is a thick, multi-layer reservoir in the Middle East, multilateral wells are adopted to improve the degree of the reservoir development. The mixed well pattern of multilateral-vertical well includes two vertical producers (P_{21} and P_{22}), one vertical injector (I_{21}), and P_{2new} is the infill multilateral producer with two branch holes.

The productivity models can be solved by VB modular programming of the object-oriented technology. The reservoir properties and wellbore geometry are listed in Tables 1 and 2. By inputting the same model parameters, the productivity evaluation of the two wells were conducted respectively with the semi-analytical models and the Eclipse simulator. The calculation results are shown in Table 3.

Table 1. Productivity calculation parameters of P_{1new}.

Parameters (Unit)	Value
Net thickness (m)	6.0
Porosity (%)	15.3
Permeability (10^{-3} μm^2)	88.5
Reservoir pressure (MPa)	38.6
Oil viscosity (mPa·s)	8.1
Horizontal length of P_{1new} (m)	335
Wellbore radius of P_{1new} (m)	0.1
Bottom hole flowing pressure (MPa)	36.6
Production rate of P_{11} (m^3/d)	18.5
Production rate of P_{12} (m^3/d)	21.6
Production rate of P_{13} (m^3/d)	24.3
Production rate of P_{14} (m^3/d)	56.8
Injection rate of I_{11} (m^3/d)	68.5
Injection rate of I_{12} (m^3/d)	79.2

Table 2. Productivity calculation parameters of P_{2new}.

Parameters (Unit)	Value
Net thickness-the 1st branch of P_{2new} (m)	5.1
Net thickness-the 2nd branch of P_{2new} (m)	9.4
Porosity (%)	22.1
Ratio of vertical to horizontal permeability (-)	0.1
Reservoir pressure (MPa)	20.3
Oil viscosity (mPa·s)	6.2
Length of the 1st branch of P_{2new} (m)	200
Length of the 2nd branch of P_{2new} (m)	140
Bottom hole flowing pressure (MPa)	10.0
Production rate of P_{21} (m^3/d)	52.9
Production rate of P_{22} (m^3/d)	67.8
Injection rate of I_{21} (m^3/d)	125.5

Table 3 indicates that the semi-analytical models have a high accuracy with the relative error of less than 15%. The results calculated by the models are consistent with those by the Eclipse numerical simulator, which verify the reliability of the productivity models. By taking into account the actual conditions, namely the interference of other wells in the well pattern, the coupling between the reservoir flow and wellbore flow, and the pressure drop along the horizontal lateral, the accuracy of the productivity prediction by the semi-analytical models is greatly improved.

Table 3. Results comparison of two methods.

Well Name		Actual Productivity (m^3/d)	The Semi-Analytical Model		The Eclipse Simulator	
			Productivity (m^3/d)	Relative Error (%)	Productivity (m^3/d)	Relative Error (%)
P_{1new}		25.0	27.3	9.1	27.2	8.8
P_{2new}	1st branch	197.7	220.7	11.7	221.3	11.9
	2nd branch	147.1	166.5	13.2	168.0	14.2
	Total	344.7	387.2	12.3	389.3	12.9

3.2. Model Application

3.2.1. Study on Seepage Mechanism of Horizontal Well

In the first case of model validation, we can further analyze the bottom hole flowing pressure, the distribution of wellbore flow rate and radial flow rate along the horizontal segment. Figures 8 and 9 show that the bottom hole flowing pressure decreases logarithmically while the wellbore flow rate increases monotonically from the toe to the heel of the horizontal well. The reservoir flow coupled with variable mass wellbore flow results in this near wellbore dynamics. On one hand, the mass flow rate of fluid from the toe to the heel increases gradually. In this case, the fluid velocity along the main flow direction also increases with an accelerating pressure drop. The radial inflow of reservoir fluid along the horizontal wellbore disturbs the boundary layer of the main stream tube and affects its velocity profile, thus altering the wall friction resistance determined by the velocity distribution. These factors lead to the bottom hole flowing pressure decreasing in the form of logarithmic function. On the other hand, the radial inflow performance affects the distribution of pressure and pressure drop, and vice versa. Due to the pseudo-hemispherical flow at each endpoint (well heel and well toe) and the pseudo-linear flow at the center of the horizontal segment, the drainage area at each endpoint is relatively larger than that at the center. The radial flow rate at each endpoint of the horizontal segment is considerably greater than that at the center, and decreases quickly toward the intermediate section, which generally presents a U-shape distribution.

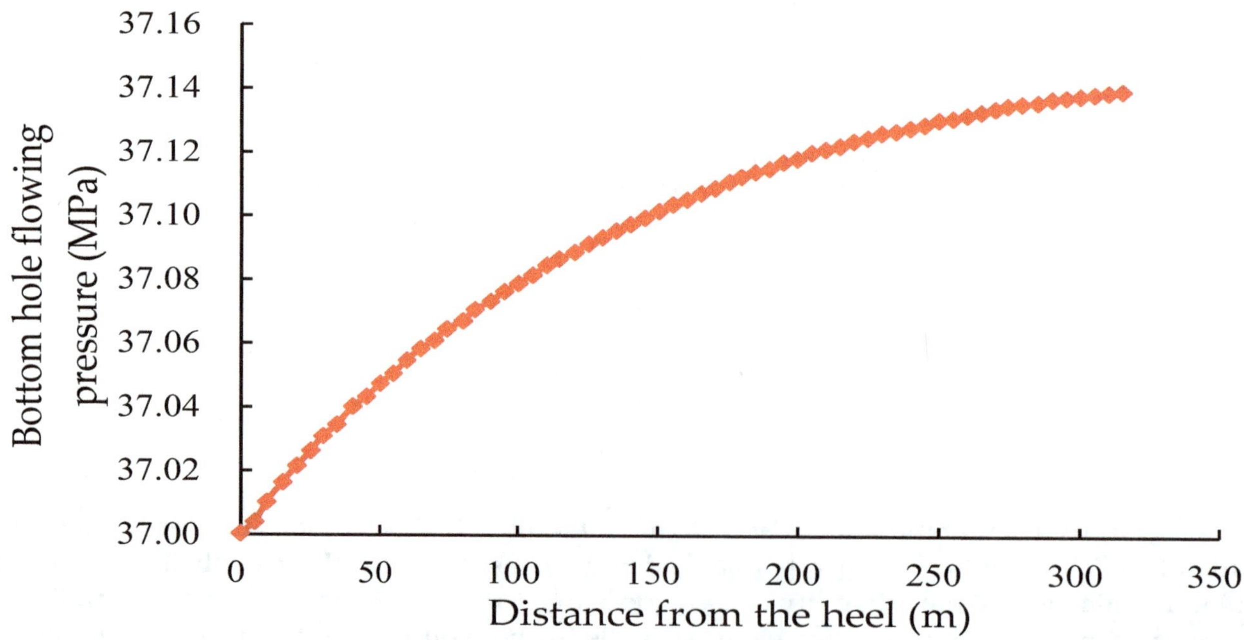

Figure 8. Distribution of bottom hole flowing pressure along the horizontal lateral.

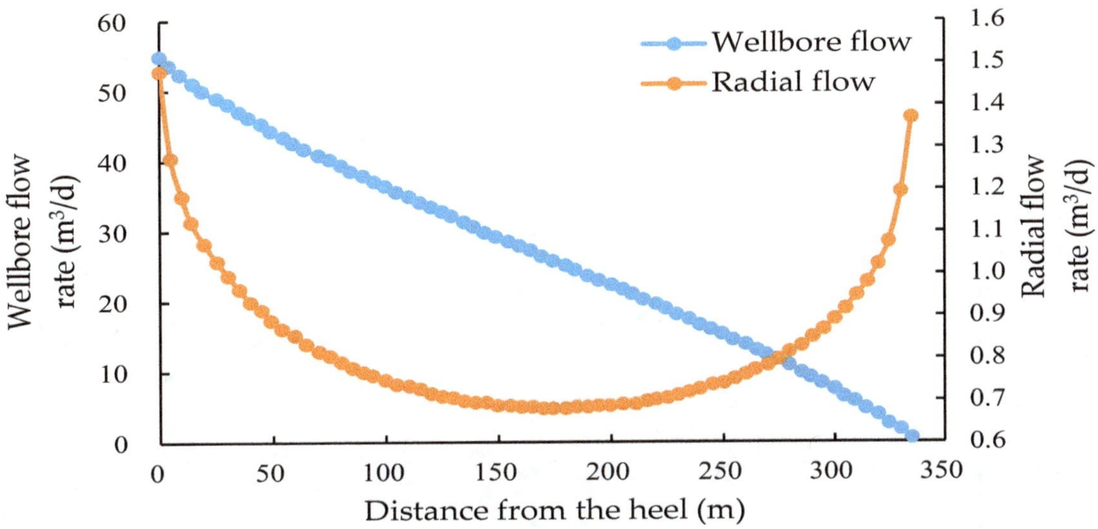

Figure 9. Distribution of wellbore flow and radial flow along the horizontal lateral.

3.2.2. Optimization of Infilling Well Location

In the intermediary and later phase of oilfield development, infill wells are usually needed to improve the well pattern and the effect of tapping potential of remaining oil. Therefore, the key is the optimization of infill well location. The productivity models provide a shortcut for this target. Based on the understanding of geological conditions and development status, the productivity of infill well can be accurately predicted, and in combination with the findings of remaining oil distribution, the infill well location could be optimized.

Taking the mixed five-spot well pattern of vertical injectors and horizontal producer as an example, we conducted a study on the productivity variation of horizontal well at different locations in the well pattern (Figure 10). The basic reservoir parameters refer to the first case of model validation. Figure 11 shows that the lateral shifting of horizontal well has some effects on the productivity, and the productivity at the center of the well pattern is relatively higher than that at other locations. However, this does not mean that the central location is the optimum selection for the horizontal well. In the adjustment of well pattern, geological condition and development factors, such as the distribution of flow field and remaining oil, should also be considered.

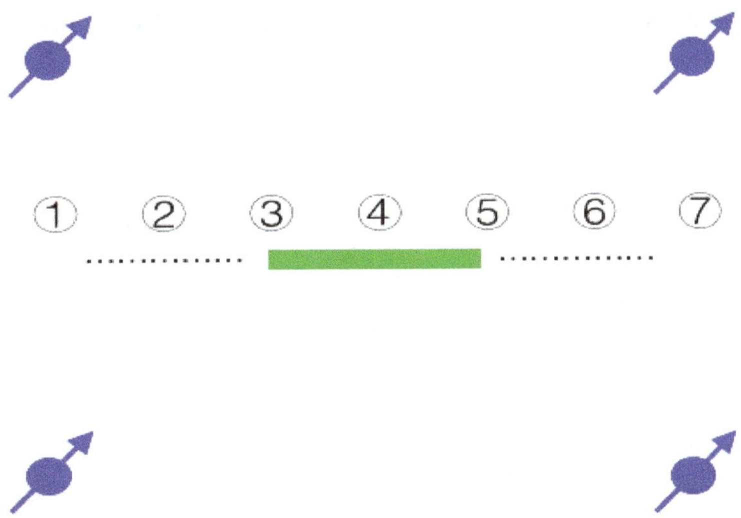

Figure 10. Mixed five-spot well pattern of vertical-injectors and horizontal producer (①–⑦ show the code of different horizontal locations).

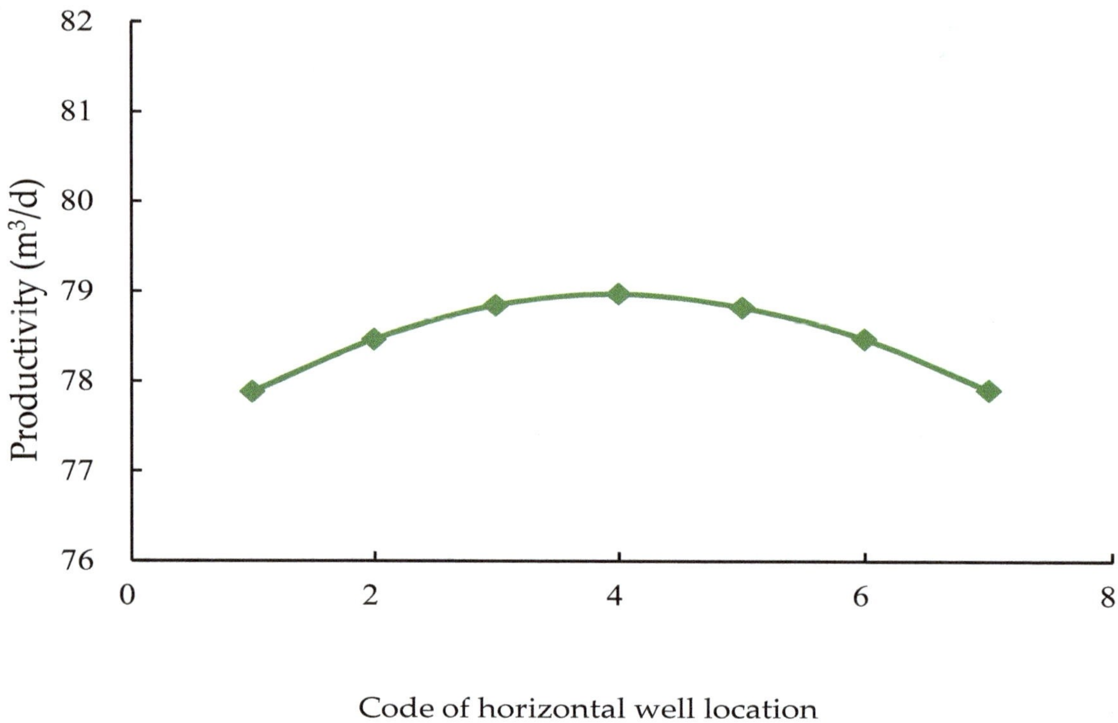

Figure 11. Productivity variation of horizontal well in mixed well pattern.

We made further study on the flow field distribution in the mixed well pattern so as to provide references for the optimization of infilling well location. Figures 12 and 13 indicate that the equipotential lines and stream lines are symmetrically distributed along the x-axis as the horizontal well moves laterally. When the horizontal well is located at the center of the well pattern, both the equipotential lines and stream lines present a uniform distribution; when the horizontal well moves toward the right side, both of the two types of the lines on the right side become dense, while those on the left side are sparsely distributed, and vice versa. The more the horizontal well location changes, the more obvious this phenomenon is. The flow pattern almost presents a linear flow as the horizontal well locates at the line of the two vertical wells on the same side, which indicates a relatively higher swept efficiency. Therefore, in combination with the findings of productivity variation and flow fluid distribution, the infilling well location could be optimized. This improves the effect of well pattern adjustment considerably.

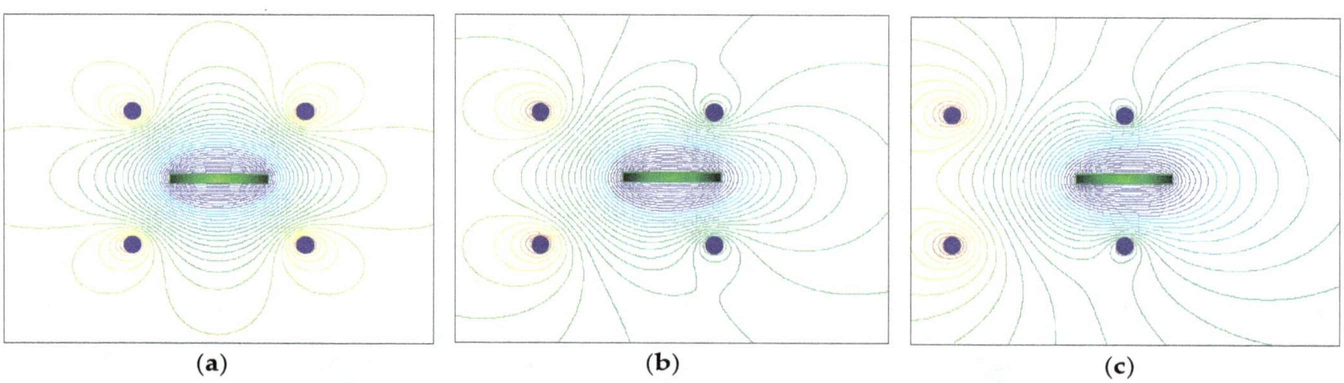

Figure 12. Equipotential line pattern for lateral shifting of the horizontal well. ((**a**–**c**) show the equipotential line pattern of different lateral location.)

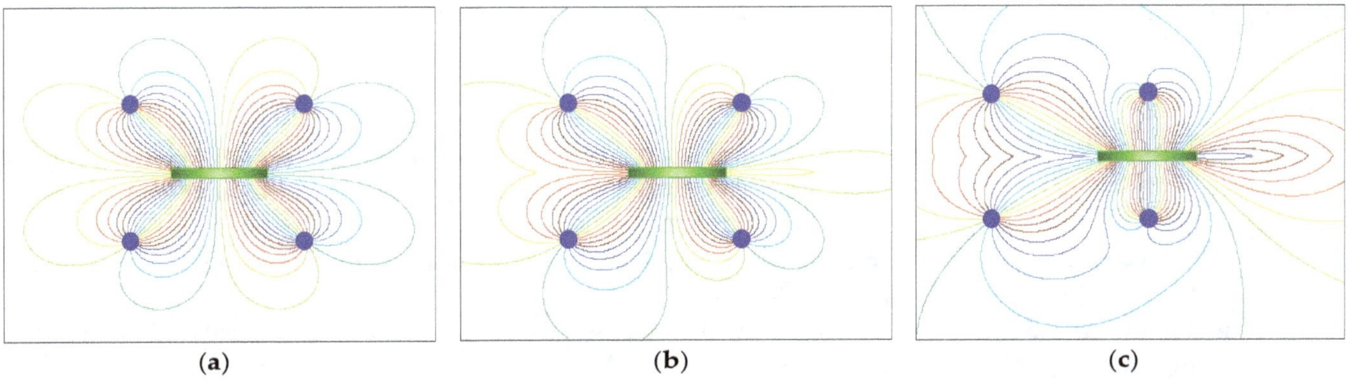

Figure 13. Stream line pattern for lateral shifting of the horizontal well. ((**a**–**c**) show the stream line pattern of different lateral location.)

3.3. Discussion

In this study, the semi-analytical models for productivity evaluation of infill complex structural wells are built with the method of images and source–sink theory. Compared with the existing models, the advantages of the semi-analytical model lie in the following three aspects. Firstly, the models are suitable for mixed well patterns of vertical wells and complex structure wells; Secondly, both the interference of other wells in the well pattern and the pressure drawdown along the horizontal segment are taken into account, which makes the models more reliable. In addition, the models provide some insight into the seepage mechanism of complex structural wells such as the distribution of wellbore flow and radial flow along the horizontal segment. However, the semi-analytical models also have some limitations, the basic assumptions of the models are steady-state flow and closed top and bottom boundaries, and the productivity evaluation of different reservoir boundaries under the condition of unsteady-state flow still needs to be further studied.

4. Conclusions

In this paper, on the basis of the mirror-image method and the source–sink theory, two semi-analytical models are established to predict the productivity of infill horizontal wells and multilateral wells in mixed well patterns respectively, in which the interference of other wells, the randomicity of well pattern, and the pressure drawdown along the horizontal lateral are taken into account.

The application of the productivity models verifies the reliability and practicability. The results calculated by the models are consistent with those by the Eclipse numerical simulator, and the semi-analytical models have a high accuracy with the relative error of less than 15%.

The results indicate that the bottom hole flowing pressure decreases logarithmically while the wellbore flow rate increases monotonically from the toe to the heel of the horizontal segment. Due to the pseudo-hemispherical flow at each endpoint and the pseudo-linear flow at the center of the horizontal segment, the drainage area at each endpoint is relatively larger than that at the center. The radial flow rate at each endpoint of the horizontal segment is considerably greater than that at the center, which generally presents a U-shape distribution.

This study also proposes a practical and efficient approach for the study on the horizontal seepage mechanism, and the optimization of infilling well locations.

Author Contributions: Conceptualization, L.S.; Data curation, L.S.; Formal analysis, L.S.; Funding acquisition, Y.L.; Methodology, L.S.; Project administration, B.L.; Supervision, Y.L.; Validation, L.S.; Writing – review & editing, L.S., Y.L.

Acknowledgments: The authors are grateful for the technical and financial support from the Research Institute of Exploration and Development, PetroChina, Beijing 100083, China.

References

1. Adesina, F.; Paul, A.; Oyinkepreye, O.; Adebowale, O. An improved model for estimating productivity of horizontal drain hole. In Proceedings of the SPE Nigeria Annual International Conference and Exhibition, Lagos, Nigeria, 2 August 2016. [CrossRef]
2. Ye, S.J. A productivity evaluation model and its application for an infill horizontal well in different types of reservoirs. *Pet. Sci. Technol.* **2012**, *30*, 1677–1691.
3. Liangbiao, O.; Arbabi, S.; Aziz, K. General wellbore flow model for horizontal, vertical and slanted well completions. *SPE J.* **1998**, *3*, 124–133.
4. Merkulov, V.P. The flow of slanted and horizontal well. *Neft. Khoz* **1958**, *6*, 51–56.
5. Borisov, J.P. *Oil Production Using Horizontal and Multiple Deviation Wells*; Nedra: Moscow, Russia, 1964.
6. Giger, F.M. Horizontal wells production techniques in heterogeneous reservoirs. In Proceedings of the Middle East Oil Technical Conference and Exhibition, Manama, Bahrain, 14–17 March 1983. [CrossRef]
7. Joshi, S.D. Augmentation of well productivity using slant and horizontal wells. In Proceedings of the SPE Annual Technical Conference and Exhibition, New Orleans, Louisiana, 5 October 1986. [CrossRef]
8. Raghavan, R.; Joshi, S.D. Productivity of multiple drainholes or fractured horizontal wells. *SPE Format. Eval.* **1993**, *8*, 11–16. [CrossRef]
9. Thomas, L.K.; Todd, B.J.; Evans, C.E.; Pierson, R.G. Horizontal well IPR calculations. *SPE Reserv. Eval. Eng.* **1998**, *1*, 392–399. [CrossRef]
10. Zhang, W.; Han, D. 3D potential distribution and precise productivity equation of horizontal well. *Pet. Explor. Dev.* **1999**, *26*, 49–52.
11. Dikken, B.J. Pressure drop in horizontal wells and its effect on production performance. *J. Pet. Technol.* **1990**, *42*, 1426–1433. [CrossRef]
12. Liu, X.; Zhang, Z.S.; Liu, X.E.; Guo, S. A model to calculate pressure drops of horizontal wellbore variable mass flow coupled with flow in a reservoir. *J. Southwest Pet. Inst.* **2000**, *22*, 36–39.
13. Wang, R.H.; Zhang, Y.Z. A segmentally numerical calculation method for estimating the productivity of perforated horizontal wells. *Pet. Explor. Dev.* **2006**, *33*, 630.
14. Li, X.P.; Guo, C.Z.; Jiang, Z.X.; Liu, X.E.; Guo, S.P. The model coupling fluid flow in the reservoir with flow in the horizontal wellbore. *Acta Pet. Sin.* **1999**, *3*, 82–86.
15. Butler, R.M. Discussion of augmentation of well productivity with slant and horizontal wells. Author's reply. *J. Pet. Technol.* **1992**, *44*, 942–943.
16. Dang, L. Analysis of productivity formulae of horizontal well. *Pet. Explor. Dev.* **1997**, *5*, 21.
17. Jiang, H.; Ye, S.; Lei, Z.; Wang, X.; Zhu, G.; Chen, M. The productivity evaluation model and its application for finite conductivity horizontal wells in fault block reservoirs. *Pet. Sci.* **2010**, *7*, 530–535. [CrossRef]
18. Huang, S.J.; Cheng, L.S.; Zhao, F.L. The productivity evaluation model of the stepped horizontal well in thin interbeded reservoirs. *J. Southwest Pet. Univ.* **2007**, *3*, 16.
19. Борисов, Ю.П.; Табаков, В.П. *Расчет взаимодействия батарей наклонных и многозабойных скважин в слоистом пласте*; НТС по добыче нефти: Оренбургскаяобл, г. Оренбург, ул., Россия, 1961.
20. Wang, W.H.; Li, D. Productivity study on branch horizontal wells. *Oil Drill. Prod. Technol.* **1997**, *4*, 12.
21. Li, C.L. Derivation of productivity formulae of a fishbone well. *J. Southwest Pet. Inst.* **2005**, *27*, 36.
22. Cheng, L.S.; Li, C.L.; Lang, Z.X.; Zhang, L.H. The productivity study of branch a horizontal well with multiple branched wells. *Acta Pet. Sin.* **1995**, *16*, 49–55.
23. Zhao, L.X.; Jiang, M.H.; Zhao, X.F. Research on deliverability relationship of complicated horizontal well. *J. Univ. Pet. China (Ed. Nat. Sci.)* **2006**, *30*, 77–80.
24. Johansen, T.E.; James, L.; Cao, J. Analytical coupled axial and radial productivity model for steady-state flow in horizontal wells. *IJPE* **2015**, *1*, 290. [CrossRef]
25. Wang, H.; Guo, J.; Zhang, L. A semi-analytical model for multilateral horizontal wells in low-permeability naturally fractured reservoirs. *J. Pet. Sci. Eng.* **2017**, *149*, 564–578. [CrossRef]
26. Simonov, M.V.; Akhmetov, A.V.; Roshchektaev, A.P. Semi-analytical model of transient fluid flow to multilateral well. In Proceedings of the SPE Annual Caspian Technical Conference and Exhibition, Baku, Azerbaijan, 1–3 November 2017. [CrossRef]

27. Hassan, A.; Abdulraheem, A.; Elkatatny, S.; Ahmed, M. New approach to quantify productivity of fishbone multilateral well. In Proceedings of the SPE Annual Technical Conference and Exhibition, San Antonio, TX, USA, 9–11 October 2017. [CrossRef]
28. Liu, G.; Meng, Z.; Cui, Y.; Wang, L.; Liang, C.; Yang, S. A semi-analytical methodology for multiwell productivity index of well-industry-production-scheme in tight oil reservoirs. *Energies* **2018**, *11*, 1054. [CrossRef]
29. Vodorezov, D.D. Estimation of horizontal-well productivity loss caused by formation damage on the basis of numerical modeling and laboratory-testing data. *SPE J.* **2018**. [CrossRef]
30. Al-Rbeawi, S.; Artun, E. Fishbone type horizontal wellbore completion: A study for pressure behavior, flow regimes, and productivity index. *J. Pet. Sci. Eng.* **2019**. [CrossRef]
31. Suprunowicz, R.; Butler, R.M. The productivity and optimum pattern shape for horizontal wells arranged in staggered rectangular arrays. *J. Can. Pet. Technol.* **1992**, *31*. [CrossRef]
32. Chunlan, L.; Lingsong, C.; Lihua, Z.; Zhaoxin, L. The Study on Productivity of Horizontal Well 9-spot Patterns. *J. Southwest Pet. Inst.* **1998**, *20*, 56–58.
33. Cheng, L.S.; Zheng, J.Q.; Li, C.L. Productivity study of horizontal wells pattern. *Oil Drill. Prod. Technol.* **2002**, *24*, 39–41.
34. Liu, Y.T.; Zhang, J.C. Stable permeating flow and productivity analysis for anisotropic reservoir in horizontal well networks. *Pet. Explor. Dev.* **2004**, *31*, 94–96.
35. Yin, G.F.; Xu, H.M.; Ye, S.J.; Li, Y.R.; Qiu, J.P. Productivity evaluation model for infill horizontal well in horizontal injection and production pattern. *J. China Univ. Pet. (Ed. Nat. Sci.)* **2011**, *4*, 18.
36. Ye, S.J.; Jiang, H.Q.; Li, J.J. Productivity calculation of infill horizontal wells in mixed well pattern. *Chin. J. Comput. Phys.* **2011**, *28*, 693–697.
37. Huang, S.J.; Cheng, L.S.; Zhao, F.L.; Li, C.L. The flow model coupling reservoir percolation and variable mass pipe flow in production section of the stepped horizontal well. *J. Hydrodyn.* **2005**, *20*, 463–471.
38. Tabatabaei, M.; Ghalambor, A. A new method to predict performance of horizontal and multilateral wells. *SPE Prod. Oper.* **2011**, *26*, 75–87. [CrossRef]

8

A Dispatching Optimization Model for Park Power Supply Systems Considering Power-to-Gas and Peak Regulation Compensation

Yunfu Qin [1], Hongyu Lin [1,2,*], Zhongfu Tan [1,2], Qingyou Yan [1], Li Li [3], Shenbo Yang [1,2], Gejirifu De [1,2] and Liwei Ju [1,2]

1 School of Economics and Management, North China Electric Power University, Beijing 102206, China; qinyunfu815@163.com (Y.Q.); tzhf@ncepu.edu.cn (Z.T.); yanqingyou@ncepu.edu.cn (Q.Y.); ysbo@ncepu.edu.cn (S.Y.); dove@ncepu.edu.cn (G.D.); hdlw_ju@ncepu.edu.cn (L.J.)
2 Beijing Key Laboratory of New Energy and Low-Carbon Development, North China Electric Power University, Beijing 102206, China
3 State Grid Jibei Electric Power Economic Research Institute, Beijing 100000, China; jbjyyll@126.com
* Correspondence: hone@ncepu.edu.cn

Abstract: To ensure the stability of park power supply systems and to promote the consumption of wind/photovoltaic generation, this paper proposes a dispatching optimization model for the park power supply system with power-to-gas (P2G) and peak regulation via gas-fired generators. Firstly, the structure of a park power system with P2G was built. Secondly, a dispatching optimization model for the park power supply system was constructed with a peak regulation compensation mechanism. Finally, the effectiveness of the model was verified by a case study. The case results show that with the integration of P2G and the marketized peak regulation compensation mechanism, preferential power energy storage followed by gas storage had the best effect on the park power supply system, which minimized the clean energy curtailment to 11.18% and the total cost by approximately $120.190 and maximized the net profit by approximately $152.005.

Keywords: park power supply system; power-to-gas; peak regulation compensation; ancillary service; wind/photovoltaic generation consumption

---

1. Introduction

Due to the fluctuation and randomness of wind/photovoltaic power generation, the issue of energy curtailment in China is still serious [1]. According to data from the National Energy Administration of China, the amount of wind power curtailment reached 10.5 billion kW·h in the first half of the year 2019 [2]. Wind/photovoltaic power curtailment is an obstacle to the development of clean energy in China and may cause greater economic losses. Therefore, research on the consumption of wind/photovoltaic power generation is still of great practical significance.

Power-to-gas (P2G) technology can, to a great extent, reduce wind/photovoltaic curtailment. Unconsumed electricity can be converted into natural gas via P2G and then into electricity in reverse through gas-fired generators (GFG), which plays a positive role in the consumption of clean energy generation [3]. Therefore, P2G can change the coupling modes of different energies [4]. In the previous studies on P2G, Gholizadeh et al. [5] evaluated and enhanced the security of P2G in networked energy hubs. Yang et al. [6] and Ye et al. [7] proposed optimal dispatching models for the multi-energy systems, which provided a new optimization method for the flexible operation of integrated energy systems. P2G technology can promote the level of renewable energy consumption in a hybrid system, as demonstrated by Marco et al. [8] and Hassan et al. [9]. Fischer al. [10] looked into the use of on-site storage and a model predictive controller and came up with possible solutions for P2G in a smart city context. All of the studies highlighted that P2G is an effective tool to promote the coupling of electricity and gas as well as clean energy consumption. Miguel et al. [11] analyzed the prospect of P2G in Portugal, while Garcia et al. [12] expanded the scope of research and extended it to Europe. Although the P2G is promising, it is still in the early stage of development [13]; thus, it is necessary to continue to study in this field.

Gas-fired power generation is the key for P2G to participate in the power-gas-power cycle. GFGs are quick to start/stop, have an easily adjustable output, and operate flexibly, which are suitable for peak regulation. Ghasemi et al. [14] looked into the importance of the gas-fired power plant location in a power-gas system with respect to peak regulation. Zhao et al. [15] studied the economy of peak regulation only via GFG, while Zhong et al. [16] analyzed the economy of peak regulation via GFG and electric vehicles under the battery-to-gas mode.

Peak regulation compensation is a method that is used to encourage the GFG to participate in peak regulation [17]. The importance of peak regulation compensation and its optimization have been discussed in the studies of Li et al. [18], Yang et al. [19], and Na et al. [20]. However, the above studies did not take P2G into consideration in their models.

In terms of the peak regulation of GFG, fuzzy clustering [21], mixed integer programming [22], and cooperative game models [16] were adopted to study its technological economy. With regard to P2G, Gil et al. [23] used the two-stage optimal power flow method to evaluate the impact of P2G technology on power and natural gas networks; the operational risks and costs of electrical systems due to wind power uncertainty were analyzed with the Conditional Value at Risk (CVaR) theory by Xu et al. [24], while Tan et al. [25] improved the CVaR theory to study the integrated energy system along with robust optimization. Weiler et al. [26] used three dimensional (3D) urban modeling tools in heat pump and co-generation systems.

Although previous research referred to integrated energy systems with P2G for energy coupling and power curtailment reduction, the role that the P2G plays in peak regulation to promote wind/photovoltaic consumption is absent, since peak regulation is so important in the current situation where China is trying to construct an electricity spot market and the production of clean energy is being encouraged to replace conventional fossil energy for sustainable development. Also, the economy of peak regulation via GFGs with the compensation mechanism was studied without the consideration of P2G's effect in prior literature. Hence, the research questions of this paper focus on three aspects. First, is the combination of P2G and peak regulation via GFG with compensation better for wind/photovoltaic consumption? Second, does this combination help with making more profit for industrial parks? Third, does the use of the energy storage device and P2G lead to different results? To answer the research questions, this study contributes in the following way, compared with the previously published papers:

First, more attention is paid to the optimization effect of the P2G on the park power supply system. In that case, whether the P2G, as an incentive, is conducive to clean energy utilization and the reduction of GFG's costs is found out. Second, the marketized peak regulation compensation mechanism, as another incentive, is introduced to better encourage GFG to cooperate with clean energy units for more clean energy consumption. Finally, with the integration of P2G and peak regulation compensation, scenarios, including different use orders of the energy storage device and the gas storage tank with the P2G, are designed in order to analyze which order is better for making the maximum profit for the system. The analysis result could serve as a reference for the practical application of these two devices.

2. Dispatching Optimization Model for the Park Power Supply System

To mitigate the clean energy curtailment without profit loss, a dispatching optimization model for the park power supply system needs to be constructed. In this section, the structure of the park power supply system is described, followed by the output model construction of different types of equipment. Finally, objective functions are established with constraints.

2.1. Structure of the Park Power Supply System with the P2G

The power supply system includes a wind turbine (WT), photovoltaic unit (PVU), energy storage device (ESD), P2G, gas storage tank, and GFG (see Figure 1). Priority is given to wind/photovoltaic power supply in the park, and the surplus power is input into the P2G or stored in the ESD. P2G is set to convert the surplus power into CH_4, and CH_4 is only used for power generation. The surplus CH_4 will be stored in the gas storage tank. Also, the WT, PVU, P2G, and gas storage tank (P2G and gas storage tank are defined as PGST in this paper) only participate in the internal transactions, and the GFG participates in the ancillary service market. When the wind/photovoltaic power generation cannot meet the power load of users in the park, the GFG provides the peak regulation service.

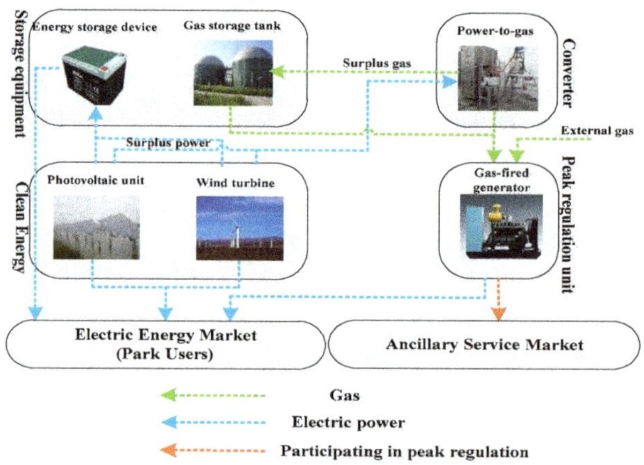

Figure 1. Park power supply system structure.

2.2. Output Models of Equipment

The available equipment in the constructed park power supply system includes clean energy units (WT and PVU), GFG, PGST, and ESD. Before building the dispatching optimization model, the output models of different kinds of equipment are given below.

The WT output is affected by the wind speed. When wind speed is within the acceptable range for the unit, the power output of the unit increases as the wind speed increases. However, if it exceeds the acceptable range, that is, the wind speed is too slow or too fast, the WT will not start in order to avoid damage to the body. The specific output model is detailed in [27]. The PVU output model generally coincides with the β distribution, which is detailed in [28].

The generation efficiency of GFG is greatly affected by its output, which decreases as the output decreases. The third-order efficiency model is adopted in this paper, which can better analyze the impact of output fluctuations on the system. The specific model [29] is

$$g_{i,g}(t) = F_{GT}(t) \cdot \eta_G(t) \tag{1}$$

where $F_{GT}(t)$ is the natural gas consumption of the GFG at time t (kW·h); $g_{i,g}(t)$ is the output of the GFG at time t (kW·h); and $\eta_G(t)$ is the generation efficiency of the GFG at time t (%).

P2G is one of the key conditions for realizing bidirectional energy coupling in the power-gas system. The reaction processes are detailed in [30]. The comprehensive energy conversion efficiency of power-to-CH_4 is about 45–60%. The power-to-CH_4 model and the state of gas storage [25] are illustrated as

$$Q_{P2G}(t) = E_{P2G}(t)\varphi_{P2G}/HCV \tag{2}$$

$$S_g(t) = S_g(T_0) + \sum_{t=1}^{T} (Q_{P2G}(t) - Q_o(t)) \tag{3}$$

where $Q_{P2G}(t)$ is the CH_4 production of the P2G at time t (m^3); $E_{P2G}(t)$ is the power consumption of P2G at time t (kW·h); φ_{P2G} is the P2G conversion efficiency (%); HCV is the high calorific value of natural gas (39 MJ/m^3) [31]; $S_g(t)$ is the gas volume in the gas storage tank at time t (m^3); $S_g(T_0)$ is the initial gas volume (m^3); and $Q_o(t)$ is the gas volume from the gas storage tank to the GFG at time t (m^3).

2.3. Peak Regulation Compensation Mechanism

Due to China's complicated electricity market, which is still developing, the peak regulation service (belonging to the ancillary service market) is an effective way to promote the consumption of clean energy generation. According to the "Market Regulation Measures for Power Peak Regulation Service" in a certain area of China, the peak regulation service is divided into basic (unpaid) and paid peak regulation based on a load rate of 52% (i.e., a peak regulation rate of 48%). The basic peak regulation service is provided when the peak regulation rate of a unit is less than 48%; the paid peak regulation service is provided when the peak regulation rate is more than 48% or a unit is started up or shut down for peak regulation.

The compensation mechanism involves units submitting the day-ahead quotations for peak regulation, and then the peak regulation service is provided according to the order from low to high. The current clearing price is taken as a settlement price, that is, the maximum quotation for the peak regulation compensation occurring in the unit in the statistical period of the same day. The peak regulation service follows the principle of "basic peak regulation taking precedence over paid peak regulation, and low-priced peak regulation taking precedence over high-priced peak regulation, on the premise of the safe operation of the power grid". The peak regulation compensation model is described as

$$P_a(t) = \begin{cases} P_1,\ 45\% < R_{load} \leq 52\% \\ P_2,\ 40\% < R_{load} \leq 45\% \\ P_3,\ R_{load} \leq 40\% \end{cases} \tag{4}$$

$$R_{load} = \frac{g(t)}{C(t)} \tag{5}$$

where $P_a(t)$ is the compensation price at time t (\$/kW·h); P_1, P_2, and P_3 are the current clearing prices under different peak regulation rates (\$/kW·h); R_{load} is the load rate of unit w at time t (%); $C(t)$ is the capacity of unit w (kW); and $g(t)$ is the available output of unit w at time t (kW·h).

2.4. Objective Functions

As illustrated above, to reduce clean energy wastage by GFG actively participating in peak regulation and gain more profit at the same time, the dispatching optimization model (including the objective functions and constraints) based on the system net profit, the wind/photovoltaic curtailment rate, and the GFG cost is constructed.

The economic benefit is the most important focus of the park power supply system; therefore, the net profits are considered. Firstly, the net profit model of WT and PVU is given by

$$Z_{wpv} = \sum_{j=1}^{J} \sum_{m=1}^{M} \sum_{t=1}^{24} [g_{j,w}(t) \times (p_{park}(t) - c_w) + g_{m,pv}(t) \times (p_{park}(t) - c_{pv})] \tag{6}$$

where Z_{wpv} is the net profit of the WT and PVU (\$); $g_{j,w}(t)$ and $g_{m,pv}(t)$ are the outputs of the WT and the PVU at time t, respectively (kW·h); $p_{park}(t)$ is the internal power price in the park (\$/kW·h); and c_w

and c_{pv} are the unit costs of the WT and the PVU ($). The unit cost covers unit power generation, operation, and maintenance costs.

The GFG costs mainly cover power generation, operation, and maintenance costs, and the income is from selling electricity and obtaining peak regulation compensation, so the net profit model is given by

$$Z_g = \sum_{i=1}^{I} \sum_{t=1}^{24} [g_{i,g}(t) \times (p_{park}(t) - c_g) + G_{i,g} \times P_a(t)] \tag{7}$$

$$c_g = F_{GT}(t) \times \rho_g + c_g^* \times g_{i,g}(t) \tag{8}$$

where Z_g is the net profit of the GFG ($); $P_a(t)$ is the peak regulation compensation price of the GFG at time t ($/kW·h); $G_{i,g}$ is the compensable electricity at time t (kW·h); c_g is the power generation cost of the GFG ($); ρ is the natural gas price ($/m^3); and c_g^* is the unit operation and maintenance cost ($).

Therefore, the net profit of the park is given by

$$f_1 = \max R = \max\left(Z_{wpv} + Z_g - c_{PGST} - c_{ESD}\right). \tag{9}$$

Therein,

$$c_{PGST} = g_{PGST}(t) \times c_{PGST}^* \tag{10}$$

$$c_{ESD} = g_{ESD}(t) \times c_{ESD}^* \tag{11}$$

where R is the system's net profit ($); c_{PGST} and c_{ESD} stand for the costs of the PGST and the ESD respectively, both of which only refer to the operation and maintenance costs ($); c_{PGST}^* and c_{ESD}^* are the unit operation and maintenance costs of the PGST and the ESD ($); and $g_{ESD}(t)$ and $g_{PGST}(t)$ are the storage capacity (kW).

Following the concern on net profits in the park, the clean energy consumption is also what this paper mainly focuses on; therefore, the objective of clean energy curtailment rate is given by

$$f_2 = \min I = \min\left(\frac{U_c}{U_t} \times 100\%\right) \tag{12}$$

where I is the clean energy curtailment rate (%); U_c is the actual clean energy curtailment (kW·h); and U_t is the total clean energy power generation (kW·h).

To ensure more electricity from clean energy units can be consumed, the GFG acts as a bridge between clean energy units and P2G; therefore, its cost is one of this paper's focuses. The objective of the minimum GFG costs is analyzed to motivate more willing cooperation with clean energy units. The corresponding objective is given by

$$f_3 = \min c_g. \tag{13}$$

2.5. Constraints

The constraints used in this paper include the supply and load balance, the running constraints of different types of equipment, and the reserve capacity of the system.

The constraint of the system power balance is stated as

$$\sum_{i=1}^{I} g_{i,g} + \sum_{j=1}^{J} g_{j,w} + \sum_{m=1}^{M} g_{m,pv} = D \tag{14}$$

where $g_{i,g}$, $g_{j,w}$, and $g_{m,pv}$ are the available outputs of GFG, WT, and PVU respectively (kW·h); and D is the load demand (kW·h).

The running constraints of GFG, WT, and PVU are detailed in [32].

The constraint of the system's reserve capacity is

$$\sum_{g=1}^{G}[g^{\max}(t)(1-\theta)](1-l) \geq D(t)+R(t) \tag{15}$$

where $D(t)$ is the load demand at time t (kW·h); $R(t)$ is the system reserve demand at time t (kW); l is the line loss rate (%); θ is the unit self-use rate (%); and $g^{\max}(t)$ is the maximum output at time t (kW·h).

Since the objective functions with the above constraints are mixed integer programming problems, the General Algebraic Modeling System (GAMS) software is a good solution tool. The mixed-integer programming models and the GAMS are detailed in [33].

3. Case Study

3.1. Scenario Settings

To verify the effectiveness of the proposed model, four scenarios based on different types of storage equipment and the different use priorities were set up (see Table 1).

The different types of storage equipment were as follows:

- In Scenario 1 (S1), the park power supply system employed an energy storage device (ESD) to store the unconsumed wind/photovoltaic power generated to be sold in the next period;
- In Scenario 2 (S2), the park power supply system employed a P2G and a gas storage tank (PGST) to convert the unconsumed wind/photovoltaic power generated into natural gas to be stored in the gas storage tank as the fuel for power generation.

The different storage equipment use priorities were as follows:

- In Scenario 3 (S3), the park power supply system employed a PGST and an ESD. The unconsumed wind/photovoltaic power generated was preferentially converted into natural gas via the P2G and stored in the gas storage tank as the fuel for power generation, and then the surplus was stored in the ESD.
- In Scenario 4 (S4), the park power supply system employed a PGST and an ESD. The unconsumed wind/photovoltaic power generated was preferentially stored in the ESD, and then the surplus was converted into natural gas via the P2G and stored in the gas storage tank as the fuel for power generation.

Table 1. Four scenarios based on different types of storage equipment and their use priorities.

Scenario	Energy Storage Device (ESD)	P2G and Gas Storage Tank (PGST)	Use Priority
S1	√	/	/
S2	/	√	/
S3	√	√	PGST > ESD
S4	√	√	ESD > PGST

3.2. Basic Data

A park power supply system used in an area of China was selected for the case study. In this park, clean energy units, gas-fired generators (GFGs), an ESD, and a PGST were employed. The clean energy units contained a wind turbine (WT) and a photovoltaic unit (PVU), and the PGST included a P2G and a gas storage tank. Besides, there were four GFGs in the park (the specific parameters are shown in Tables 2 and 3). The capacity of the ESD was 1000 kW, the maximum charging/discharging power was 200 kW, the charging/discharging loss was 0.4, and the initial energy storage was set to be 0; the unit cost of operation and maintenance was 0.071 $/kW·h.

Table 2. Equipment parameters of the wind turbine (WT), the photovoltaic unit (PVU), and the P2G and gas storage tank (PGST).

Equipment	Capacity	Unit Cost ($/kW·h)	Current Limitation (km^3/h)	Conversion Efficiency	Unit Operation and Maintenance Cost ($/kW·h)
Wind turbine (WT)	10 (MW)	0.049	/	/	/
Photovoltaic unit (PVU)	500 (kW)	0.085	/	/	/
P2G	400 (kW)	/	/	60%	0.078
Gas store tank	50 (km^3)	/	20~50	/	

Table 3. Equipment parameters of the gas-fired generators (GFGs)

GFG	*a*	*b*	*c*	*d*
Capacity(kW)	1000	500	500	500
Generation efficiency	60%	60%	60%	60%
Unit operation and maintenance cost ($/kW·h)	0.098	0.098	0.098	0.098

The power load demand and the output of clean energy units in this park are illustrated in Figure 2. Given the uncertainty of wind and photovoltaic power generation, the output data of the clean energy units on a typical day were selected for this research. GFGs satisfied the load demand that WT and PVU could not cope with.

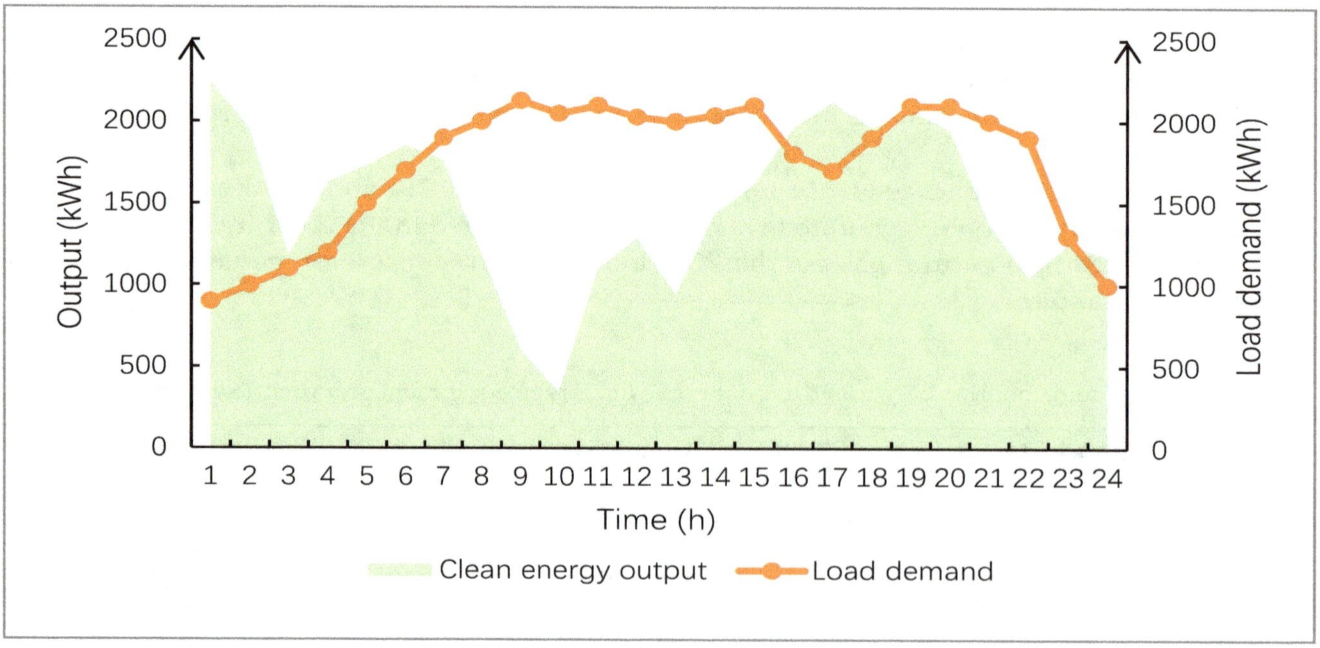

Figure 2. Power load demand of the park and output of clean energy units.

Based on the load demand in the park, a day was divided into three periods: 8:00–15:00 and 19:00–22:00 were the peak periods, 23:00–4:00 was the valley period, and 5:00–7:00 and 16:00–18:00 were the flat periods. For the given periods, the power price data are presented in Table 4, according to the actual prices of a province in China. In addition to power prices, the natural gas price was 0.488 $/m^3 (for convenient calculation, it was converted into the unit calorific value price, 0.049 $/kW·h).

Table 4. Price data.

Category	Item	Price ($/kW·h)
Park power price	Peak period	0.170
	Valley period	0.113
	Flat period	0.141
External power price	Peak period	0.204
	Valley period	0.157
	Flat period	0.172

Besides selling electricity, GFGs participate in peak regulation to gain profit, which increases clean energy consumption. In the light of peak regulation compensation mechanism, GFGs propose their quotations the day before operation. The quotation standards under different peak regulation rates are shown in Table 5.

Table 5. Quotation intervals for peak regulation.

Peak Regulation Rate (PRR)	Lower Limit ($/kW)	Upper Limit ($/kW·h)
48% < PRR ≤ 55%	0.042	0.071
55% < PRR ≤ 60%	0.071	0.113
PRR > 60%	0.113	0.141

According to the load demand and peak regulation rate of each unit, the day-ahead peak regulation quotations of all GFGs during each period were generated randomly, and these are displayed in Table 6.

Table 6. gas-fired generators' (GFGs') day-ahead quotations for peak regulation (Unit: $/kW·h).

Times \ Unit	GFG *a*	GFG *b*	GFG *c*	GFG *d*
1	0.061	0.056	0.062	0.069
2	0.066	0.055	0.063	0.069
3	0.065	0.054	0.060	0.058
4	0.060	0.062	0.059	0.066
5	0.082	0.094	0.085	0.097
6	0.108	0.107	0.078	0.113
7	0.091	0.078	0.109	0.110
8	0.139	0.126	0.135	0.132
9	0.113	0.131	0.118	0.121
10	0.125	0.113	0.114	0.139
11	0.138	0.140	0.137	0.124
12	0.131	0.123	0.140	0.117
13	0.121	0.132	0.123	0.129
14	0.123	0.124	0.115	0.122
15	0.130	0.117	0.135	0.132
16	0.089	0.103	0.107	0.092
17	0.106	0.105	0.085	0.098
18	0.096	0.082	0.097	0.084
19	0.124	0.130	0.129	0.139
20	0.126	0.119	0.130	0.132
21	0.123	0.136	0.137	0.130
22	0.135	0.117	0.138	0.122
23	0.062	0.055	0.066	0.058
24	0.054	0.063	0.066	0.065

3.3. Results Analysis

3.3.1. Wind/Photovoltaic Curtailment

WT, PVU, GFG, and ESD were the units providing power. The power output of PGST was in the form of gas-fired power generation. Figure 3 shows the output of different types of equipment in each scenario, and Table 7 shows the clean energy curtailment rates of different scenarios. The results indicate that the available outputs of the WT and PVU were basically the same in different scenarios, but the wind/photovoltaic curtailment differed accordingly.

The curtailment rate in the S1 was the highest, reaching 16.70%, whereas when the PGST was employed, the curtailment rate decreased (12.37% in S2), as did the GFG output. Moreover, in the scenario where both the PGST and ESD were employed, because of the different priorities of using the two devices, different results occurred for the wind/photovoltaic curtailment and the GFG output. The curtailment rate in S3 decreased to 11.59%compared with that in S2, but that in S4 decreased more—11.18%. Thus, the method of 'energy storage preferentially, then gas storage' has the best effect on reducing wind/photovoltaic curtailment.

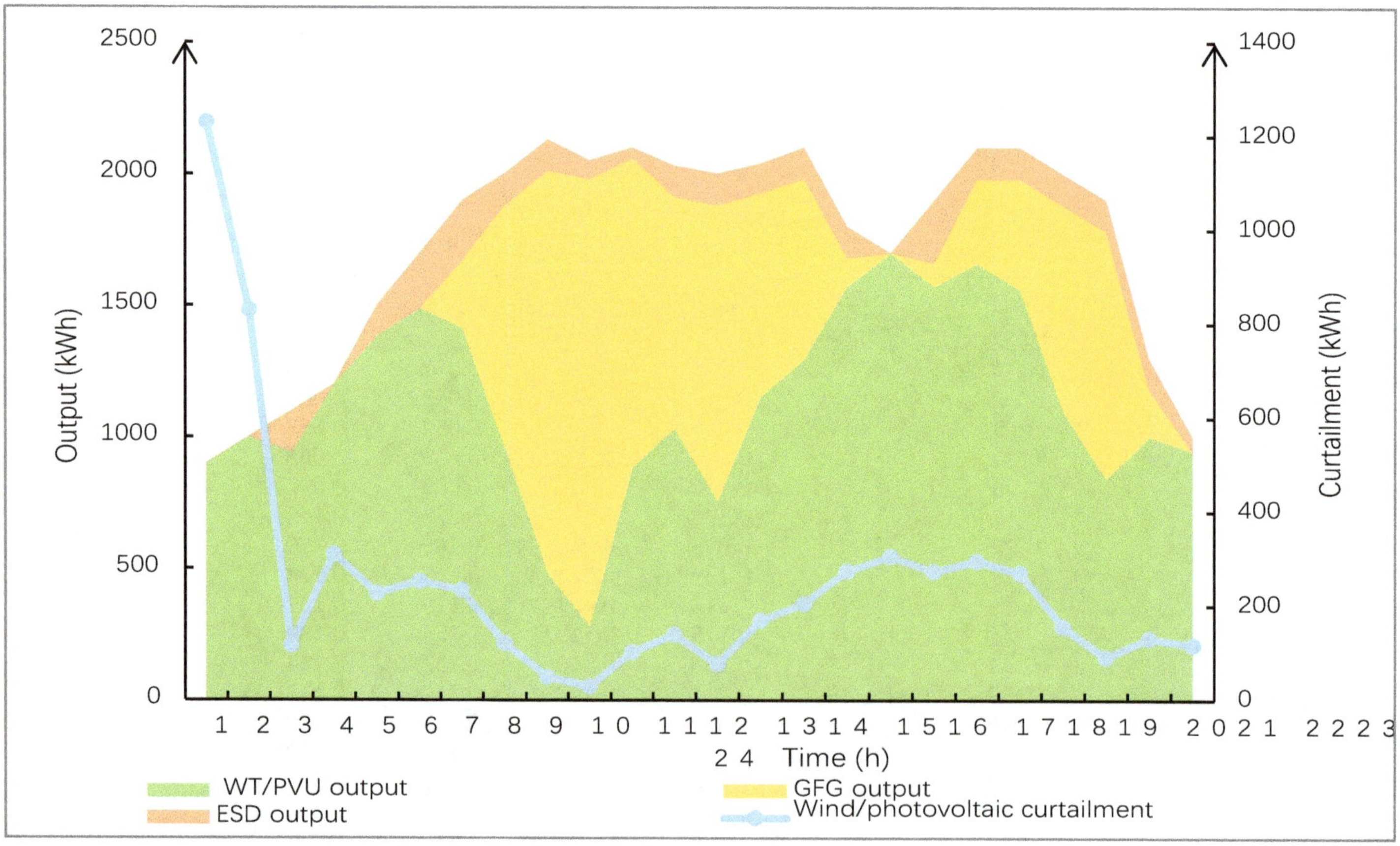

(a)

Figure 3. *Cont*

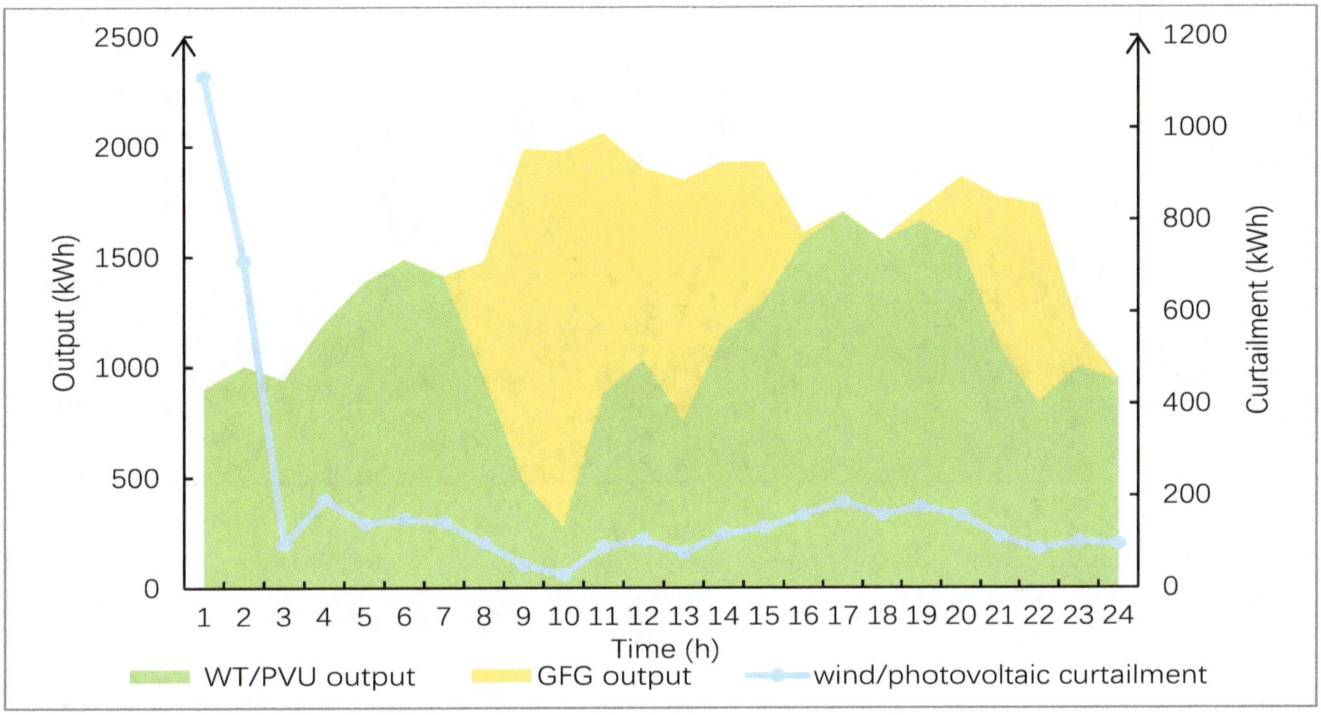

(**b**)

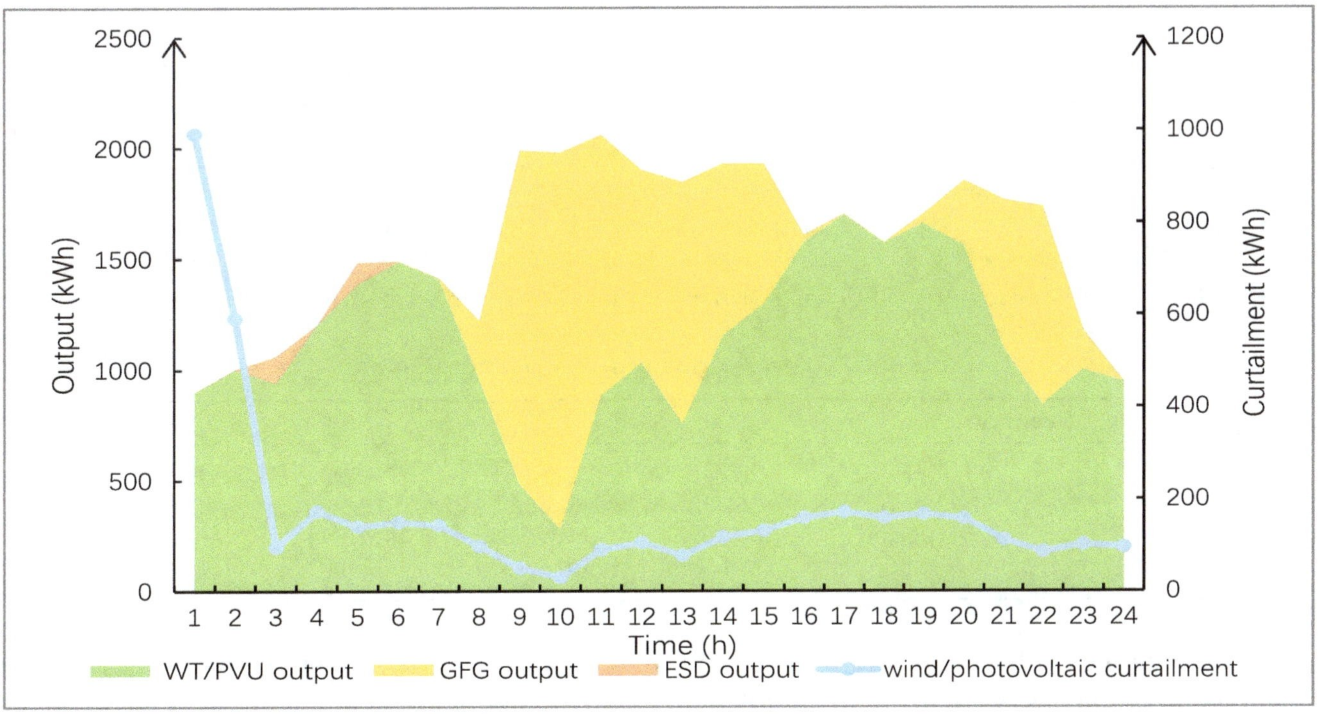

(**c**)

Figure 3. *Cont.*

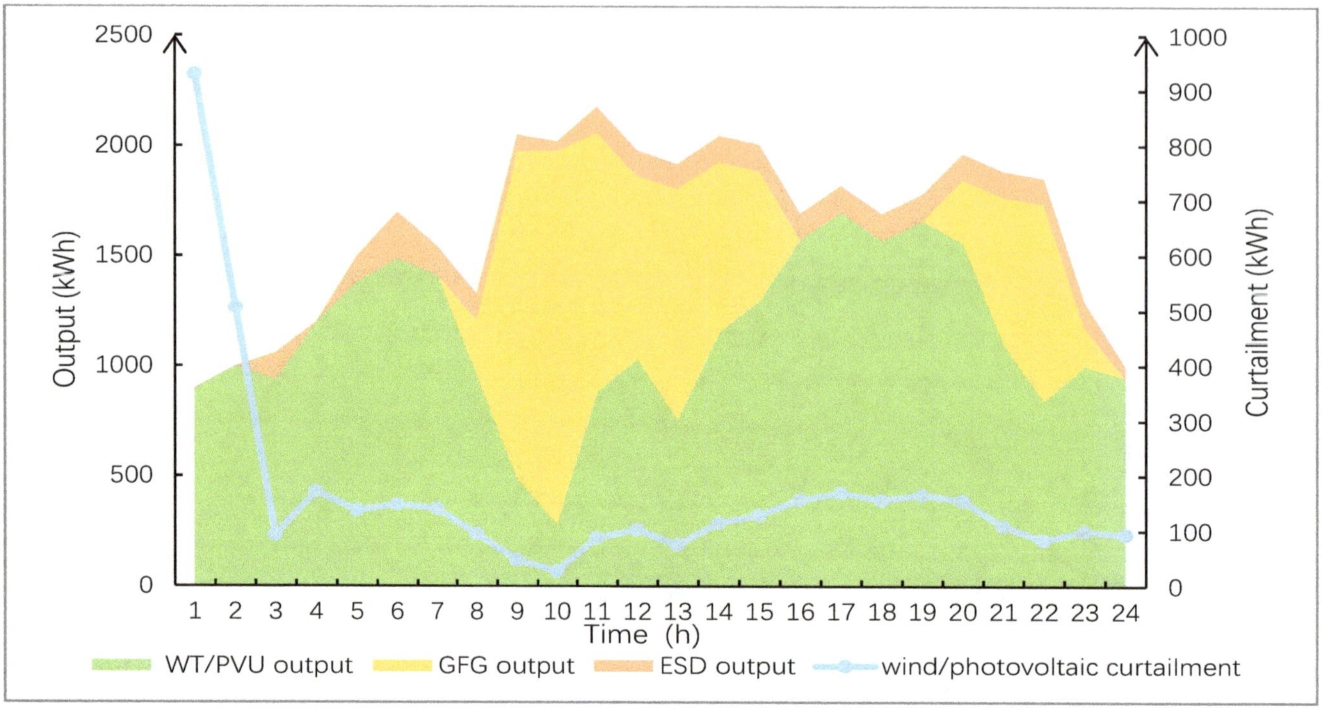

(**d**)

Figure 3. Unit output and wind/photovoltaic curtailment in different scenarios. (**a**) Unit output and wind/photovoltaic curtailment in S1. (**b**) Unit output and wind/photovoltaic curtailment in S2. (**c**) Unit output and wind/photovoltaic curtailment in S3. (**d**) Unit output and wind/photovoltaic curtailment in S4.

Table 7. Clean energy curtailment rates in different scenarios.

Scenario	1	2	3	4
Clean energy curtailment rate	16.70%	12.37%	11.59%	11.18%

3.3.2. GFG Costs

For different scenarios, the GFG costs during different periods are shown in Table 8.

Table 8. GFG costs during different periods (Unit: $).

Scenario		S1	S2	S3	S4
Period	Valley period	26.442	25.593	25.593	25.593
	Flat period	66.458	4.666	4.666	0
	Peak period	1651.693	1494.174	1452.744	1426.302
	Total	1744.593	1524.433	1483.003	1451.754

Table 8 indicates that, with the participation of the PGST, the GFG costs decreased significantly, especially in flat and peak periods. Compared with S1, the GFG costs in S2 decreased by $220.160, because the GFG used gas from the P2G instead of purchasing it from the outside; thus, its generation cost decreased. The GFG costs in S3 and S4 were both lower than that of S2 by $41.430 and $72.680, respectively, which indicates that with the combination of the ESD and the GPST, the costs were further reduced. Additionally, compared with S3, the costs in S4 decreased by $31.249, so the different priorities of using the PGST and the ESD affect the GFG costs, and the method of 'energy storage preferentially, then gas storage' has a better effort.

3.3.3. System Economic Benefits

In view of the above four scenarios, taking the transactions in the electric energy market and ancillary service market into account, the results of system economic benefits in different scenarios are shown in Table 9.

Table 9. System total costs and net profit in different scenarios (Unit: $).

Scenario	S1	S2	S3	S4
Total costs	3728.577	3662.543	3641.050	3608.387
Net profit	2203.436	2319.526	2336.069	2355.441

As presented in Table 9, the scenario with PGST (namely, S2) earned $116.090 more and cost $66.034 less than S1 where the PGST was not considered. This demonstrates that the P2G can effectively reduce the total cost and increase the net profit. Under the joint operation of the PGST and the ESD, the method of 'energy storage preferentially, then gas storage' (namely, S4) minimized the total cost of the system and maximized its net profit.

4. Conclusions

To promote clean energy consumption to mitigate resource wastage and encourage gas-fired generators to collaborate with clean energy units, the construction of a dispatching optimization model with P2G and the peak regulation compensation mechanism for GFG was carried out in this paper. The model tends to pursue the maximum net profit of the park power supply system and the minimum clean energy curtailment rate and GFG costs. After conducting a case study, it was found out that systems with P2G can effectively promote clean energy consumption in comparison with ones involving no P2G. Additionally, with the joint operation of the PGST and the ESD, the method of 'energy storage preferentially, then gas storage' can maximize benefits.

Due to the high installation capacity of clean energy units but insufficient peak regulation sources and immature market mechanism development in China, marketized compensation is still an important means to encourage conventional generators to participate in peak regulation for clean energy consumption, and P2G not only increases the consumption but also reduces GFG costs, being a novel means in the power industry. As P2G technology develops (lower investment and higher conversion efficiency), it will be more practical and popular in the future.

In future research, the effect of P2G on CO_2 emission reduction (environmentally) will be taken into account and how much it can gain for the park power supply system in terms of green certificate trading and carbon trading (economically) will be investigated. Also, more marketized peak regulation compensation means and more types of peak regulation units based on local resource advantages will be discussed, respectively.

Author Contributions: Supervision, Z.T.; Writing—Original draft preparation, Y.Q. and H.L.; Writing—Review and editing, Q.Y. and L.J.; Formal analysis, L.L. and S.Y.; Software, G.D.

References

1. Pu, L.; Wang, X.; Tan, Z.; Wu, J.; Long, C.; Kong, W. Feasible electricity price calculation and environmental benefits analysis of the regional nighttime wind power utilization in electric heating in Beijing. *J. Clean. Prod.* **2019**, *212*, 1434–1445. [CrossRef]
2. Renewable Energy Grid Operation in the First Half of 2019. Available online: http://www.nea.gov.cn/2019-07/25/c_138257185.htm (accessed on 25 July 2019).

3. Stephen, C.; Pierluigi, M. Integrated modeling and assessment of the operational impact of power-to-Gas (P2G) on electrical and gas transmission networks. *IEEE Trans. Sustain. Energy* **2015**, *6*, 1234–1244.
4. Ju, L.; Zhao, R.; Tan, Q.; Lu, Y.; Tan, Q.; Wang, W. A multi-objective robust scheduling model and solution algorithm for a novel virtual power plant connected with power-to-gas and gas storage tank considering uncertainty and demand response. *Appl. Energy* **2019**, *250*, 1336–1355. [CrossRef]
5. Gholizadeh, N.; Vahid-Pakdel, M.J.; Mohammadi-ivatloo, B. Enhancement of demand supply's security using power to gas technology in networked energy hubs. *Int. J. Electr. Power Energy Syst.* **2019**, *109*, 83–94. [CrossRef]
6. Yu, J.; Ma, M.; Guo, L.; Zhang, S. Reliability evaluation of integrated electrical and natural-gas system with power-to-gas. *Proc. CSEE* **2018**, *38*, 708–715.
7. Yang, D.; Xi, Y.; Cai, G. Day-Ahead Dispatch Model of Electro-Thermal Integrated Energy System with Power to Gas Function. *Appl. Sci.* **2017**, *7*, 1326. [CrossRef]
8. Marco, B.; Gabriele, F. Optimising energy flows and synergies between energy networks. *Energy* **2019**, *173*, 400–412.
9. Hassan, A.; Patel, M.K.; Parra, D. An assessment of the impacts of renewable and conventional electricity supply on the cost and value of power-to-gas. *Int. J. Hydrogen Energy* **2019**, *44*, 9577–9593. [CrossRef]
10. Fischer, D.; Kaufmann, F.; Oliver, S.L.; Christoper, V. Power-to-gas in a smart city context—Influence of network restrictions and possible solutions using on-site storage and model predictive controls. *Int. J. Hydrogen Energy* **2018**, *43*, 9483–9494. [CrossRef]
11. Miguel, C.V.; Mendes, A.; Madeira, L.M. An Overview of the Portuguese Energy Sector and Perspectives for Power-to-Gas Implementation. *Energies* **2018**, *11*, 3259. [CrossRef]
12. Garcia, D.A.; Barbanera, F.; Cumo, F.; di Matteo, U.; Nastasi, B. Expert Opinion Analysis on Renewable Hydrogen Storage Systems Potential in Europe. *Energies* **2016**, *9*, 963. [CrossRef]
13. Liu, W.; Wen, F.; Xue, Y.; Zhao, J.; Dong, Z.; Zheng, Y. Cost characteristics and economic analysis of power-to-gas technology. *Autom. Electr. Power Syst.* **2016**, *40*, 1–11.
14. Masoud, G.; Rajabi, M.H.; Amin, H. Importance of gas-fired power plants location in integrated operation of power and natural gas systems: Peak load condition analysis. In Proceedings of the 26th Iranian Conference on Electrical Engineering, Mashhad, Iran, 8–10 May 2018; pp. 1227–1232.
15. Zhao, C.; Zhang, H.; Li, F.; Yuan, J. Economic research on gas CHP and peaking generation in Beijing. *Int. Pet. Econ.* **2017**, *25*, 99–105.
16. Zhong, K.; Wang, D.; Yang, G.; Fu, J. Peak shaving analysis of electric vehicle and gas turbine under B2G mode. *Electrotech. Appl.* **2018**, *37*, 47–52.
17. Shen, W.; Zhen, W.; Zhou, D.; Gao, Y. Wind power peak Regulation pricing model under wind and fire alternative trading mechanism—A case study of wind Power Integration, Gansu Province. In Proceedings of the 2018 China International Conference on Electricity Distribution, Tianjin, China, 17–19 September 2018.
18. Li, Y.; Li, H. Improvement of compensation mechanism on peak-regulating auxiliary service for thermal power unit with energy saving dispatch. *Heilongjiang Electr. Power* **2014**, *26*, 194–197.
19. Yang, Z.; Li, K.; Wang, N.; Jin, Z.; Song, S.; Guo, X. A Model of Considering the Economic Analysis and Environmental Protection for Thermal Power Compensation on Peak Regulation. *J. Eng. Thermophys.* **2018**, *39*, 2124–2130.
20. Na, C.; Yuan, J.; Zhu, Y.; Xue, L. Economic Decision-Making for Coal Power Flexibility Retrofitting and Compensation in China. *Sustainability* **2018**, *10*, 348. [CrossRef]
21. Wang, Y.; Tian, Y.; Wu, M.; Geng, J. Two-part electricity price model for peak load regulation of natural gas power based on fuzzy clustering. *Proc. CSEE* **2017**, *6*, 38–46.
22. Kong, D.; Long, H.; Li, G.; Liu, Y. Research on equivalent smoothing strategy of PV output based on micro energy system. *Acta Energy Sol. Sin.* **2017**, *9*, 33–41.
23. Gil, M.; Duenas, P.; Reneses, J. Electricity and natural gas interdependency: Comparison of two methodologies for coupling large market models within the European regulatory framework. *IEEE Trans. Power Syst.* **2016**, *31*, 361–369. [CrossRef]
24. Xu, Z.; Zhang, Y.; Chen, Z.; Lin, X.; Chen, B. Bi-level optimal capacity configuration for power to gas facilities considering operation strategy and investment subject benefit. *Autom. Electr. Power Syst.* **2018**, *42*, 76–84.
25. Tan, Z.; Tan, Q.; Yang, S.; Ju, L.; De, G. A Robust Scheduling Optimization Model for an Integrated Energy System with P2G Based on Improved CVaR. *Energies* **2018**, *11*, 3437. [CrossRef]

26. Weiler, V.; Stave, J.; Eicker, U. Renewable Energy Generation Scenarios Using 3D Urban Modeling Tools Methodology for Heat Pump and Co-Generation Systems with Case Study Application †. *Energies* **2019**, *12*, 403. [CrossRef]
27. Cheng, J.; Choobineh, F. A Novel Wind Energy Conversion System with Storage for Spillage Recovery. *J. Power Energy Eng.* **2015**, *3*, 33–38. [CrossRef]
28. Wang, Y.; Lu, Y.; Ju, L.; Wang, T.; Tan, Q.; Wang, J.; Tan, Z. A Multi-objective Scheduling Optimization Model for Hybrid Energy System Connected with Wind-Photovoltaic-Conventional Gas Turbines, CHP Considering Heating Storage Mechanism. *Energies* **2019**, *12*, 425. [CrossRef]
29. Ju, L.; Zhang, Q.; Tan, Z.; Wang, W.; Xin, H.; Zhang, Z. Multi-agent-system-based coupling control optimization model for micro-grid group intelligent scheduling considering autonomy-cooperative operation strategy. *Energy* **2018**, *157*, 1035–1052. [CrossRef]
30. Andrea, M.; Ettore, B.; Gianfranco, C. Applications of power to gas technologies in emerging electrical systems. *Renew. Sustain. Energy Rev.* **2018**, *92*, 794–806.
31. Wei, Z.; Zhang, S.; Sun, G.; Zang, H.; Chen, S.; Chen, S. Power-to-gas Considered Peak Load Shifting Research for Integrated Electricity and Natural-gas Energy Systems. *Proc. CSEE* **2017**, *37*, 4601–4609.
32. Yang, S.; Tan, Z.; Ju, L.; Lin, H.; De, G.; Tan, Q.; Zhou, F. An Income Distributing Optimization Model for Cooperative Operation among Different Types of Power Sellers Considering Different Scenarios. *Energies* **2018**, *11*, 2895. [CrossRef]
33. Wang, J. Optimization of Power System Operation Based on GAMS. Master's Thesis, South China University of Technology, Guangzhou, China, 2014.

Energy Model for Long-Term Scenarios in Power Sector under Energy Transition Laws

Gabriela Hernández-Luna [1], Rosenberg J. Romero [1], Antonio Rodríguez-Martínez [1,*], José María Ponce-Ortega [2], Jesús Cerezo Román [1] and Guadalupe Diocelina Toledo Vázquez [1]

1 Engineering and Applied Sciences Center Research—CIICAp, Autonomous University of Morelos State—UAEM, Av. Universidad 1001, Chamilpa, Cuernavaca 62210, Morelos, Mexico; gabriela.hernandez@uaem.mx (G.H.-L.); rosenberg@uaem.mx (R.J.R.); jesus.cerezo@uaem.mx (J.C.R.); diocelina.toledo@gmail.com (G.D.T.V.)

2 Faculty of Chemical Engineering, Universidad Michoacana de San Nicolás de Hidalgo, Av. Francisco J. Múgica S/N Ciudad Universitaria, Morelia 58030, Michoacán, Mexico; jmponce@umich.mx

* Correspondence: antonio_rodriguez@uaem.mx

Abstract: High electricity demand, as well as emissions generated from this activity impact directly to global warming. Mexico is paying attention to this world difficulty and it is convinced that sustainable economic growth is possible. For this reason, it has made actions to face this problem like as launching constitutional reforms in the power sector. This paper presents an energy model to optimize the grid of power plants in the Mexican electricity sector (MES). The energy model considers indicators and parameters from Mexican Energy Reforms. Electricity demand is defined as a function of two population models and three electricity consumption per capita. Prospectives are presented as a function of total annual cost of electricity generation, an optimal number of power plants—fossil and clean—as well as CO_2eq emissions. By mean of the energy model, optimized grid scenarios are identified to meet the governmental goals (energy and environment) to 2050. In addition, this model could be used as a base to identify optimal scenarios which contribute to sustainable economic growth, as well as evaluate the social and environmental impacts of employed technologies.

Keywords: electricity model; power plants prospectives; Mexican prospectives

1. Introduction

Power in any region of the world is essential for development and economic growth. The electricity sector has historically developed under uncertainty and constant changes, especially since the end of the 1980s when the damage to the ozone layer was evidenced [1]. The magnitude of the event gave a global social concern, which is reflected in the Montreal Protocol in 1987 [2], the multilateral treaty on the environment has had the most success in all of history. This treaty subsequently gives rise to international agreements such as the Kyoto Protocol in 1998 [3] until reaching the Sustainable Development Goals of the United Nations launched in 2015 [4], also considering the Paris Agreement of 2015 [5] which established keeping the global temperature rise below 2 °C. However, despite the international effort, electric consumption continues, increasing incessantly, as reported by the World Bank in 2017 [6], as can be seen in Figure 1.

This high electricity demand brings an environmental impact, which has reached 30% of the total global emissions of greenhouse gases (GHG) as Figure 2 shows.

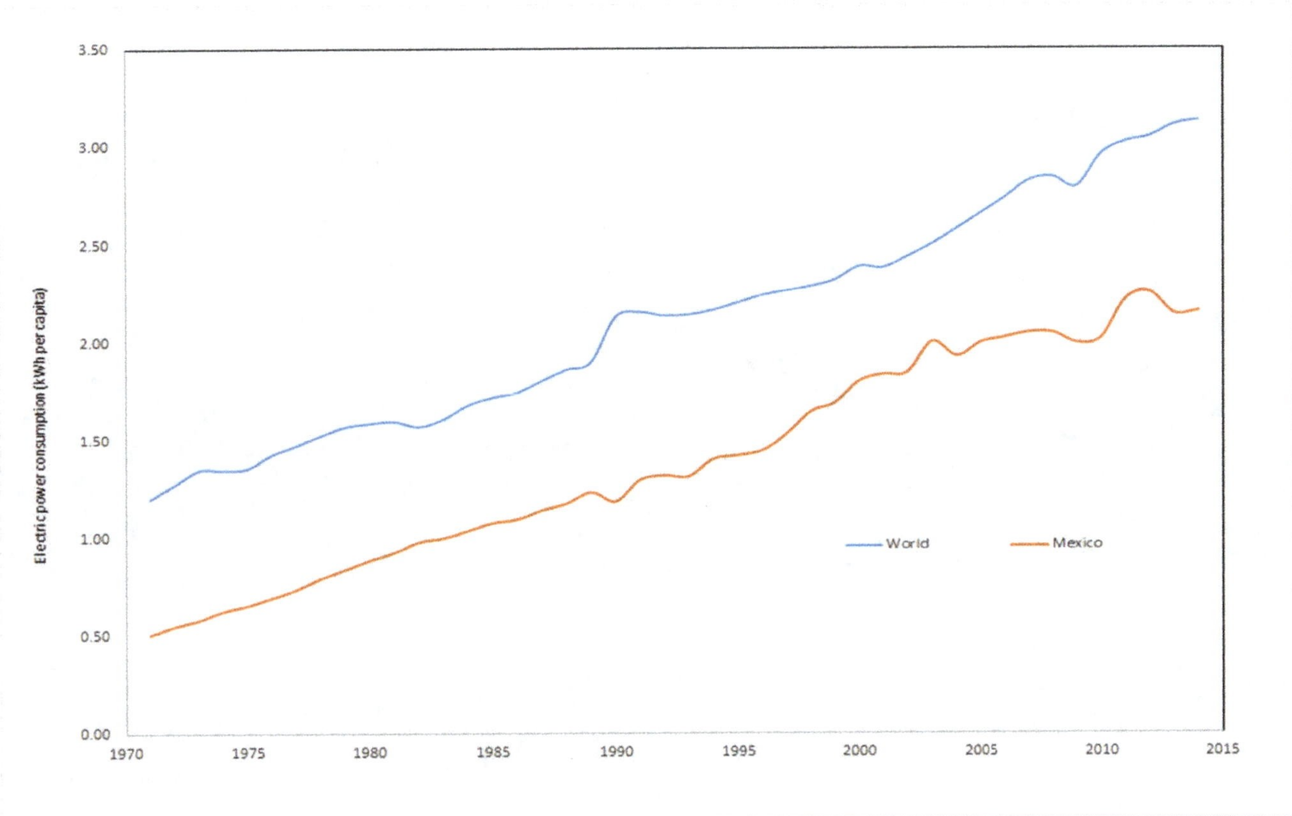

Figure 1. World and Mexico electric power consumptions. Prepared by the authors based on data from cited Reference [6].

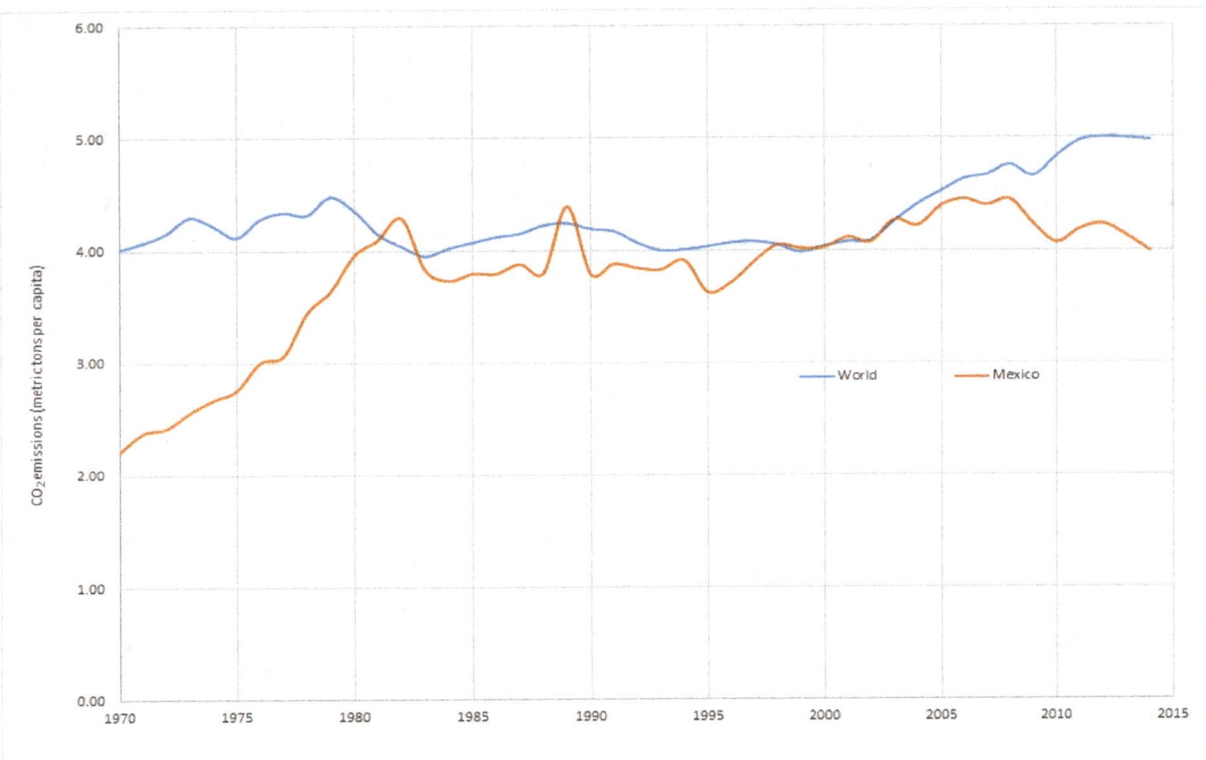

(a)

Figure 2. *Cont.*

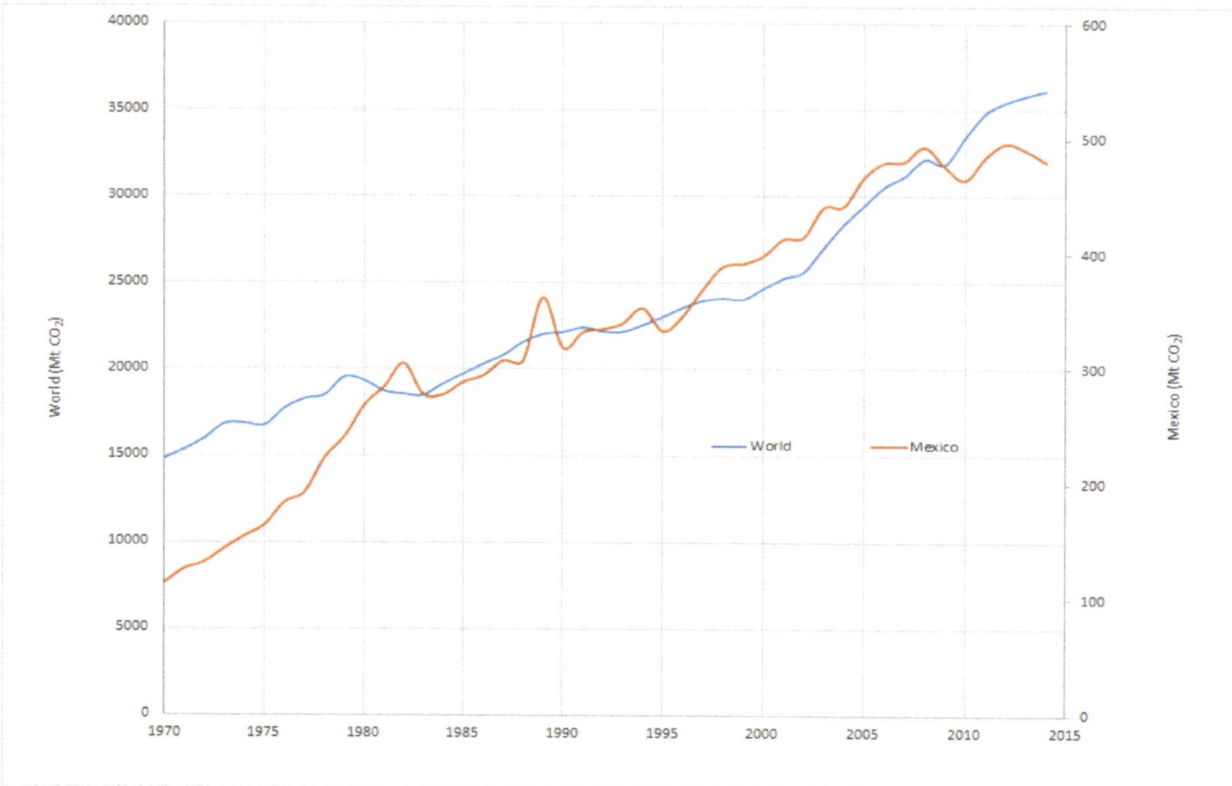

(**b**)

Figure 2. World and Mexico (**a**) CO_2 emissions per capita and (**b**) CO_2 total emissions. Prepared by the authors based on data from cited Reference [6].

The world faces a challenge between meeting the world's electricity demand and reducing greenhouse gas emissions produced by this sector. Mexico is not only sensitive to this global reality but it is also convinced that economic progress can and should be parallel to environmental protection. Under this premise, the Mexican government has taken actions to face this challenge as issuing reforms related to the energy sector, specifically in the Mexican Electricity System (MES) since this is the second largest emitter of GHG in the country, only after the transport sector. The GHG emissions produced by electricity sector has reached approximately 25% of total national emissions in 2015, as reported by National Institute of Climate Change (INECC) in 2015 [7] (see Figure 3) due to electricity generated in Mexico which is around 80% from fossil energy sources as Energy State Secretary (SENER) indicated in 2017 [8].

The amendments or reforms in energy affairs are also a product of the influence of international organizations and financial institutions as well as of changes in the price of fuels [9]. The main reforms launched in this field are "*General Climate Change Law*" (GCCL) in 2012 [10] and the "*National Climate Change Strategy*" (NCCS) in 2013 [11], whose main objective is minimizing GHG emissions produced by electricity generation and also modernize the Mexican Electricity System. The strategy considers the planning and sustainable growth, low carbon emissions of the MES in long term, reflected in an increase of the so-called "clean technologies" (According to the Energy Transition Law, 2015 [12]) for electricity generation. The goal is to achieve at least 35% of electricity generated from clean sources for the year 2024, in addition in achieving a specific GHG emission reduction of 30% and 50% for the years 2020 and 2050, respectively, with respect to the 2000 year emissions [10,11].

In response to this questioning, the use of renewable technologies is promoted to replace the main conventional technologies and achieve the proposed goals, however, this option can lead to risks such as potential incidences in the planning and construction of plants generating power, in addition to potential situations that would increase the start-up time of the electricity generating plants. This is one of the reasons why the use of simulations and numerical models can be a useful tool that allows strategically planning and visualizing the sector to suggest scenarios where proposed objectives by reforms are achieved.

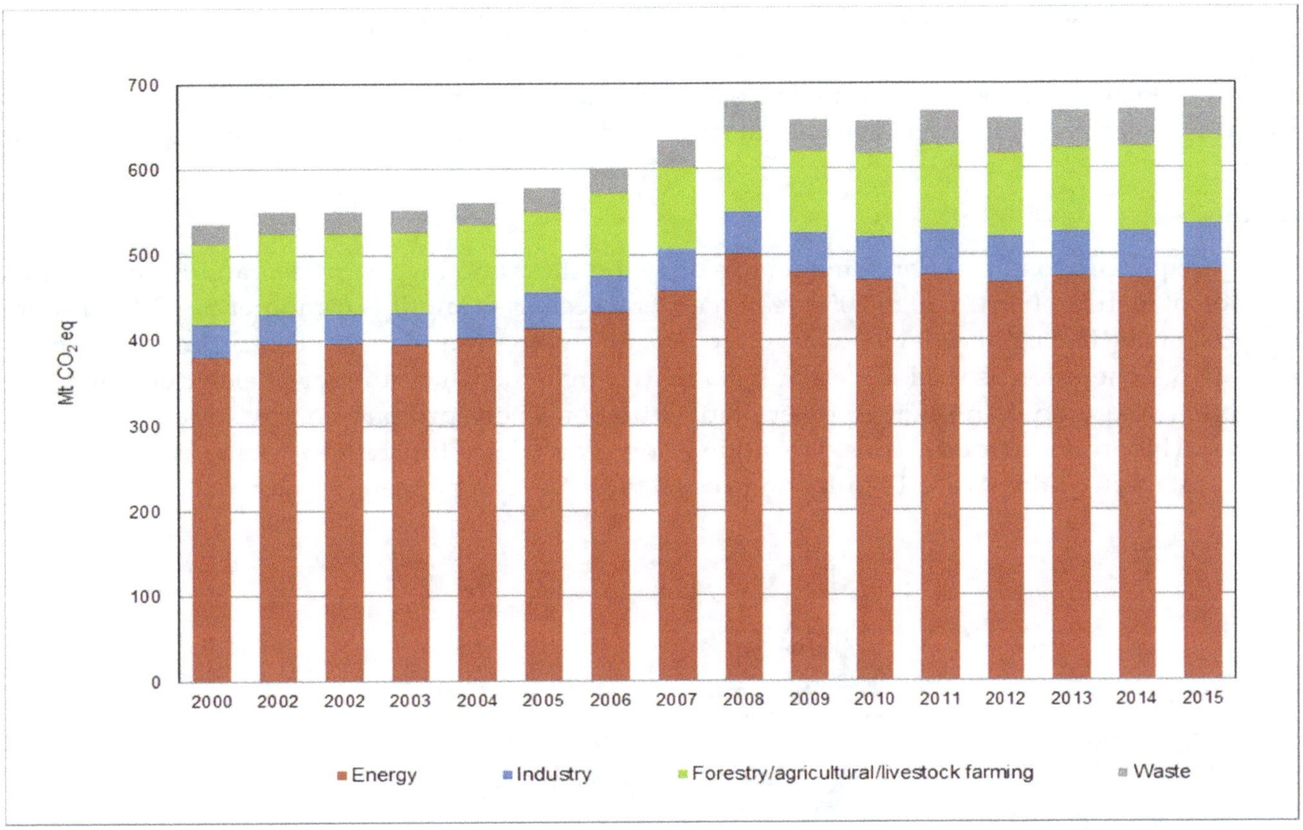

Figure 3. Mexican CO_2 emissions by sector. Prepared by the authors based on data from cited Reference [7].

The energy sector has been the object of study from various points of view; from the economic, technical as well as environmental. This section presents a brief review of the research.

Manzini et al. [13] analyzed the environmental impacts of the use of renewable sources in electricity generation. They basically identified CO_2, CH_4, NOx and SOx emissions for conditions where renewable energy reaches a contribution of 31 and 43% of installed capacity.

The World Bank [14] has also studied this sector and in 2009 they conducted an analysis to identify areas of influence in the reduction of GHG; to identify the most important to the electricity sector. In this study, the authors also identify the minimum costs available to be implemented using the cost-benefit methodology.

This cost-benefit methodology has been used by Lund and Mathiesen [15] for the study of the electrical system in Denmark, analyzing transition scenarios emphasizing areas such as energy efficiency, reduction of emissions of CO_2eq and industrial development, whose objective was is to reach up to 100% of the demand for electrical energy using renewable sources.

In 2014, Santoyo-Castelazo et al. [16] analyzed the environmental implications of decarbonization of the MES, by proposing scenarios with diverse configurations of the energy matrix of technologies used for combined-cycle power plant with life cycle analysis studies. The authors concluded that it is evident the possibility of reducing the environmental impact with the proposals made.

The MES energy transition has been developing also studied by Vidal-Amaro et al. [17]. The authors made a proposal to determine an optimal configuration of the energy matrix consisting of fossil and renewable sources for one moment in time, 2024 with the goal of reaching 35% of electricity from renewable sources, besides the authors managed to identify an optimal configuration as well as quasi-optimal configurations as an alternative.

The review focuses its efforts on analyzing GHG emissions as well as on identifying or evaluating the contributions of renewable energy sources to the MES in a specific scenario; however, no methodologies have been identified that propose identifying the number of power plants generating electricity to meet the demand neither short nor long term. This paper presents a proposal for an alternative electricity generation model to optimize the number of power plants to satisfy demand under Mexico's government policies with at short and at long term environmental impact scenarios.

2. Methods

This paper proposes an energy model to satisfy the short and long term demand for power plants in periods of lustrum from 2020 to 2050 year at several scenarios and it is compared with International Energy Agency (IEA) [18] data. The objective function is to minimize the Total Annual Cost (TAC) of electricity generation as well as determine a matrix made up of a number of electricity generating power plants required, minimizing power plants number whose primary energy source is from fossil resources. The proposed model considers population growth as a function to determine the electricity demand, taken from the World Population Prospects [19] as can be seen in Figure 4.

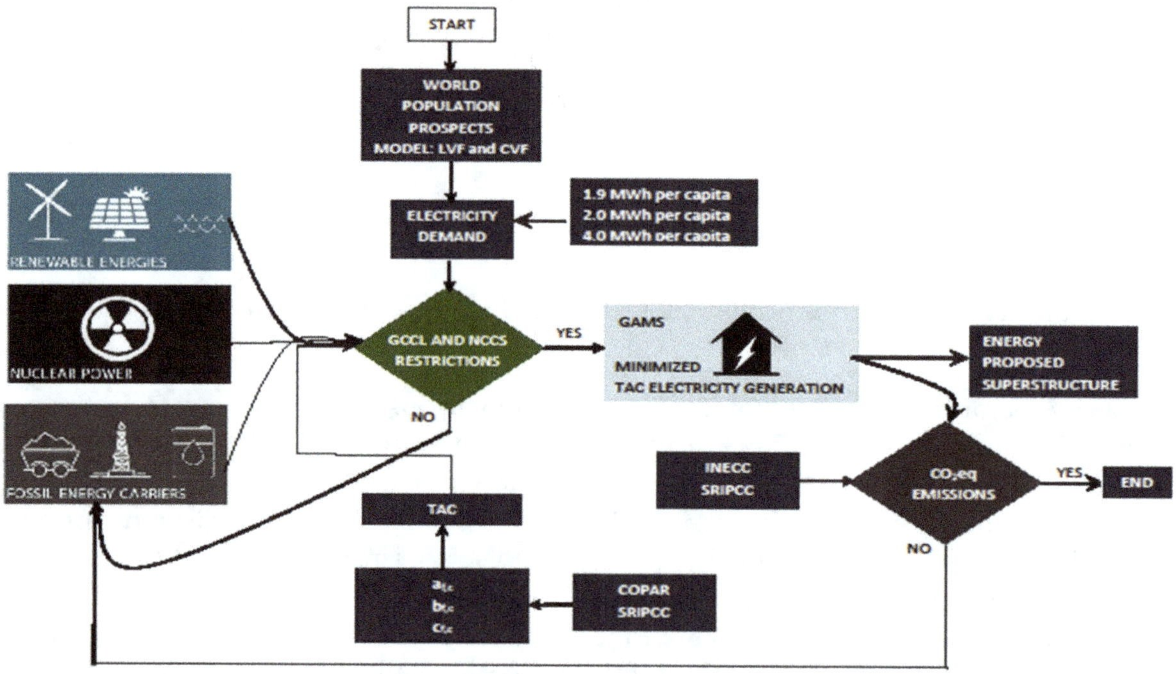

Figure 4. Schematic representation of Mexican Electricity System (MES) optimization.

The proposed matrix (composed by the number of power plants) is based on a superstructure which considers technologies based on fossil and clean energy sources for electricity generation, presented in Figure 4 and the code is in Appendix A. In addition, the software GAMS © identifies the optimal number of power plants generation to satisfy a period demand and CO_2eq emissions. The CO_2eq emissions are compared with NCCS and GCCL, if the value is lower than the constraint then the superstructure is shown as a result.

Electricity demand is determined using two population models from the United Nations (2017) [19], which are detailed next.

2.1. Electricity Demand

The proposed mathematical model assumes that the annual electricity demand is satisfied with electricity generation from two main energy sources: from fossil and clean sources, which is stated as follows:

$$E_d = \sum_i E_i \quad (1)$$

E_d is the annual electricity demand to be met, E_i the total electricity generated for each analyzed period corresponding to the total amount of the electricity coming from fossil sources, E_f, as well as from clean sources, E_c, in other words, i = f for fossil sources, while i = c for clean sources.

Electricity demand, E_d, is defined with two population models used in the realization of World Population Prospects of the Department of Economics and Social Affairs of the United Nations [19]. The population models are the Low Variability of Fertility (LVF) and life expectancy at birth and Constant Variability of Fertility (CVF) and life expectancy at birth. Three demand conditions per capita are assumed in each model: 1.9, 2.0 and 4.0 MWh. The demand of 1.9 MWh per capita is taken from the historical demand of Mexico in the period of 2000 to 2010; the demand of 2.0 MWh per capita considers an increase of 10% in the population demand and finally, the demand of 4.0 MWh per capita is the typical demand of the population consumption from developed countries. Table 1 presents a matrix with the described conditions.

Table 1. Matrix of described conditions.

	Population Model					
	LVF			CVF		
	Electricity Demand [MWh per capita]					
Year	1.9	2.0	4.0	1.9	2.0	4.0
2015	LVF-1.9-2010	LVF-2.0-2010	LVF-4.0-2010	CVF-1.9-2010	LVF-2.0-2010	LVF-4.0-2010
2020	LVF-1.9-2010	LVF-2.0-2010	LVF-4.0-2010	CVF-1.9-2010	LVF-2.0-2010	LVF-4.0-2010
2025	LVF-1.9-2025	LVF-2.0-2025	LVF-4.0-2025	CVF-1.9-2025	LVF-2.0-2025	LVF-4.0-2025
2030	LVF-1.9-2030	LVF-2.0-2030	LVF-4.0-2030	CVF-1.9-2030	LVF-2.0-2030	LVF-4.0-2030
2035	LVF-1.9-2035	LVF-2.0-2035	LVF-4.0-2035	CVF-1.9-2035	LVF-2.0-2035	LVF-4.0-2035
2040	LVF-1.9-2040	LVF-2.0-2040	LVF-4.0-2040	CVF-1.9-2040	LVF-2.0-2040	LVF-4.0-2040
2045	LVF-1.9-2045	LVF-2.0-2045	LVF-4.0-2045	CVF-1.9-2045	LVF-2.0-2045	LVF-4.0-2045
2050	LVF-1.9-2050	LVF-2.0-2050	LVF-4.0-2050	CVF-1.9-2050	LVF-2.0-2050	LVF-4.0-2050

2.2. Model Formulation

Electricity generation from each type of fuel is determined with Equation (2).

$$\sum E_i = \sum_f \sum_c E_{f,c} \leq \sum_f \sum_c \left(E_{f,c_{inst}} + x_{f,c} \cdot E_{f,c_{cap}} \cdot fc\right) \quad (2)$$

The fossil energy sources, $i = f$, consider electricity generation from coal, *co*; diesel, *d*; gas *g* and fuel oil *o*, while electricity generation from clean energy sources, i = c, considers the use of biomass, *bm*; carbon capture and storage, *ccs*; eolic, *eo*; photovoltaic, *phv*; geothermal, *gtr*; hydraulic, *hdr*; nuclear, *nc* and solar concentration, *sc*.

The proposed model determines superstructure electricity generation in each analyzed period taking into account electricity contribution generated in a period prior to the analyzed one, $E_{f,c_{ins}}$, the number of power plants generation $x_{f,c}$ and the capacity of each power plant, $E_{f,c_{cap}}$ as well as a coefficient fc to homogenize units.

Total country annual cost of electricity generation from each energy plant as well as the emissions due to this process is determined by Equations (3) and (4), respectively.

$$TAC(E_i) = \sum_f \sum_c \left(E_{f,c} \cdot Cost\left(a_{f,c} + b_{f,c} + c_{f,c}\right)\right) \quad (3)$$

where $a_{f,c}$ represents investment costs, $b_{f,c}$ is the associated cost from fuel type used and finally $c_{f,c}$ represents operation and maintenance cost, considering fixed and variable costs, taken from Generation, Costs and Reference Parameters for the Formulation of Investment Projects for the Mexican Electricity Sector (COPAR) [20] and Special Report of the Intergovernmental Panel of Climate Change (SRIPCC) of 2012 [21].

CO_2eq emissions are determined with the Equation (4).

$$CO_2eqEm(E_i) = \sum_f \sum_c \left(E_{f,c} \cdot CO_2eq_{f,c}\right) \tag{4}$$

$COeq_{f,c}$ factor is particular for each type of fuel, taken from the INECC [7] and SRIPCC [21].

In an integrated manner, for each analyzed electricity demand condition, TAC of optimized electricity generation, matrix formed by the number of power plants required to satisfy the demand, integrated by power plants that use fossil fuel technologies and clean, as well as the emissions generated by each type of technology are calculated.

2.3. Solution Strategy

To determine the set of optimal solutions that satisfy governmental criteria in each electrical demand condition, epsilon constraint method [22] is implemented in the proposed model. The Pareto chart is constructed using that demand condition and optimized TAC. The epsilon constraint method satisfies governmental objectives proposed in the GCCL and NCCS, (2012) and (2013), respectively. These restrictions have implications to be considered in the application of the model, presented in Equations (5) and (6).

$$E_f < \%f \cdot E_d \tag{5}$$

$$E_c < \%c \cdot E_d \tag{6}$$

E_f is the amount of electric energy generated from fossil sources, E_c is the amount of electric energy generated from clean sources, described in Table 2 and in Figure 5.

Table 2. Objectives and goals to electricity generation by source energy defined in the General Climate Change Law (GCCL) Prepared by the authors based on public data from cited Reference [10].

Electricity Generation Percentage by Energy Source, %		
Year	**Fossil, f**	**Clean, c**
2024	65	35
2035	60	40
2050	50	50

The model considers linear tendency and it was formulated in the General Algebraic Modelling System (GAMS) software used for modelling and mathematical optimization. The results obtained for each condition is shown in Table 1 and under the restrictions presented in Table 2 are detailed in the next section.

The restrictions are used in MES planning scenarios throughout the analysis period.

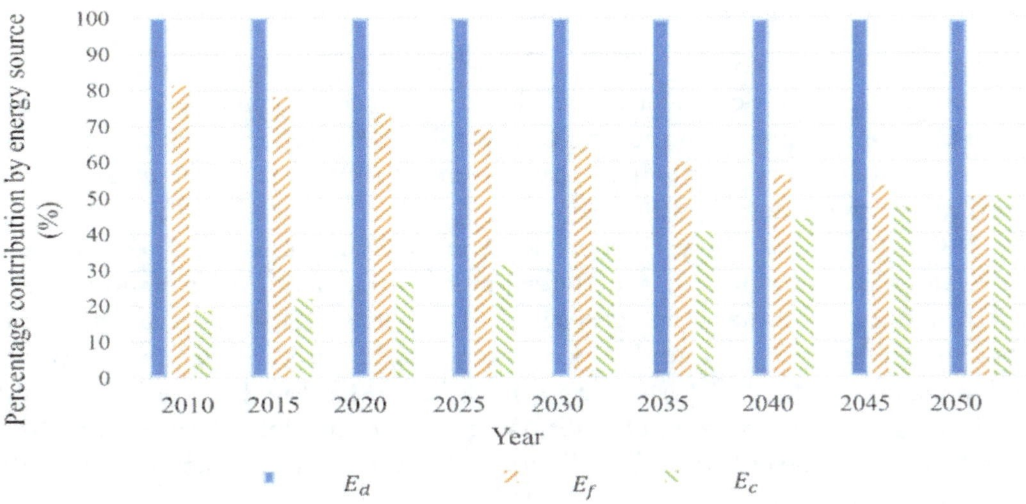

Figure 5. Governmental objectives to satisfy in General Climate Change Law (GCCL). Prepared by the authors based on data from cited reference [10].

3. Results

The modelling scenarios are presented in each electricity demand condition at three different conditions: electricity generation costs, optimized power plants number and generated emissions by this activity.

3.1. Electricity Demand

Electricity demand conditions determined with two population models throughout the analyzed period is presented in Figure 6. Three analyzed demand conditions using LVF presents a growth rate of 13% during all period, meanwhile the corresponding three analyzed demand conditions growth rate using CVF model is 50% are shown.

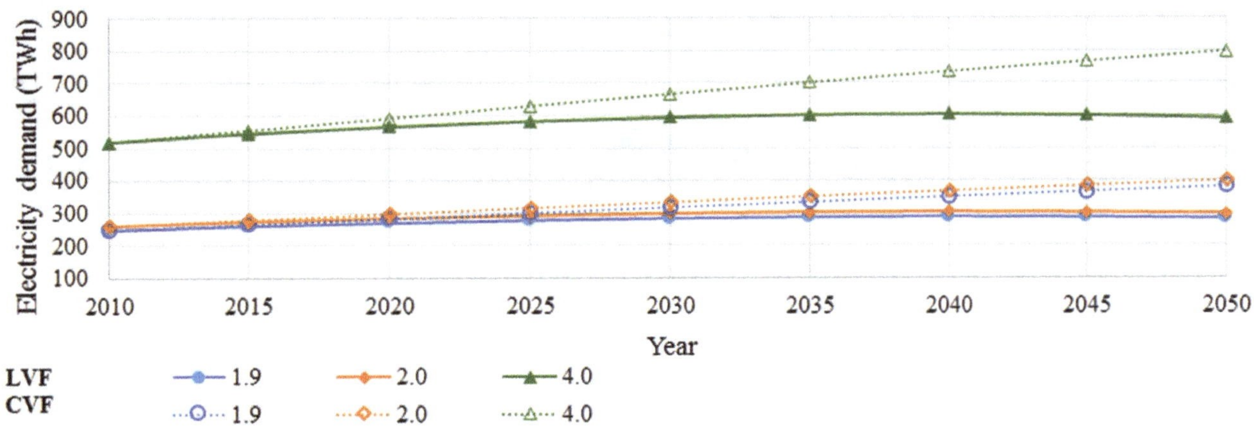

Figure 6. The analyzed electricity demand conditions determined with two population models and three demand conditions per capita. Prepared by the authors based on data from cited reference [19].

Electricity demand for both population models shows similar behavior, at first sight, both vary directly with time, however, it is possible to identify differences. Electricity demand determined with the LVF model presents three different stages throughout the period of analysis. From the beginning of the analysis until the year 2030 a constant growth is observed, followed by a stage without growth until the year 2045; finally, in the last stage of the period, the modelling shows a demand deceleration. The electric demand with the CVF model shows a constant behavior throughout the analysis period, basically an incessant increase.

3.2. *TAC of Electricity Power Generation and Optimized Centrals Number*

Once the energy demand to be supplied has been defined throughout the analyzed period, the modelling takes the TAC of electric power generation and number of plants for takes two population models (LVF and CVF) and three demand conditions (1.9, 2.0 and 4.0 MWh per capita). In order to determine the number of power plants, priority is given to those that use clean technologies and minimize those of fossil technologies.

Optimized TACs of electricity power generation compute differences between the two population models. The corresponding one from demand determined with LVF population model shows an almost constant growth, however, the demand decreases in the last period of the analysis, specifically in the last two decades. Maximum electricity demand is located in years 2035 and 2045, after this period, electricity demand has a decreasing tendency. This behavior is observed for the three demand conditions.

Optimized TACs present a different behavior using the CVF model—it grows directly proportional to electricity demand during the entire analyzed period. Details for the scenarios using both population models are presented next.

3.2.1. LVF Population Model

Details for the electricity demand determined using LVF population model as a function of TAC and optimized power plants number are analyzed and classified as fossil fuel and clean fuels of energy supply in Figure 7. Total electricity demand is presented by (T), electricity demand from fossil fuels are represented by (F) and electricity demand from clean fuels are represented by (C). Total electricity demand exhibit positive growth behavior meanwhile electricity demand from fossil technologies show an opposite trend as a function of time. Clean technologies have a constant growth meaning the main energy technologies to be employed are from clean energy sources.

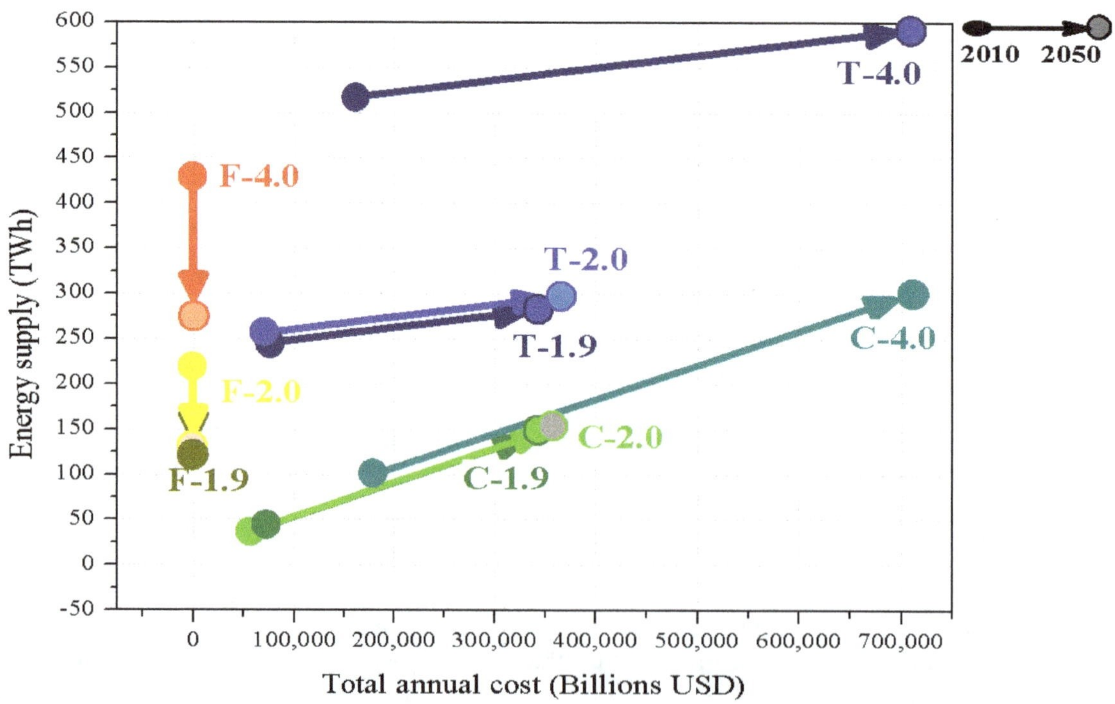

Figure 7. Scheme of the optimized power plants generation centrals number minimizing Total Annual Cost (TAC) of electricity generation for Low Variability of Fertility (LVF) population model [19] and three electricity demand conditions per capita. Prepared by the authors based on data from cited reference.

Clean energy electricity demand shows a growth rate of 204.37%, meanwhile electricity demand from fossil energy sources has a decrease of 30.04% at the end of analyzed period respect year 2010.

If fossil fuel energy (F) does not have new investment, just maintenance, then the value in Figure 7 is relatively low, compared with that of clean energy, so F looks like 0, until the value is 2.1E + 08.

The case of the electricity demand of 1.9 KWh per capita shows total electricity supply from 2010 to 2050 initializing at 2.46 TWh until it rises to 2.80 TWh at the end of the analyzed period. This energy requirement impacts directly in the number of electric power plants necessary to satisfy this demand, presenting a growth rate from 11.27% and 99.51% from fossil and clean energy sources, respectively, at the end of the analyzed period.

Electricity demand of 2.0 KWh per capita, shows that the electricity supply in the analyzed period starts at 2.60 TWh until it raises 2.95 TWh. This energy requirement impacts directly in the number of electric power plants necessary to satisfy this demand, which presents a growth rate from 11.06% and 75.69% from fossil and clean energy sources.

The Electricity demand of 4.0 KWh per capita, shows electricity to supply from from 2010 to 2050 initializing with 5.18 TWh until it raises 5.90 TWh. It is important to highlight that even though this is a necessity, electricity demand presents a growth rate of 9.38% and 74.98%, from fossil and clean energy sources, respectively.

Optimized matrix of number of electricity generating power plants from fossil and clean energy sources to satisfy the demand of 1.9 MWh per capita are presented in Figures 8 and 9.

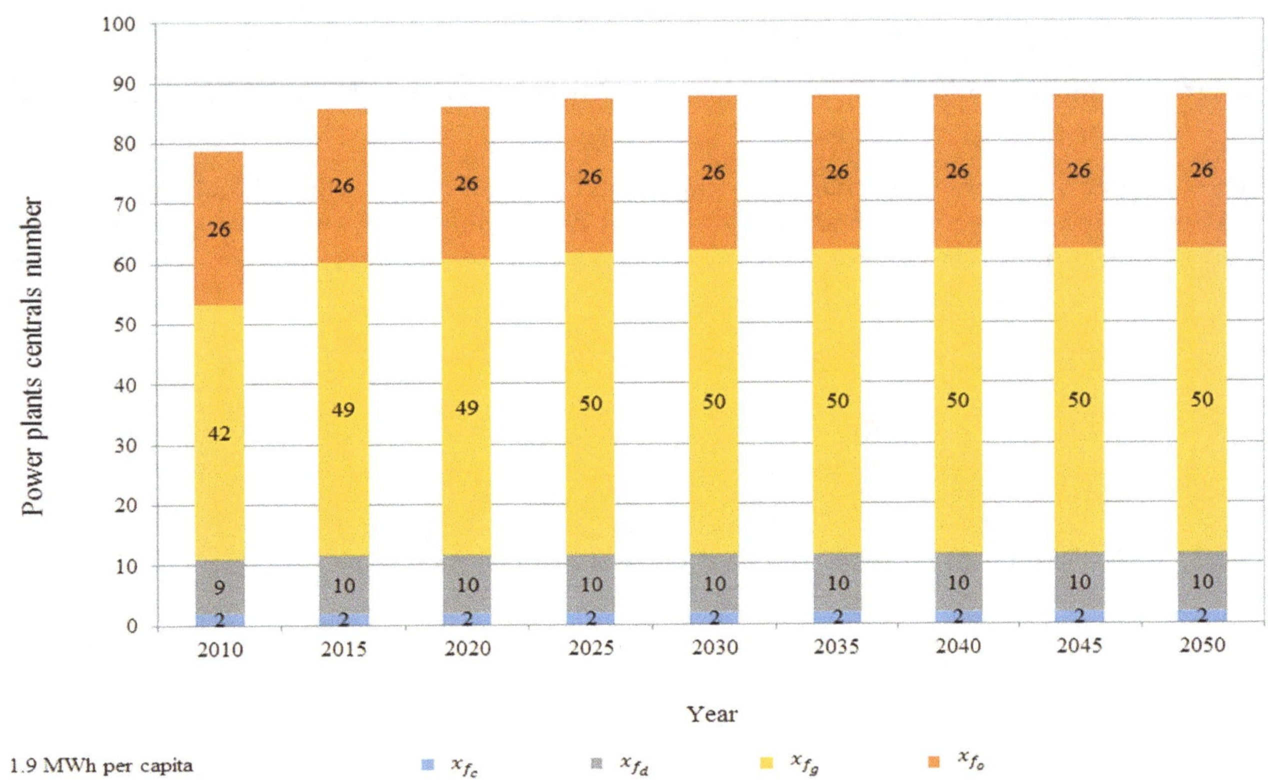

Figure 8. Optimized power plants number from fossil energy resource for 1.9 MWh per capita demand with the LVF model.

As can be appreciated from Figure 8, optimized fossil power plants' number for demand electricity condition of 1.9 MWh per capita, presents a quasi-constant behavior during the analyzed period; only a minimum growth is observed from gas power plants number.

The optimized power plants' number from clean sources presents discrete but constant growth during the whole analyzed period. Power plants which use technology from bioenergy source present higher growth, although their contribution is modest. Hydraulics is a technology source with the greatest contribution within the energy matrix.

Figure 9. Optimized power plants number from clean energy resource for 1.9 MWh per capita demand with LVF model.

The number of electricity demand per capita of 2.0 MWh of optimized fossil and clean power plants is presented in Figures 10 and 11. The model shows that the optimized fossil power plant's number remains almost constant during the whole analyzed period, meanwhile the optimized clean power plant's number present a discrete growth.

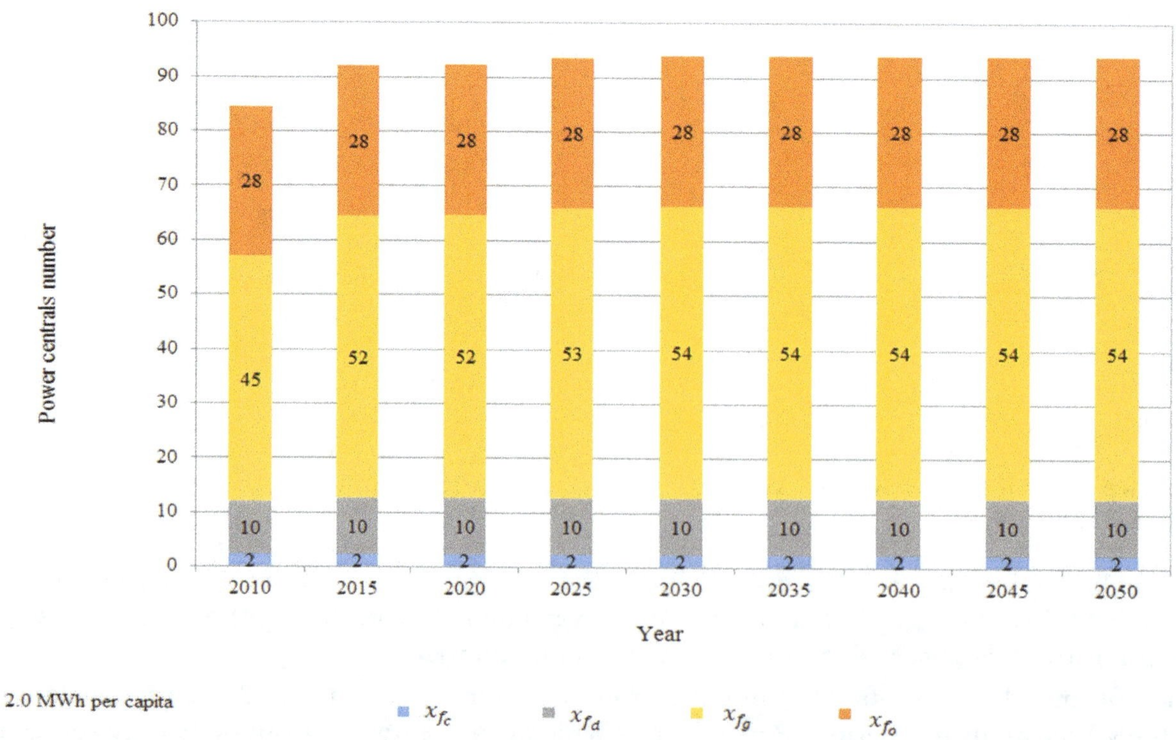

Figure 10. Optimized power plants number from fossil energy resource for 2.0 MWh per capita demand with the LVF model.

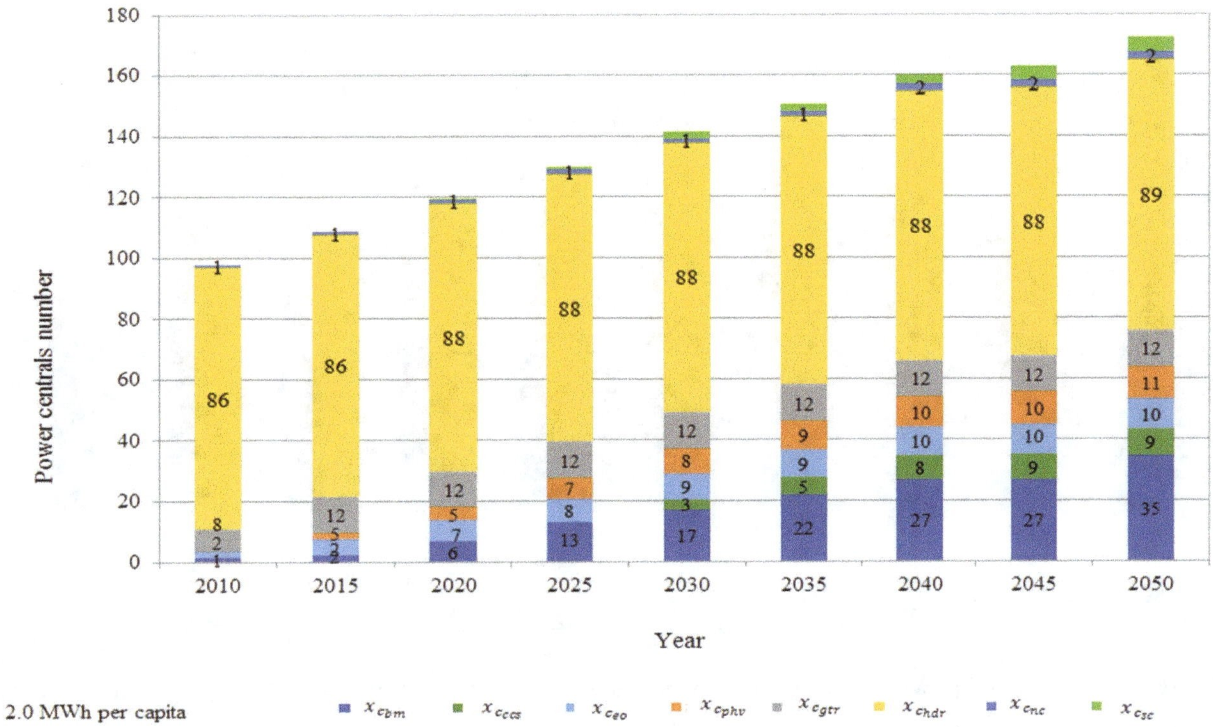

Figure 11. Optimized power plants number from clean energy resource for 2.0 MWh per capita demand with the LVF model.

Optimized fossil and clean power plants number of electricity demand per capita of 4.0 MWh are presented in Figures 12 and 13.

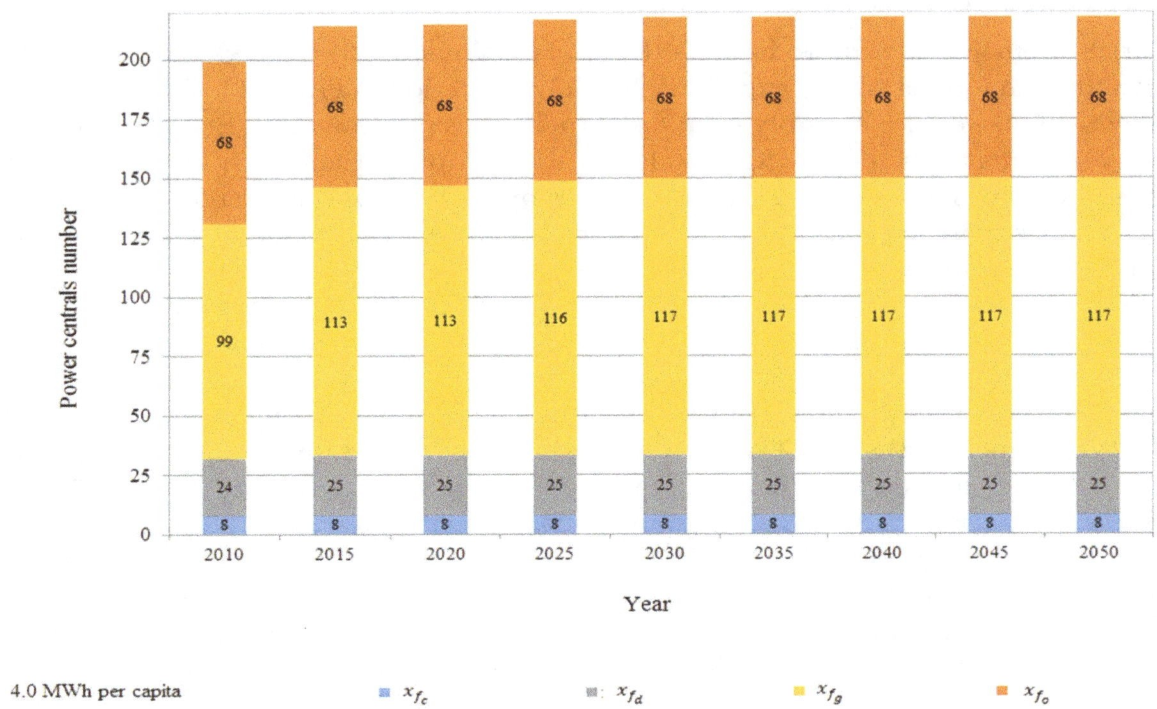

Figure 12. Optimized power plants number from fossil energy resource for 4.0 MWh per capita demand with the LVF model.

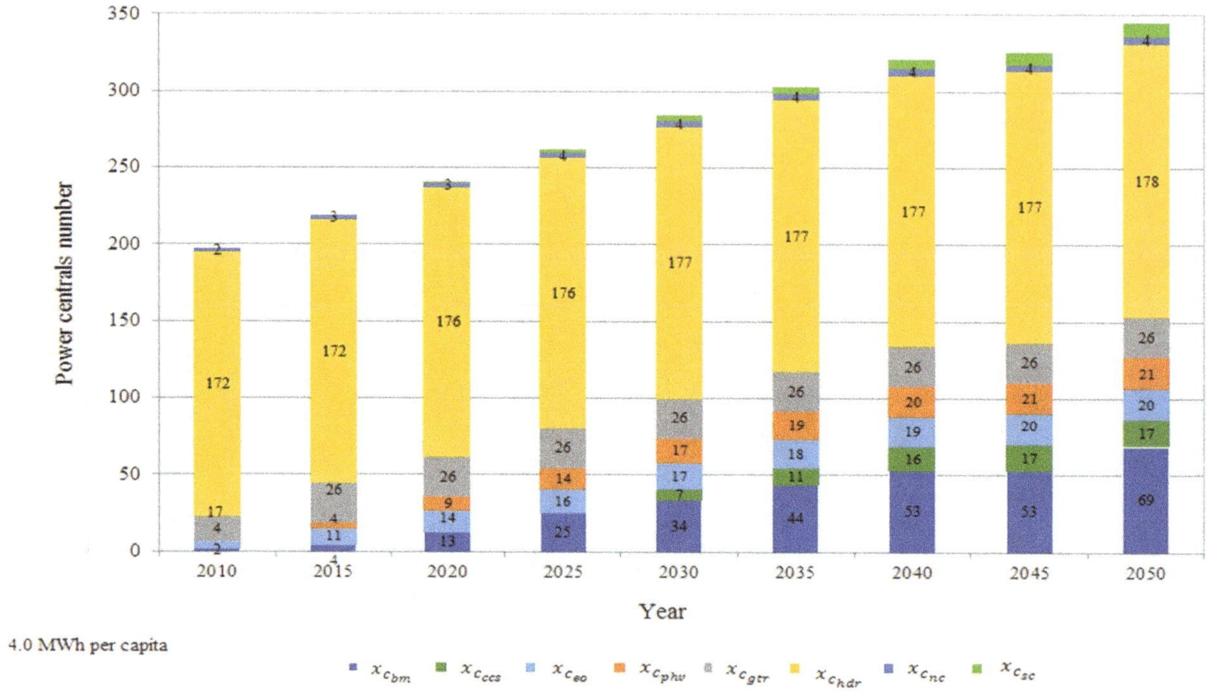

Figure 13. Optimized power plants number from clean energy resource for 4.0 MWh per capita demand with the LVF model.

In this electricity demand case, fossil power plants present an almost constant behavior, contrary to clean energy sources for which growth is constant, discrete but constant. Power plants which use technology from bioenergy and hydraulic sources present higher growth, although their contribution is modest.

3.2.2. CVF Population Model

Electricity demand determined using CVF population model is presented disaggregated between fossil fuel and clean fuels of energy supply the in analyzed period in Figure 14. Total electricity supply is presented by (T), electricity supply from fossil fuels are represented by (F) and electricity supply from clean fuels are represented by (C). Total electricity supply presents positive a growth behavior meanwhile electricity supply from fossil technologies present an opposite trend and at the same time, clean technologies present a constant growth meaning main energy technologies to be employed are from clean energy sources.

Electricity supply for demand from clean energy sources presents a growth rate of 310.24%, meanwhile electricity demand from fossil energy sources presents a decrease of 5.70% at the end of analyzed period respect year 2010. Same aspect for F values close to 0 were explained on Figure 7.

Electricity demand of 2.0 KWh per capita, shows electricity to supply in analyzed period starts with 2.60 TWh until it raises 3.98 TWh impacting directly in the number of electric power plants generation necessary to satisfy this demand, presenting a growth rate from 25.21% and 149.78% from fossil and clean energy sources, respectively, in the end of analyzed period.

The case of electricity demand of 4.0 KWh per capita, shows electricity to supply from year 2010 to year 2050 initializing with 5.18 TWh until it raises 7.95 TWh. This energy increment implies a growth rate of optimized power plants of 21.40% and 148.36%, from fossil and clean energy sources, respectively.

Optimized fossil and clean power plants number of electricity demand per capita of 1.9 MWh are presented in Figures 15 and 16. For this condition, optimized total power plants are calculated to be 313, from which 99 are planned to be from fossil technology and the rest from clean technology.

The optimized power plants' number from fossil fuel sources behaves quite similar to results obtained when the LVF population model is used, that means, power plant number present an almost constant behavior during all analyzed period, only gas power plants slightly increase. Clean power plants number presents a discrete but constant growth being bioenergy power plants the technology which presents higher growth, although their contribution is modest. Hydraulic is also the technology which presents the greatest contribution in the energy matrix.

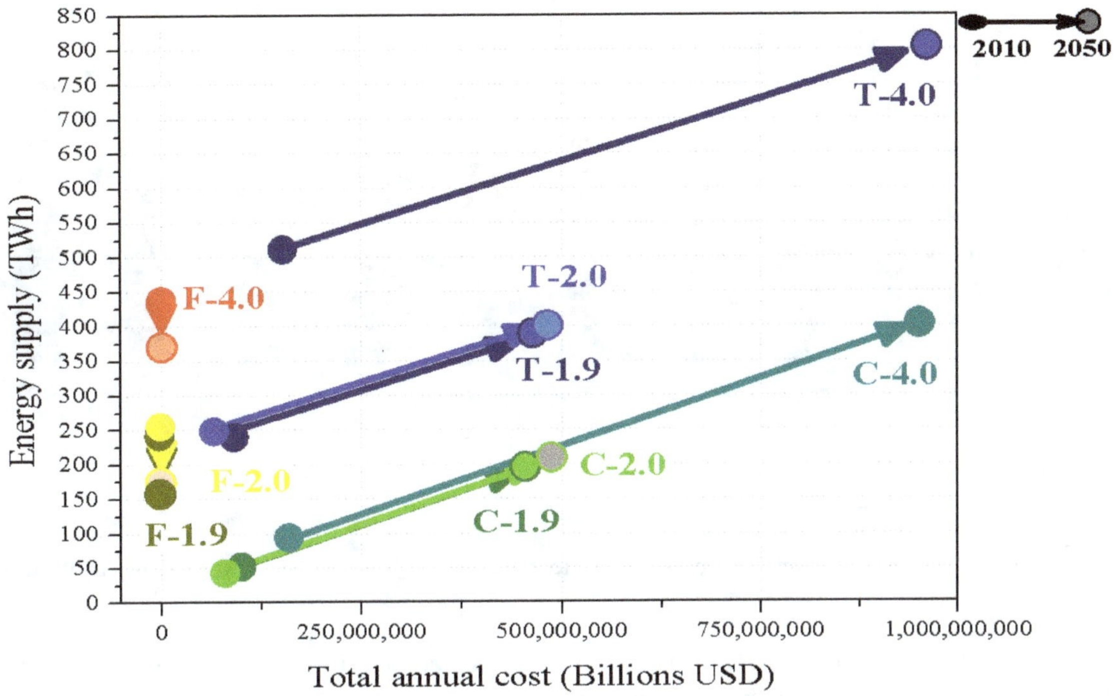

Figure 14. Schematic optimized power plants generation centrals number minimizing total annual cost (TAC) of electricity generation for the CVF population model [19] and three electricity demand conditions per capita. Prepared by the authors based on data from own results.

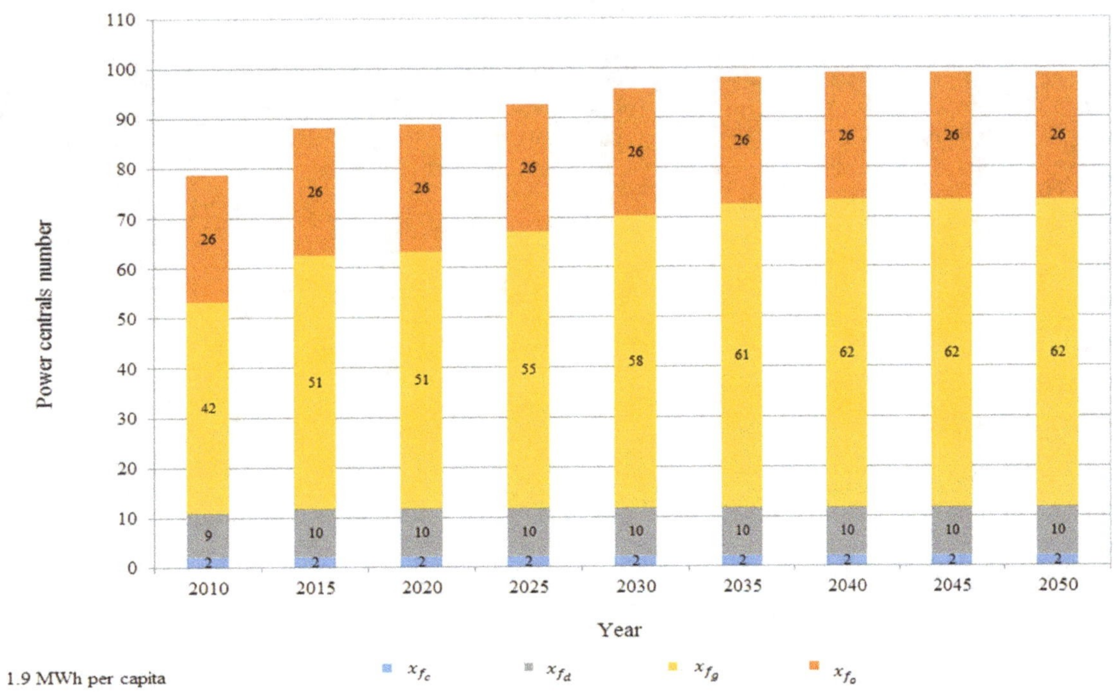

Figure 15. Optimized power plants number from fossil energy resource for 1.9 MWh per capita demand with the CVF model.

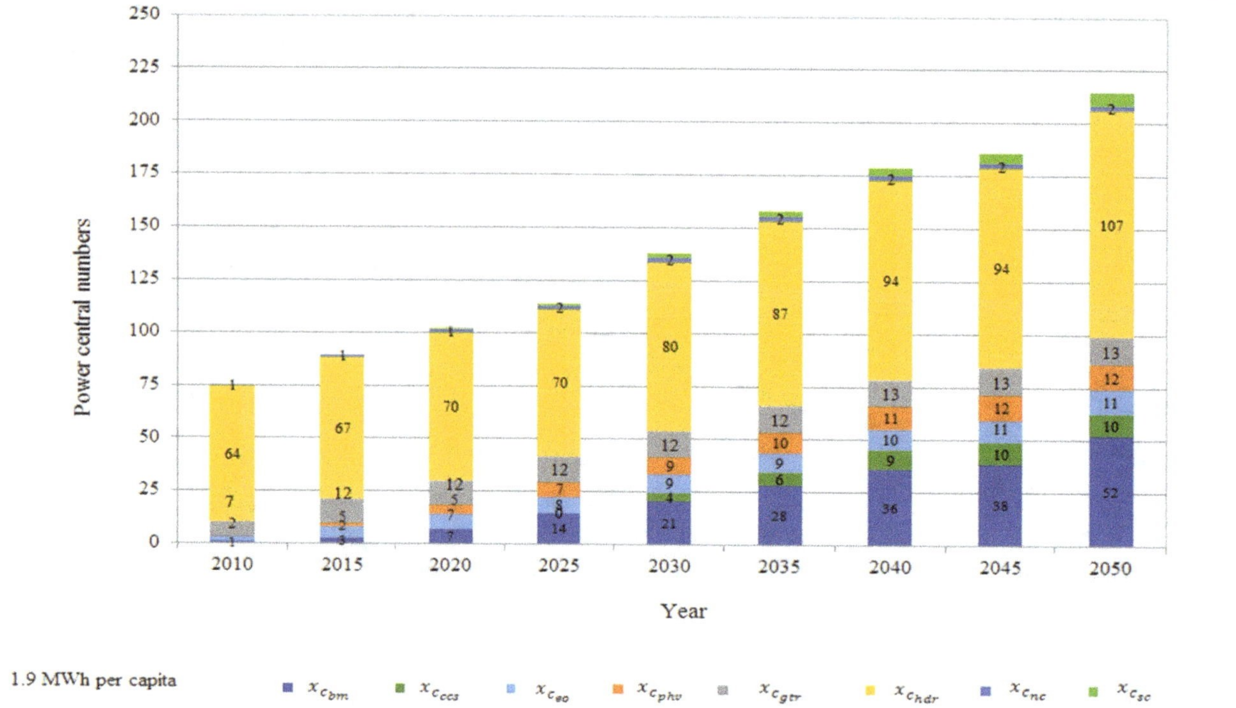

Figure 16. Optimized power plants number from clean energy resource for 1.9 MWh per capita demand with the CVF model.

To satisfy electricity demand of 2.0 MWh per capita, required optimized fossil and clean power plants number are presented in Figures 17 and 18.

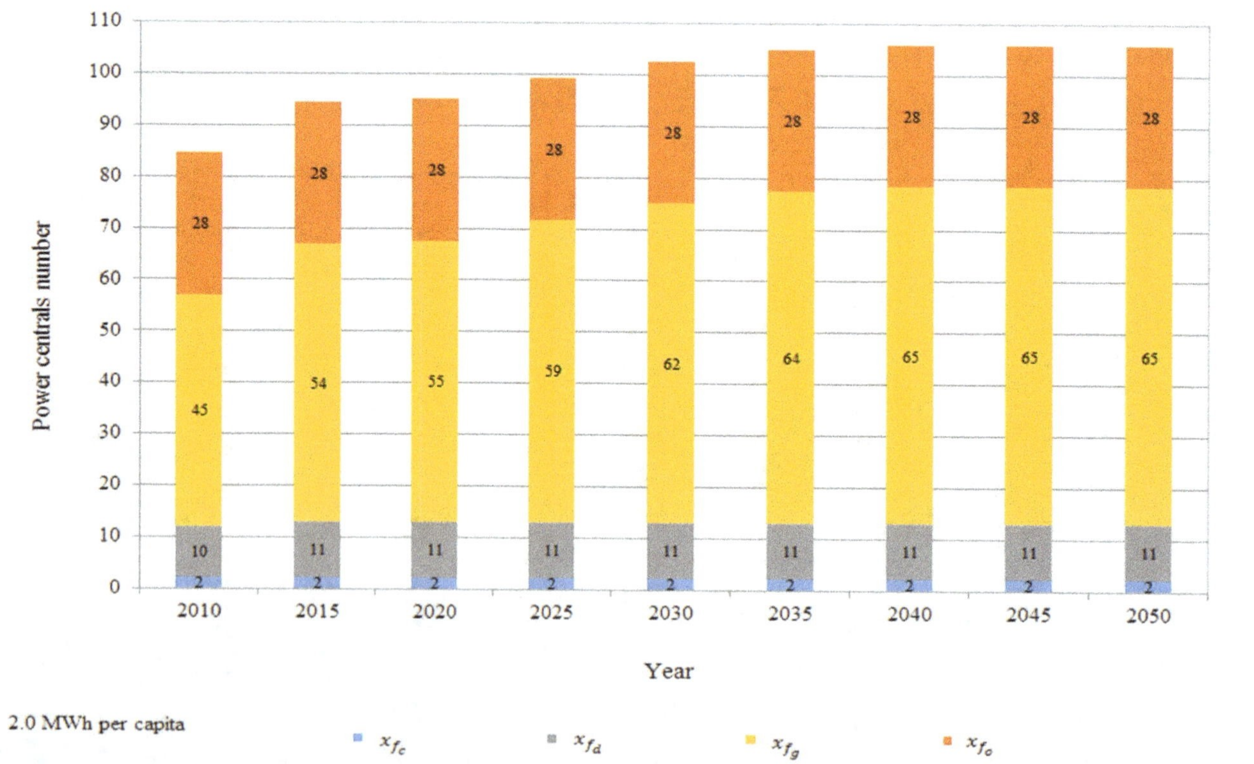

Figure 17. Optimized power plants number from fossil energy resource for 2.0 MWh per capita demand with the CVF model.

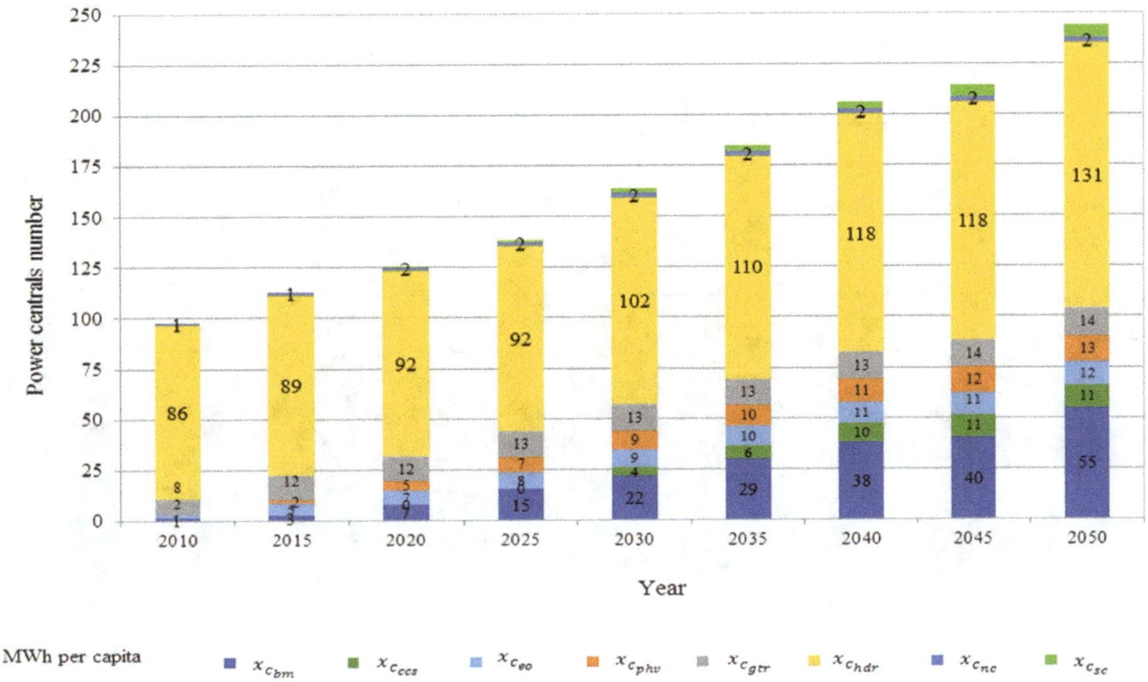

Figure 18. Optimized power plants number from clean energy resource for 2.0 MWh per capita demand with the CVF model.

Electricity demand of 4.0 MWh per capita, required optimized fossil and clean power plants number are presented in Figures 19 and 20.

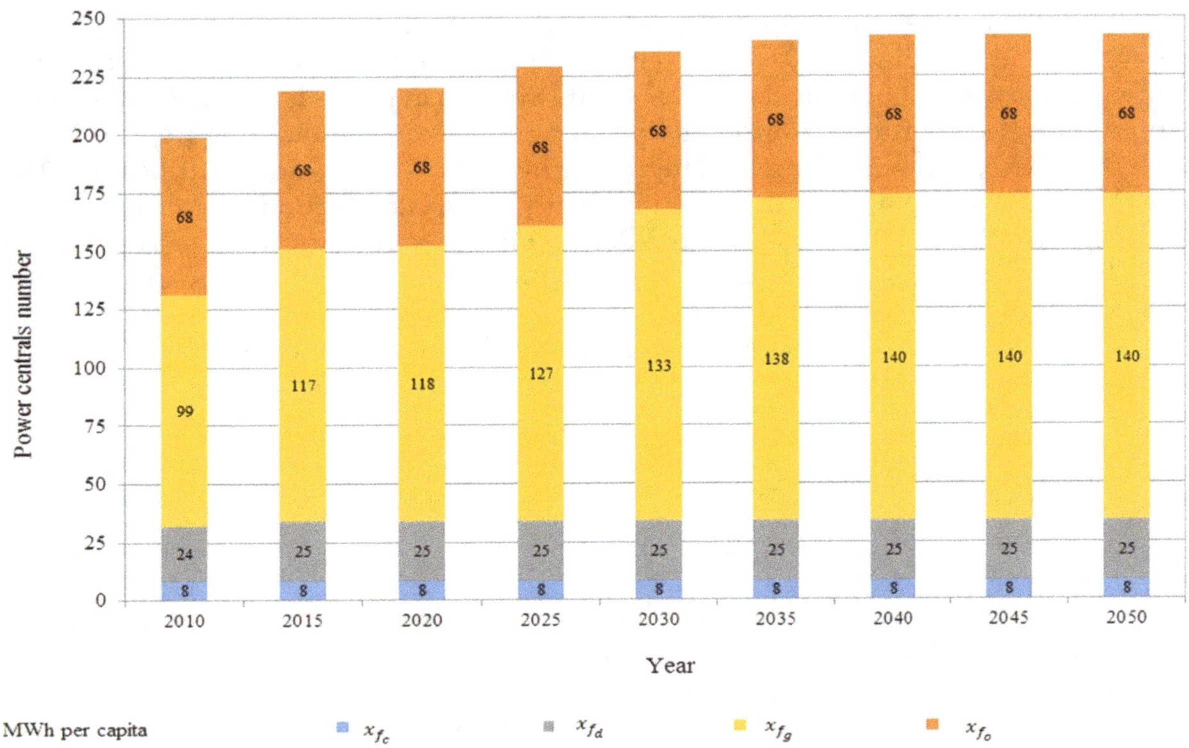

Figure 19. Optimized power plants number from fossil energy resource for 4.0 MWh per capita demand with the CVF model.

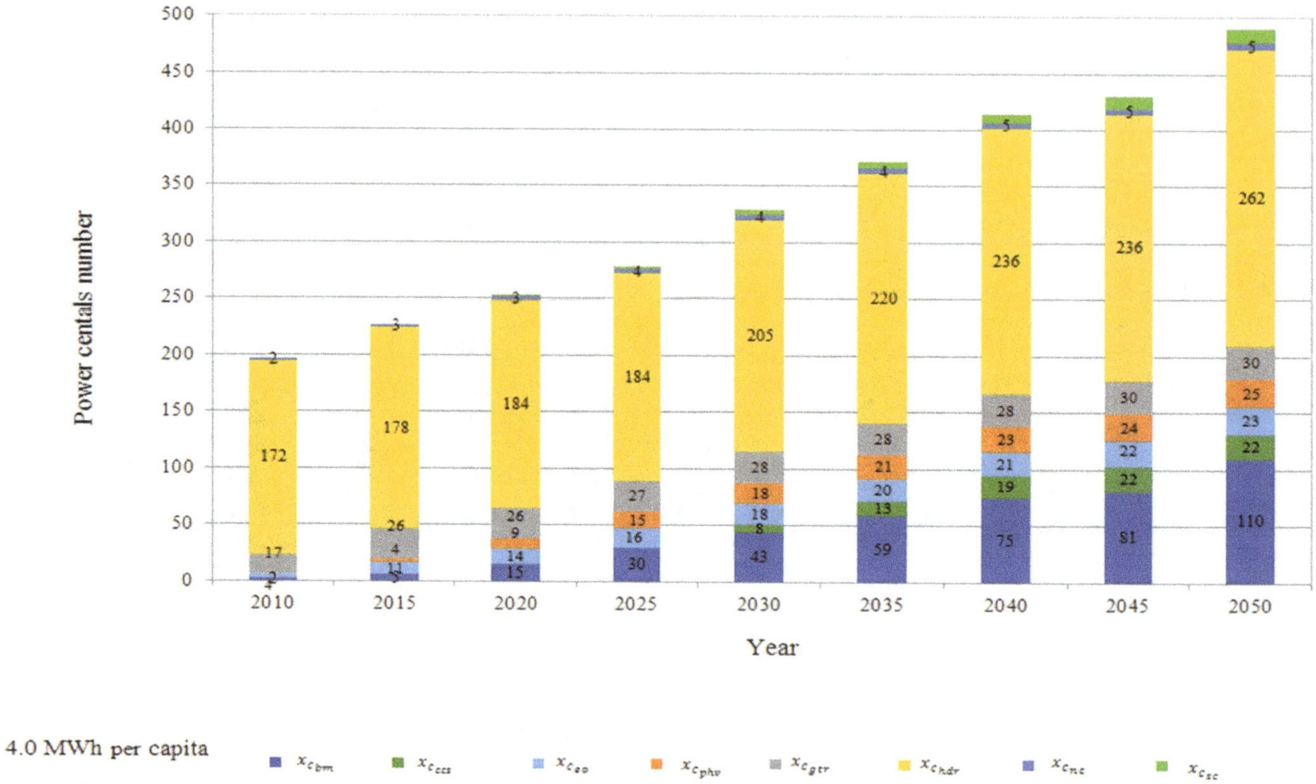

Figure 20. Optimized power plants number from clean energy resource for 4.0 MWh per capita demand with the CVF model.

From Figure 19 a behavior similar to the LVF model is observed which means quasi-constant growth during the analysis period for all technologies except gas. The number of plants that use gas presents an increase almost 50% at the end of the period compared to the initial number. The complement of the matrix of optimized power plants from clean plants is presented in Figure 20. As it is observed, clean power number presents constant growth during analyzed period. Power plants which use bioenergy technology present highest growth, with a modest contribution. The energy source with the greatest contribution within the energy matrix corresponds to the hydraulic resource.

It is important to bear in mind that the determined power plant number corresponds to the required plants in each analyzed period, taking as references the number of plants installed at the beginning of the analysis.

Each optimized TAC of electricity power generation condition, implies the energy matrix definition to satisfy the electricity demand as well as CO_2eq emissions intrinsic to this process. CO_2eq generated due this process are presented below and complement the proposed scenarios in each analyzed period.

3.2.3. CO_2eq Emissions

CO_2eq emissions generated by optimized number of fossil and clean power plants are presented by energy type energy source—fossil and clean.

Produced emissions using fossil sources as primary energy in electrical energy generation for three population demands are presented in Figure 21.

As shown, CO_2eq emissions present a decrease as the analysis period increases, expected behavior given that contribution from this energy type source decreases, satisfying objectives proposed in the General Climate Change Law (GCCL) [10].

The corresponding emissions produced by clean sources as primary energy in electricity generation are presented in Figure 22.

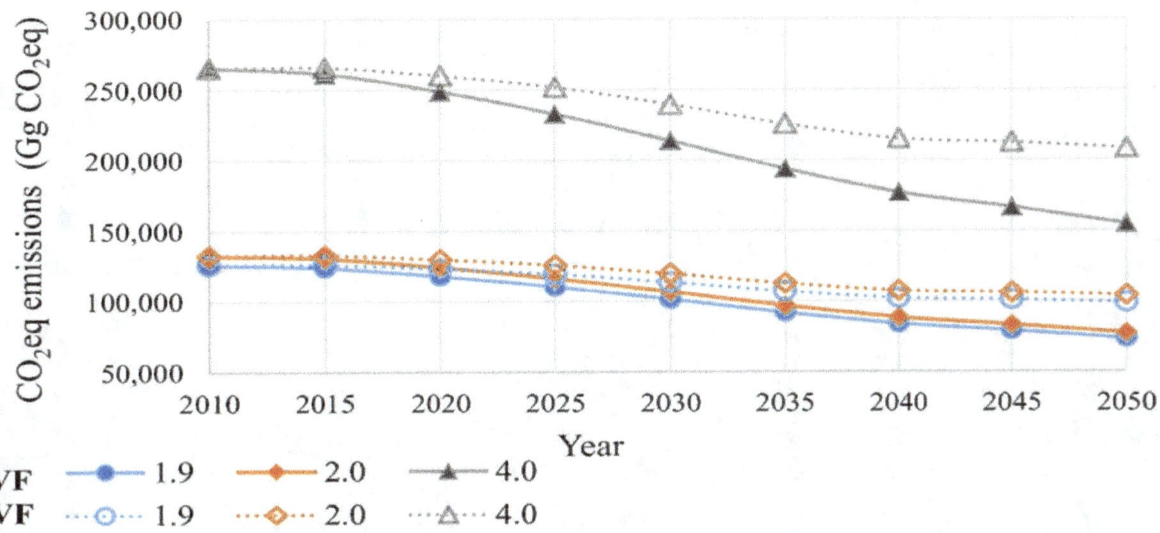

Figure 21. CO_2eq emissions generated by fossil central power plants for three electricity demand per capita and two population models. Prepared by the authors based on own results.

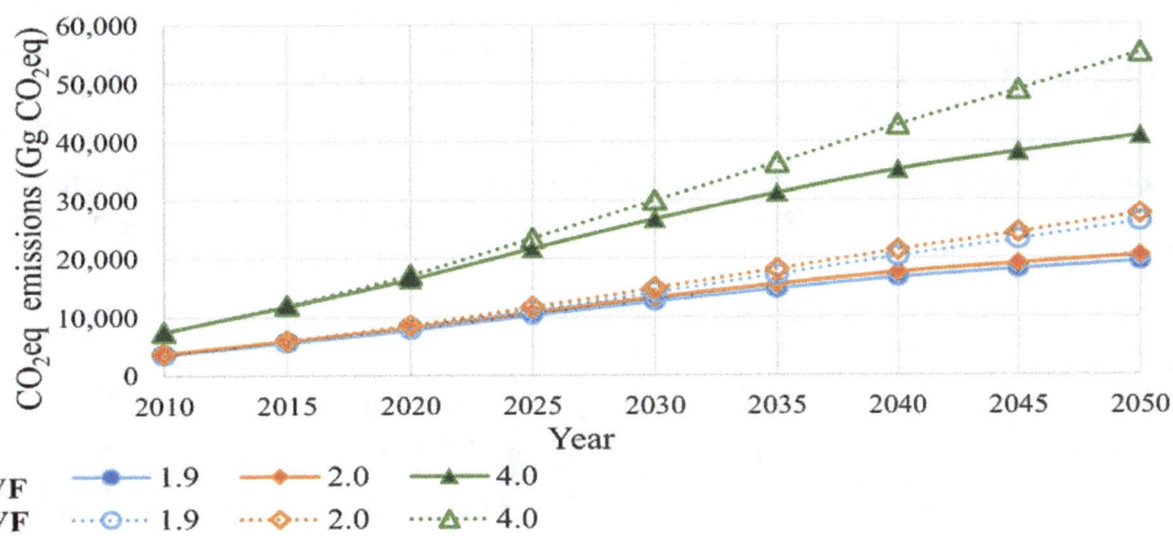

Figure 22. CO_2eq emissions generated by clean central power plants for three electricity demand per capita and two population models. Prepared by the authors based on own results.

CO_2eq emissions generated with clean technologies show an inverse behavior of emissions generated from fossil fuels, that means increasing directly with the period of analysis, however, it is very important to note the amount of emissions generated with this type of energy is, at least, three magnitude orders lower than emissions from fossil sources, satisfying the objective proposed in the General Climate Change Law, GCCL [11].

4. Discussion

Energy planning to explore possible alternative scenarios is a necessary tool for the economic development of the country in the face of global, regional events and international commitments acquired in the face of climate change. This tool serves to guide those responsible for energy policy and regulators in the development of policies to visualize scenarios of energy development that allow to contribute effectively to the sustainable growth of the region as well as to demonstrate the impact of policies and plans established in the short and long term. For two grow population model the Figures 23 and 24 are plotted as follow. The data is obtained from the detailed model in Section 2.2.

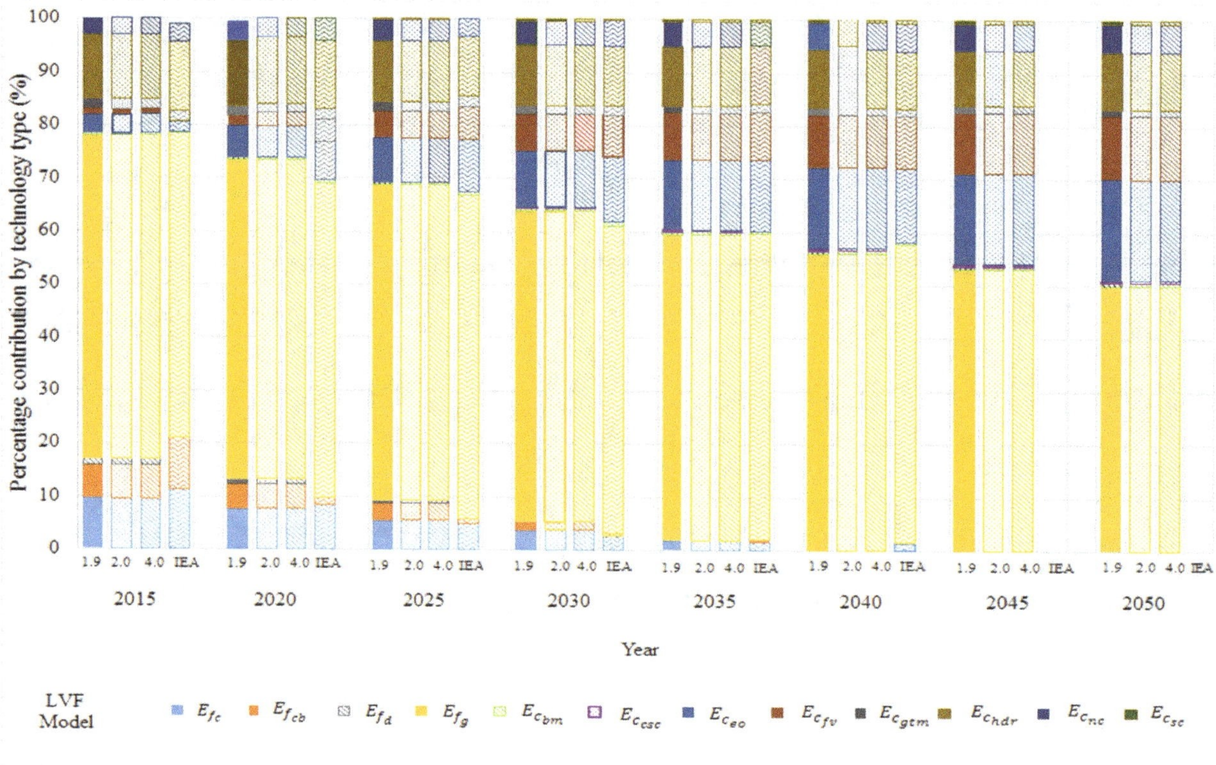

Figure 23. LVF Energy superstructure scenario satisfying governmental objectives from GCCL and National Climate Change Strategy (NCCS) [11,12] compared with IEA scenarios. Prepared by the authors based on own results and public data from cited Reference [18].

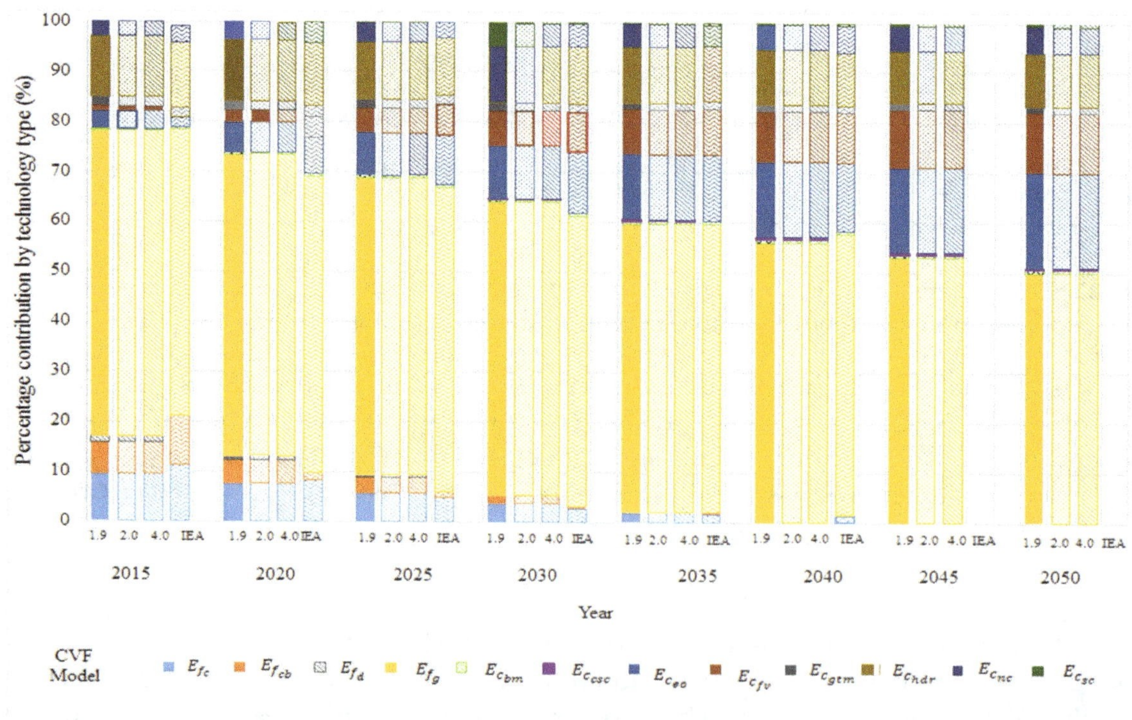

Figure 24. CVF Energy superstructure scenario satisfying governmental objectives from GCCL and NCCS [11,12] compared with IEA scenarios. Prepared by the authors based on own results and public data from cited Reference [18].

In these Figures 23 and 24, for the 2015 year, all scenarios take 10% approximately for carbon transformation in the bottom of the columns. The oil plants are close 5% for 1.9, 2.0 and 4.0% except IEA. For IEA in 2015 about 10% was reported. Diesel plants are very low, close to 1% and for IEA it is not reported. The biggest contribution for this scenarios, including IEA is gas process plant, this model has a 61.2% value and IEA is 57.3%. For clean source the values are 21.9 calculated for this model and 20.7% for IEA data.

The maximum deviations for the gas process plant are from 6.3 to 6.8% based on this modelling and IEA reports in 2015. For that year the clean energy variation is from 5.4 to 5.7% variation.

On 2020 calculation, the fossil energy is 73.5%, 73.5% and 73.8% for 1.9, 2.0 and 4.0 scenarios respectively. For IEA projection, fossil energy is 69.0%. The maximum deviation goes from 6.1 to 6.5%. For clean energy, the values are 26.5, 26.5 and 26.2 for 1.9, 2.0 and 4.0 scenarios respectively. But IEA the predicted value is 31.0%. This mean the largest deviation value for our modelling: 14.5% to 16.9% variation.

On 2025, the fossil energy is computed as 68.7% for any scenario and for IEA is 66.8%. This is a variation close to 2.75%. Clean energy values are 31.3% from this work and 33.2% from IEA. This is a variation from 5.7 to 6.0%.

On 2030 predictions, the values for fossil energy the value is constant at 63.8% and for IEA 61.1%. This variation is just from 4.2 to 4.4% for all scenarios. Clean energy for that year is 36.2% compared with IEA 38.9%. Those values have been 6.9% to 7.4% lower.

For 2035 compared values, fossil energy is 59.5% from our model, compared with 59.6% for IEA prediction. They are almost the same value (0.1% variation). This is similar behavior for clean energy—our modelling reports 40.5 and IEA 40.4 (0.2% variation).

Predictions for 2040 for fossil energy must be 55.9% compared with 57.6% from IEA. These values represent 2.95% variation. Clean energy are 44.1% and 42.4% for this work and IEA, respectively. The variation is close to 3.9%.

The IEA do not report for 2045 and 2050. The values for fossil energy are 52.9% and 49.9% for these years. Then the clean energy 47.1 and 50.1% for same years, respectively. This is the goal for NCCS.

There are considerations in each model described in Section 3.2 and a consequence of these, the resulting data are not the same.

Like other optimized models, there are disadvantages, for example dependency of macroeconomic circumstances such as technology cost trends, as well as demographic dependency population but this dependency is inherent to the modelling process.

5. Conclusions

This paper has proposed a model that satisfactorily defines energy scenarios integrated by the optimized matrix of power plants required to satisfy the electricity demand, by optimized costs of electricity generation as well as CO_2eq emissions produced by this process. In each analysis scenario, governmental objectives with impact on energy security, economy and environmental sustainability are reached satisfactory.

Electricity demand presents a constant growth with both models and three conditions for almost all the period of analysis. This growth is reflected in the gradual increase in the number of plants needed to meet demand. Power plants which use fossil technologies, specifically gas, are those that increase, the rest kept constant. Power plants that use clean source technologies, show a gradual growth, highlighting hydraulics and biomass. The results indicate, in each scenario analyzed, CO_2eq emissions satisfy objectives proposed in the General Climate Change Law [11], thus, despite the increase in electricity demand, CO_2eq emissions of a reverse behavior.

With the application of the proposed model to the MES, scenarios obtained from applying restricted public policies are visualized. In this scenarios each optimal conditions to achieve goals and plans established in the short and long term are visualized. Given the results obtained with this proposed model, bases are laid with possibility of using the model to identify alternative scenarios to the optimal ones as well as to evaluate costs of environmental impacts of the technologies in different stages of modelling.

This modelling would be implemented for another country's conditions with few data to predict good agreement with IEA long-term predictions. For the Mexico case, the energy superstructures percentage contribution for 2025 and forward years shows a 7.4% variation or lower compared with this work and IEA data.

Author Contributions: G.H.-L., J.M.P.-O. and G.D.T.V. were working in GAMS programming for this Project. R.J.R., A.R.-M., J.C.R. analyzed the data from the Mexican electricity system.

Appendix A

The GAMS © code is shown as follow.

VARIABLES EFdemand, TAC; POSITIVE VARIABLES	
Energy contributions (MWh)	
Ffos	FOSSIL
Frenv	CLEAN
Ffoscarb	CARBON
Ffoscomb	FUEL OIL
Ffosdis	DIESEL
Ffosgas	GAS
Frenvbio	BIOMASS
Frenvcscrb	CARBON CAPTURE AND STORAGE
Frenveol	EOLIC
Frenvftv	PHOTOVOLTAIC
Frenvgtm	GEOTHERMIC
Frenvhdr	HYDRAULIC
Frenvnucl	NUCLEAR
Frenvsc	SOLAR CONCENTRATION
COSTS	
TAC	TOTAL ANNUAL COST FROM FOSSIL AND CLEAN
Costfos	FOSSIL
Costrenv	CLEAN
Costfoscarb	CARBON
Costfoscomb	FUEL OIL
Costfosdis	DIESEL
Costfosgas	GAS

Costrenvbio	BIOMASS
Costrenvcscrb	CARBON CAPTURE AND STORAGE
Costrenveol	EOLIC
Costrenvftv	PHOTOVOLTAIC
Costrenvgtm	GEOTHERMIC
Costrenvhdr	HYDRAULIC
Costrenvnucl	NUCLEAR
Costrenvftrm	SOLAR CONCENTRATION
******************************CO_2 Emission (kg/MWh)**	
EFdeman	TOTAL EMISSION FROM FOSSIL AND CLEAN
EFfos	FOSSIL
EFrenv	CLEAN
EFfoscarb	CARBON
EFfoscomb	OIL FUEL
EFfosdis	DIESEL
EFfosgas	GAS
EFrenvbio	BIOMASS
EFrenvcscrb	CARBON CAPTURE AND STORAGE
EFrenveol	EOLIC
EFrenvftv	PHOTOVOLTAIC
EFrenvgtm	GEOTHERMIC
EFrenvhdr	HYDRAULIC
EFrenvnucl	NUCLEAR
EFrenvftrm	SOLAR CONCENTRATION
********************* NUMBER OF POWER PLANTS GENERATION **********************************	
xtot	TOTAL
xf	FOSSILS
xr	CLEAN
xfoscarb	CARBON
xfoscomb	FUEL OIL
xfosdis	DIESEL
xfosgas	GAS
xrenvbio	BIOMASA
xrenvcscrb	CARBON CAPTURE AND STORAGE
xrenveol	EOLICA
xrenvftv	PHOTOVOLTAIC
xrenvgtm	GEOTHERMIC
xrenvhdr	HYDRAULIC
xrenvnucl	NUCLEAR
xrenvftrm	SOLAR CONCENTRATION

PARAMETERS		
COSTS (USD/MWh neto)*		
****************************** INVESTMENT COSTS BY ENERGY FUEL**		
afoscarb	CARBON	/37.39/
afoscomb	OIL FUEL	/41.891/
afosdis	DIESEL	/64.16/
afosgas	GAS	/72.53/
arenvbio	BIOMASS	/150/
arenvcscrb	CARBON CAPTURE AND STORAGE	/47.39/
arenveol	EOLIC	/1.65E6/
arenvftv	PHOTOVOLTAIC	/5.11E6/
arenvgtm	GEOTHERMIC	/3.18E6/
arenvhdr	HYDRAULIC	/2.0E6/
arenvnucl	NUCLEAR	/87.93/
arenvftrm	SOLAR CONCENTRATION	/180/
***************************** ASSOCIATED COST FROM FUEL TYPE *********************************		
bfoscarb	CARBON	/27.57/
bfoscomb	OIL FUEL	/93.89/
bfosdis	DIESEL	/147.06/
bfosgas	GAS	/49.37/
brenvbio	BIOMASS	/7.9/
brenvcscrb	CARBON CAPTURE AND STORAGE	/27.67/
brenveol	EOLICA	/0/
brenvftv	PHOTOVOLTAIC	/0/
brenvgtm	GEOTHERMIC	/50.12/
brenvhdr	HYDRAULIC	/6.38/
brenvftrm	SOLAR CONCENTRATION	/0/
OPERATION AND MAINTENANCE COST, INCLUDING FIXED AND VARIABLE COSTS *		
cfoscarb	CARBON	/8.13/
cfoscomb	OIL FUEL	/9.21/
cfosdis	DIESEL	/12.88/
cfosgas	GAS	/11.33/
crenvbio	BIOMAS	/4.33/
crenvcscrb	CARBON CAPTURE AND STORAGE	/8.13/
crenveol	EOLIC	/1.75/
crenvftv	PHOTOVOLTAIC	/7.72/
crenvgtm	GEOTHERMIC	/19.41/
crenvhdr	HYDRAULICA	/5.71/
crenvftrm	SOLAR CONCENTRATION	/20/

************************ CO_2eq BY FUEL (Kg CO_2/MWh)************************		
ECO2eqfoscarb	CARBON	/1089.81/
ECO2eqfoscomb	OIL FUEL	/822/
ECO2eqfosdis	DIESEL	/274.44/
ECO2eqfosgas	GAS	/524/
ECO2eqrenvbio	BIOMAS	/1403.75/
ECO2eqrenvcscrb	CARBON CAPTURE AND STORAGE	/217.2/
ECO2eqrenveol	EOLIC	/210/
ECO2eqrenvftv	PHOTOVOLTAIC	/106/
ECO2eqrenvgtm	GEOTHERMIC	/372/
ECO2eqrenvhdr	HYDRAULIC	/15/
ECO2eqrenvftrm	SOLAR CONCENTRATION	/14/
Fdemand	ANNUAL NATIONAL ELECTRICITY DEMAND MWh	/246381740/

* ELECTRICITY CONTRIBUTION GENERATED IN A PERIOD PRIOR TO THE ANALYZED (MWh)*		
Ffcarbins	CARBON	/18380000/
Ffcombins	OIL FUEL	/6451200/
Ffdisins	DIESEL	/780000/
Ffgasins	GAS	/27166000/
Frbioins	BIOMAS	/0/
Frcscrbins	CARBON CAPTURE AND STORAGE	/0/
Freolins	EOLIC	/5000/
Frftvins	PHOTOVOLTAIC	/00/
Frgtmins	GEOTHERMIC	/729900/
Frhdrins	HYDRAULIC	/26851000/
Frnuclins	NUCLEAR	/1080500/
Frftrmins	SOLAR CONCENTRATION	/0/

***************************** CAPACITY OF POWER PLANT (MW)**		
Fcapfoscarb	CARBON	/2600/
Fcapfoscomb	OIL FUEL	/12711/
Fcapfosdis	DIESEL	/182/
Fcapfosgas	GAS	/7230/
Fcaprenvbio	BIOMASA	/1500/
Fcaprenvcscrb	CARBON CAPTURE AND STORAGE	/1000/
Fcaprenveol	EOLIC	/20000/
Fcaprenvftv	PHOTOVOLTAIC	/1000/
Fcaprenvgtm	GEOTHERMIC	/1000/
Fcaprenvhdr	HYDRAULIC	/10270/
Fcaprenvnucl	NUCLEAR	/1365/
Fcaprenvftrm	SOLAR CONCENTRATION	/960/

EQUATIONS	
R1	ANNUAL ELECTRICY DEMAND
R2	FOSSIL FUEL CONTRUIBUTION TO ELECTRICITY DEMAND
R3	CLEAN FUEL CONTRUIBUTION TO ELECTRICITY DEMAND
R4	ELECTRICITY GENERATION FROM CARBON
R5	ELECTRICITY GENERATION FROM FUEL OIL
R6	ELECTRICITY GENERATION FROM DIESEL
R7	ELECTRICITY GENERATION FROM GAS
R8	ELECTRICITY GENERATION FROM BIOMAS
R9	ELECTRICITY GENERATION FROM CARBON CAPTURE AND STORAGE
R10	ELECTRICITY GENERATION FROM EOLIC
R11	ELECTRICITY GENERATION FROM PHOTOVOLTAIC
R12	ELECTRICITY GENERATION FROM GEOTHERMAL
R13	ELECTRICITY GENERATION FROM HYDRAULIC
R14	ELECTRICITY GENERATION FROM NUCLEAR
R15	ELECTRICITY GENERATION FROM SOLAR CONCENTRATION
COSTS	
R16	TOTAL ANNUAL COST OF ELECTRICITY GENERATION
R17	TOTAL ANNUAL COST OF ELECTRICITY GENERATION FROM FOSSIL FUELS
R18	TOTAL ANNUAL COST OF ELECTRICITY GENERATION FROM CLEAN FUELS
R19	TOTAL ANNUAL COST OF ELECTRICITY GENERATION FROM CARBON
R20	TOTAL ANNUAL COST OF ELECTRICITY GENERATION FROM OIL FUEL
R21	TOTAL ANNUAL COST OF ELECTRICITY GENERATION FROM DIESEL
R22	TOTAL ANNUAL COST OF ELECTRICITY GENERATION FROM GAS
R23	TOTAL ANNUAL COST OF ELECTRICITY GENERATION FROM BIOMAS
R24	TOTAL ANNUAL COST OF ELECTRICITY GENERATION FROM CARBON CAPTURE AND STORAGE
R25	TOTAL ANNUAL COST OF ELECTRICITY GENERATION FROM EOLIC
R26	TOTAL ANNUAL COST OF ELECTRICITY GENERATION FROM PHOTOVOLTAIC
R27	TOTAL ANNUAL COST OF ELECTRICITY GENERATION FROM GEOTHERMAL
R28	TOTAL ANNUAL COST OF ELECTRICITY GENERATION FROM HYDRAULIC
R29	TOTAL ANNUAL COST OF ELECTRICITY GENERATION FROM NUCLEAR
R30	TOTAL ANNUAL COST OF ELECTRICITY GENERATION FROM SOLAR CONCENTRATION
CO_2eq EMISSIONS*	
R31	CO2eq EMISSION FROM TOTAL ELECTRICITY GENERATION
R32	CO2eq EMISSION FROM ELECTRICITY GENERATION FROM FOSSIL FUELS
R33	CO2eq EMISSION FROM ELECTRICITY GENERATION FROM CLEAN FUELS
R34	CO2eq EMISSION FROM ELECTRICITY GENERATION FROM CARBON
R35	CO2eq EMISSION FROM ELECTRICITY GENERATION FROM FUEL OIL
R36	CO2eq EMISSION FROM ELECTRICITY GENERATION FROM DIESEL

R37	CO2eq EMISSION FROM ELECTRICITY GENERATION FROM GAS
R38	CO2eq EMISSION FROM ELECTRICITY GENERATION FROM BIOMAS
R39	CO2eq EMISSION FROM ELECTRICITY GENERATION FROM CARBON CAPTURE AND STORAGE
R40	CO2eq EMISSION FROM ELECTRICITY GENERATION EOLIC
R41	CO2eq EMISSION FROM ELECTRICITY GENERATION PHOTOVOLTAIC
R42	CO2eq EMISSION FROM ELECTRICITY GENERATION GEOTHERMIC
R43	CO2eq EMISSION FROM ELECTRICITY GENERATION HYDRAULIC
R44	CO2eq EMISSION FROM ELECTRICITY GENERATION NUCLEAR
R45	CO2eq EMISSION FROM ELECTRICITY GENERATION SOLAR CONCENTRATION
RESTRICTIONS	
R46	ELECTRICITY CONTRIBUTION FROM CARBON
R47	ELECTRICITY CONTRIBUTION FROM FUEL OIL
R48	ELECTRICITY CONTRIBUTION FROM DIESEL
R49	ELECTRICITY CONTRIBUTION FROM GAS
R50	ELECTRICITY CONTRIBUTION FROM BIOMAS
R51	ELECTRICITY CONTRIBUTION FROM CARBON CAPTURE AND STORAGE
R52	ELECTRICITY CONTRIBUTION FROM EOLIC
R53	ELECTRICITY CONTRIBUTION FROM PHOTOVOLTAIC
R54	ELECTRICITY CONTRIBUTION FROM GEOTHERMIC
R55	ELECTRICITY CONTRIBUTION FROM HYDRAULIC
R56	ELECTRICITY CONTRIBUTION FROM NUCLEAR
R57	ELECTRICITY CONTRIBUTION FROM SOLAR CONCENTRATION
R58	ELECTRICITY CONTRIBUTION FROM FOSSIL FUELS
R59	TOTAL NUMBER OF POWER PLANTS GENERATION
R60	FOSSIL NUMBER OF POWER PLANTS GENERATION
R61	CLEAN NUMBER OF POWER PLANTS GENERATION

MODEL ***ELECTRICIY CALCULUS ***	
R1..	Fdemand = E = Ffos + Frenv;
R2..	Ffos = E = Ffoscarb + Ffoscomb + Ffosdis + Ffosgas;
R3..	Frenv = E = Frenvbio + Frenvcscrb + Frenveol + Frenvftv + Frenvftrm + Frenvgtm + Frenvhdr + Frenvnucl;
*************************FÓSILES***	
R4..	Ffoscarb =L= Ffcarbins + (Fcapfoscarb * xfoscarb * 8760 * fafoscarb);
R5..	Ffoscomb =L= Ffcombins + (Fcapfoscomb * xfoscomb * 8760 * fafoscomb);
R6..	Ffosdis =L= Ffdisins + (Fcapfosdis * xfosdis * 8760 * fafosdis);
R7..	Ffosgas =L= Ffgasins + (Fcapfosgas * xfosgas * 8760 * fafosgas);

**************************RENOVABLES**	
R8..	Frenvbio =L= Frbioins + (Fcaprenvbio * xrenvbio * 8760 * farenvbio) ;
R9..	Frenvcscrb =L= Frcscrbins + (Fcaprenvcscrb * xrenvcscrb * 8760 * farenvcscrb) ;
R10..	Frenveol =L= Freolins + (Fcaprenveol * xrenveol * 8760 * farenveol) ;
R11..	Frenvftv =L= Frftvins + (Fcaprenvftv * xrenvftv * 8760 * farenvftv) ;
R12..	Frenvftrm =L= Frftrmins + (Fcaprenvftrm * xrenvftrm * 8760 * farenvftrm) ;
R13..	Frenvgtm =L= Frgtmins + (Fcaprenvgtm * xrenvgtm * 8760 * farenvgtm) ;
R14..	Frenvhdr =L= Frhdrins + (Fcaprenvhdr * xrenvhdr * 8760 * farenvhdr) ;
R15..	Frenvnucl =L= Frnuclins + (Fcaprenvnucl * xrenvnucl * 8760 * farenvnucl) ;
***	
********************************COST CALCULUS**	
R16..	TAC =E= Costfos + Costrenv ;
R17..	Costfos =E= Costfoscarb + Costfoscomb + Costfosdis + Costfosgas ;
R18..	Costrenv =E= Costrenvbio + Costrenvcscrb + Costrenveol + Costrenvftv + Costrenvftrm + Costrenvgtm + Costrenvhdr + Costrenvnucl;
R19..	Costfoscarb =E= Ffoscarb * (afoscarb + bfoscarb + cfoscarb) * 1000 ;
R20..	Costfoscomb =E= Ffoscomb * (afoscomb + bfoscomb + cfoscomb) * 1000 ;
R21..	Costfosdis =E= Ffosdis * (afosdis + bfosdis + cfosdis) * 1000 ;
R22..	Costfosgas =E= Ffosgas * (afosgas + bfosgas + cfosgas) * 1000 ;
R23..	Costrenvbio =E= Frenvbio * (arenvbio + brenvbio + crenvbio) * 1000 ;
R24..	Costrenvcscrb =E= Frenvcscrb * (arenvcscrb + brenvcscrb + crenvcscrb) * 1000 ;
R25..	Costrenveol =E= Frenveol * (arenveol + brenveol + crenveol) * 1000 ;
R26..	Costrenvftv =E= Frenvftv * (arenvftv + brenvftv + crenvftv) * 1000 ;
R27..	Costrenvgtm =E= Frenvgtm * (arenvgtm + brenvgtm + crenvgtm) * 1000 ;
R28..	Costrenvhdr =E= Frenvhdr * (arenvhdr + brenvhdr + crenvhdr) * 1000 ;
R29..	Costrenvnucl =E= Frenvnucl * (arenvnucl + brenvnucl + crenvnucl) * 1000
R30..	Costrenvftrm =E= Frenvftrm * (arenvftrm + brenvftrm + crenvftrm) * 1000 ;
**	
R31..	EFdemand =E= EFfos + EFrenv;
**	
***************************** EMISSIONS CALCULUS**	
R32..	EFfos =E= EFfoscarb + EFfoscomb + EFfosdis + EFfosgas ;
R33..	EFrenv =E= EFrenvbio + EFrenvcscrb + EFrenveol + EFrenvftv + EFrenvftrm + EFrenvgtm + EFrenvhdr;
****************************DE FUENTES FÓSILES***	
R34..	EFfoscarb =E= ECO2eqfoscarb * Ffoscarb ;
R35..	EFfoscomb =E= ECO2eqfoscomb * Ffoscomb ;
R36..	EFfosdis =E= ECO2eqfosdis * Ffosdis ;
R37..	EFfosgas =E= ECO2eqfosgas * Ffosgas ;

***************************DE FUENTES RENOVABLES**	
R38..	EFrenvbio =E= ECO2eqrenvbio * Frenvbio ;
R39..	EFrenvcscrb =E= ECO2eqrenvcscrb * Frenvcscrb ;
R40..	EFrenveol =E= ECO2eqrenveol * Frenveol ;
R41..	EFrenvftv =E= ECO2eqrenvftv * Frenvftv ;
R42..	EFrenvftrm =E= ECO2eqrenvftrm * Frenvftrm ;
R43..	EFrenvgtm =E= ECO2eqrenvgtm * Frenvgtm ;
R44..	EFrenvhdr =E= ECO2eqrenvhdr * Frenvhdr ;
R45..	EFrenvnucl =E= ECO2eqrenvnucl * Frenvnucl ;
**************************PERCENTAGE TECHNOLOGY CONTRIBUTIONS*************************	
R46..	Ffoscarb =L= Ffos ;
R47..	Ffoscomb =L= Ffos ;
R48..	Ffosdis =L= Ffos ;
R49..	Ffosgas =L= Ffos ;
R50..	Frenvbio =L= Fdemand - Ffos ;
R51..	Frenvcscrb =L= Fdemand - Ffos ;
R52..	Frenveol =L= Fdemand - Ffos ;
R53..	Frenvftv =L= Fdemand - Ffos ;
R54..	Frenvftrm =L= Fdemand - Ffos ;
R55..	Frenvgtm =L= Fdemand - Ffos ;
R56..	Frenvhdr =L= Fdemand - Ffos ;
R57..	Frenvnucl =L= Fdemand - Ffos ;
************FOSSIL PERCENTAGE CONTRIBUTION ***********	
R58..	Ffos =L= Fdemand * 0.8132 ;
******************************CENTTRAL NUMERO DE CENTRALES*******************************	
R59..	xtot =E= xf + xr ;
R60..	Xf =E= xfoscarb + xfoscomb + xfosdis + xfosgas ;
R61..	xr =E= xrenvbio + xrenvcscrb + xrenveol + xrenvftv + xrenvftrm + xrenvgtm + xrenvhdr + xrenvnucl;
**********************NUMBER OF POWER PLANTS GENERATION ***	
	Ffoscarb.LO = Fdemand * 0.11 ;
	Ffoscarb.UP = Fdemand * 0.115 ;
	Ffoscomb.LO = Fdemand * 0.075 ;
	Ffoscomb.UP = Fdemand * 0.0781 ;
	Ffosdis.LO = Fdemand * 0.008 ;
	Ffosdis.UP = Fdemand * 0.01 ;
	Ffosgas.LO = Fdemand * 0.45 ;
	Ffosgas.UP = Fdemand * 0.61 ;
	Frenvbio.LO = Fdemand * 0.00 ;

	Frenvbio.UP = Fdemand * 0.003 ;
	Frenvcscrb.LO = Fdemand * 0.00 ;
	Frenvcscrb.UP = Fdemand * 0.00 ;
	Frenveol.LO = Fdemand * 0.190 ;
	Frenveol.UP = Fdemand * 0.013 ;
	Frenvftrm.LO = Fdemand * 0.00 ;
	Frenvftrm.UP = Fdemand * 0.00 ;
	Frenvftv.LO = Fdemand * 0.0055 ;
	Frenvftv.UP = Fdemand * 0.0115 ;
	Frenvgtm.LO = Fdemand * 0.0145 ;
	Frenvgtm.UP = Fdemand * 0.015 ;
	Frenvhdr.LO = Fdemand * 0.11 ;
	Frenvhdr.UP = Fdemand * 0.1194 ;
	Frenvnucl.LO = Fdemand * 0.05 ;
	Frenvnucl.UP = Fdemand * 0.027 ;
MODEL PCL05 /ALL/; SOLVE PCL05 USING MIP MINIMIZING TAC;	

References

1. Farman, J.C.; Gardiner, B.G.; Shanklin, J.D. Large losses of total ozone in Antarctica reveal seasonal ClOx/NOx interaction. *Nature* **1985**, *315*, 207–210. [CrossRef]
2. *Montreal Protocol of Substances that Deplete the Ozone Layer*. United Nations Environment Programme, Ozone Secretariat; Montreal, Canada. 1987. Available online: https://treaties.un.org/doc/Treaties/1989/01/19890101%2003-25%20AM/Ch_XXVII_02_ap.pdf (accessed on 13 August 2019).
3. *Kyoto Protocol to the United Nations Framework Convention on Climate Change*. United Nations Framework Convention on Climate Change; Kyoto, Japan. 1998. Available online: https://unfccc.int/resource/docs/convkp/kpeng.pdf (accessed on 13 August 2019).
4. Sustainable Development Goals. United Nations. Available online: https://www.un.org/sustainabledevelopment/ (accessed on 7 July 2018).
5. Paris Agreement. In Proceedings of the Conference of the Parties of the United Nation Framework Convention on Climate Change, United Nations Framework Convention on Climate Change, Le Bourget, France, 15 December 2015.
6. The World Bank Open Data. World Development Indicators. 2017. Available online: http://datos.bancomundial.org/indicador/EG.USE.ELEC.KH.PC?view=chart (accessed on 26 August 2019).
7. National Ecology and Climate Change Institute (INECC). National Inventory of Emissions of Gases and Compounds of Greenhouse Gases. 2015. Available online: www.gob.mx/inecc/acciones-y-programas/emisiones-80133 (accessed on 26 August 2019).
8. SENER; Mexican Energy State Secretariat; Secretaría de Energía. *Electricity Sector Prospectives 2018–2032*. Mexico City, Mexico. 2018. Available online: https://www.gob.mx/cms/uploads/attachment/file/284345/Prospectiva_del_Sector_El_ctrico_2017.pdf (accessed on 13 August 2019).
9. Ruiz-Mendoza, B.J.; Sheinbaum-Pardo, C. Electricity sector reforms in four Latin-American countries and their impact on carbon dioxide emissions and renewable energy. *Energy Policy* **2010**, *38*, 6755–6766. [CrossRef]
10. SEGOB; Mexican Governing Secretariat; Secretaría de Gobernación. General Climate Change Law. Mexico City, Mexico. 2012. Available online: www.diputados.gob.mx/LeyesBiblio/pdf/LGCC.pdf (accessed on 7 September 2018).

11. SEMARNAT; Mexican Ministry of the Environment and Natural Resources; Undersecretary of Planning and Environmental Policy. *National Climate Change Strategy*; Vision 10-20-40; Secretaría del Medio Ambiente y recusos Naturales, Subsecretaría de Planeación y Política Ambiental; Dirección General de Políticas para el Cambio Climático: Mexico City, Mexico, 2013.
12. SEGOB; Mexican Governing Secretariat. Secretaría de Gobernación. *Energy Transition Law*. Energy Transition Law; Tenth-Five Transitory from the Third Article. Ley de Transición Energética. Transitorio Décimo quinto del Artículo Tercero; Mexico City, Mexico. 2015. Available online: www.diputados.gob.mx/LeyesBiblio/pdf/LTE.pdf (accessed on 5 September 2018).
13. Manzini, F.; Islas, J.; Martínez, M. Reduction of greenhouse gases using renewable energies in México in 2025. *Int. J. Hidrogen Energy* **2001**, *26*, 145–149. [CrossRef]
14. Johnson, T.M.; Alatorre, C.; Romo, Z.; Lui, F. *Low Carbon Development for Mexico*; World Bank: Washington, DC, USA, 2009.
15. Lund, H.; Mathiesen, B.V. Energy system analysis of 100% renewable energy systems—The case of Denmark in years 2030 and 2050. *Energy* **2009**, *34*, 524–531. [CrossRef]
16. Santoyo-Castelazo, E.; Stamford, L.; Azapagic, A. Environmental implications of decarbonizing electricity supply in large economies: The case of Mexico. *Energy Convers. Manag.* **2014**, *85*, 272–291. [CrossRef]
17. Vidal-Amaro, J.J.; Østergar, P.A.; Sheinbaum-Pardo, C. Optimal energy mix for transition from fossil fuel to renewable energy sources—The case of the Mexican electricity system. *Appl. Energy* **2015**, *150*, 80–96. [CrossRef]
18. International Energy Agency (IEA). *CO_2 Emissions from Fuel Combustion*, Highlights: Paris, France. November 2017.
19. Department of Economic and Social Affairs; United Nations. World Population Prospects. 2017. Available online: https://esa.un.org/unpd/wpp/Graphs/Probabilistic/POP/TOT/ (accessed on 29 March 2018).
20. Federal Electricity Commission (CFE); COPAR. Costos y Parámetros de Referencia para la Formulación de Proyectos de Inversión en el Sector Eléctrico. *Generation, Costs and Reference Parameters for the Formulation of Investment Projects in the Electricity Sector. Mexico*, Programmation Subdirection; Evaluation Coordination; México, 2012.
21. Intergovernmental Panel of Climate Change; Renewable Energy Sources and Climate Change Mitigation. *Special Report of the Intergovernmental Panel of Climate Change*, 1th ed.; Cambridge University: Cambridge, MA, USA, 2012.
22. Diwekar, U.M. *Introduction to Applied Optimization*, 2nd ed.; Springer: New York, NY, USA, 2008.

A Model for Optimizing Location Selection for Biomass Energy Power Plants

Chia-Nan Wang, Tsang-Ta Tsai * and Ying-Fang Huang

National Kaohsiung University of Science and Technology, Kaohsiung 80778, Taiwan; cn.wang@newfancy.com (C.-N.W.); winner@nkust.edu.tw (Y.-F.H.)

* Correspondence: tsangtatsai2019@gmail.com

Abstract: In addition to its potential for wave power, wind power, hydropower, and solar power, it can be said that Vietnam is a country with great potential for biomass energy derived from agricultural waste, garbage, and urban wastewater, which are resources widely available across the country. This huge amount of biomass, however, if left untreated, could become a major source of pollution and cause serious impacts on ecosystems (soil, water, and air), as well as on human health. In this research, the authors present a fuzzy multicriteria decision-making model (FMCDM) for optimizing the site selection process for biomass power plants. All of the criteria affecting location selection are identified by experts and literature reviews; in addition, the fuzzy analytic hierarchy process (FAHP) method was utilized so as to identify the weight of all of the criteria in the second stage. Furthermore, the Technique for Order Preference by Similarity to an Ideal Solution (TOPSIS) is applied for ranking potential locations in the final stage of this research. As a result, Long An (DMU/005) was found to be the best location for building biomass energy in Vietnam. The main contributions of this work include modeling the site selection decision process under fuzzy environment conditions. The proposed approaches also can address the complex problems in site selection; it is also a flexible design model for considering the evaluation criteria, and is applicable to location selection for other industries.

Keywords: biomass energy; site selection; optimization; MCDM; FMCDM; FAHP; TOPSIS

1. Introduction

In the context of increasingly depleting domestic fossil fuels, rising world oil prices, and increasing reliance on world energy prices, the ability to meet energy requirements for domestic demand is increasingly becoming difficult, as well as a major challenge. Thus, considering that the exploitation of clean renewable energy (RE) has important meaning in terms of economy, society, food security, and sustainable development, the energy demand in Vietnam has increased at twice the rate of the gross domestic product (GDP) growth, while this rate is only approximately 1% in developed countries. Vietnam's energy consumption has increased four times since 2005. From 1998 to 2008, the total electricity consumption demand in Vietnam increased by about 400% compared with the prior period. If this trend continues, Vietnam will become an energy importer in the future [1]. Vietnam's power transmission system includes voltage levels of 500, 220, and 110 kV. A power transmission system of 500 KV with a total length of 4670 km from north to south facilitates the transmission of electricity exchange among the north, central, and south of Vietnam. Circuit 1 of the 500-kV line was put into operation in September 1994, while Circuit 2 was put into operation in late 2005. Although the power distribution system is in relatively good condition, it still has high power loss. Overloaded lines, transformers operating with low efficiency, and poor-quality cables are the main causes of high losses [2].

In addition to the potential for wave power, wind power, hydropower, and solar power, it can be said that Vietnam is a country with great potential for creating biomass energy from agricultural waste, garbage, and urban wastewater, which are widely distributed across the country. This huge amount of biomass, however, if left untreated, will become a major source of pollution and continue to cause serious impacts on ecosystems (soil, water, and air), as well as on human health [1].

In recent years, interest in developing renewable energy technologies to replace fossil energy sources has increased globally, as fossil energy sources are in danger of depletion and fuel costs are increasing. Vietnam is a long-term agricultural country; furthermore, raw materials needed for producing biomass energy are about 118 million tons/year [3]. Vietnam has favorable natural conditions, such as a hot and humid climate, heavy rain, and fertile land; as a result, biomass grows quickly. Byproducts from agriculture and forestry are abundant and constantly increasing. However, these byproducts are typically considered natural waste, thus becoming more dangerous and causing environmental pollution. Therefore, Vietnam has many favorable conditions for developing biomass energy. Taking advantage of biomass energy will simultaneously provide energy for economic development and ensure environmental protection [3].

Vietnam's impressive economic reform over the past two decades has been accompanied by a sharp increase in energy demand. The Vietnamese government's strategy is to ensure sustainable economic growth in the future. Therefore, the Vietnamese government has set renewable energy development targets, including those for solar energy, wind energy, and biomass energy. Biomass energy is targeted to achieve 2.1% of the total electricity output by 2030 [4].

As an agricultural country, Vietnam has great potential to develop biomass energy. Accordingly, the ability to sustainably exploit biomass for energy production in Vietnam is about 150 million tons/year [5]. The main types of energy biomass include wood, waste byproducts from crops, livestock waste, municipal waste, and other organic waste. Sources of renewable energy can be burned directly or be used to create biomass fuel pellets [5]. Currently, on the global scale, biomass is the fourth-largest energy resource, accounting for 14%–15% of the world's total energy consumption. In developing countries, biomass is often the largest energy resource, averaging about 35% of the total energy supply [6].

One of the most important activities of the biomass energy project is site selection for building plants. A good location of a biomass plant is of vital importance for the project's economic survival, and it includes solving complex location and transportation problems. In this research, we propose an MCDM model for biomass power plant location selection in Vietnam. All of the criteria affecting the location selection are identified by experts and literature reviews; furthermore, the fuzzy analytic hierarchy process (FAHP) method was utilized in order to identify the weight of all of the criteria in the second stage. TOPSIS was applied for ranking the potential locations in the final stage of this research. The general flow of the MCDM model is shown in Figure 1.

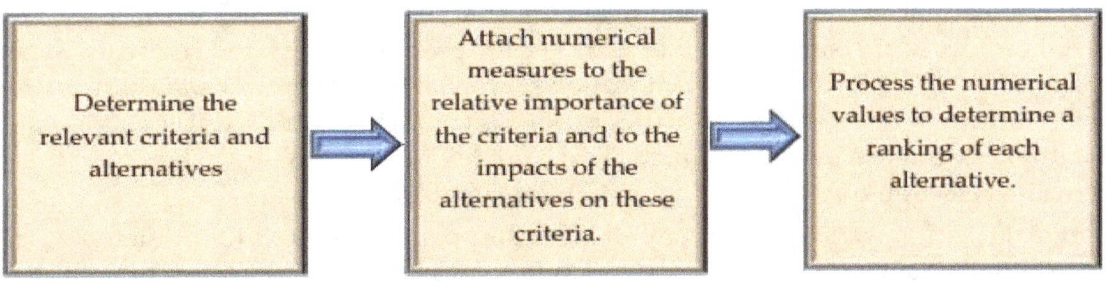

Figure 1. General flow of multicriteria decision-making model (MCDM) model [7,8].

The primary goal of this research is to propose a useful fuzzy MCDM model for biomass power plant location selection. The proposed approaches can also address different complex problems in site selection; in addition, it is also a flexible design model for considering the evaluation criteria, and it is applicable to location selection for other industries.

The remainder of the research is to provide background to assist the authors in building the fuzzy MCDM model. Then, an integrated model using the FAHP–TOPSIS approach is introduced so as to select the best site for biomass power plant from eight potential locations in Vietnam. The results and contributions will be presented at the end of this paper.

2. Literature Review

Nowadays, there are many studies that have applied the MCDM model to select locations in the energy sector and some other industries, such as Roghayeh Ghasempour et al. [9], who have reviewed an application of the MCDM model for solar plant location selection and technology. In this paper, a wide variety MCDM methods, investigated by various researchers, are presented so as to obtain effective criteria for selecting solar plant sites and solar plant technologies. C. Chen et al. [10] proposed an optimization model for biomass location selection. In this research, the number of biomass power plants, optimum site, and transportation schemes were obtained with minimum cost, minimum energy consumption, and the lowest effect on the environment. Jakkawan Patomtummakan et al. [11] presented a mixed-integer linear programming for biomass power plant location selection. The objective of the optimization model is to minimize the total cost at selected bio-power plants by considering four cost components, namely, the fixed opening of the proposed bio-power plants, the material cost from purchasing biomass, the transportation cost between suppliers and bio-power plants, and the inventory holding cost.

Massimiliano Cattafi et al. [12] proposed an integer linear programming approach for defining the energy and cost-efficient biomass plant location, along with the corresponding provisioning basin. This optimization tool is just a small part of a wider perspective that is aimed at defining the decision support tools for the improvement of regional planning and its precise strategic environmental assessment. Heesung Woo et al. [13] integrated a multi-criteria analysis (MCA) and geographical information systems (GIS) for optimizing the site of the biomass energy plants. The research results show that land use, resource availability, and supply chain cost data can be integrated and mapped using GIS, in order to facilitate the determination of different sustainable factors weightings, and to ultimately generate optimal candidate sites for biomass energy plants.

Hao Lv et al. [14] proposed a multi-objective mixed-integer programming approach to solve the location selection issue for a straw-based power generation plant with CO_2 emissions. It is anticipated that this paper will make a contribution to the current scientific knowledge by presenting innovative approaches for the sustainable utilization of forest harvest residues as a resource for the generation of bioenergy in Tasmania. Wang et al. [15] applied an MCDM model for solid waste to energy plant location selection. Jin Su Jeong et al. [16] proposed a multicriteria GIS assessment with a weighted linear combination (WLC) to various disciplines, using suitable criteria to optimize a biomass facility location. This assessment could be used in studies to verify suitable biomass plant sites with corresponding geographical and spatial circumstances and available spatial data necessary in various governmental and industrial sectors. Ali Asghar Isalou et al. [17] proposed an integrated fuzzy logic and analytic network process (FANP) to locate a suitable location for the landfilling municipal solid waste generated in Iran. Their findings revealed that the integration of fuzzy logic and ANP can give a better idea compared with other models like AHP, fuzzy logic, and ANP (individually). Therefore, this model can be applied in site selection for landfill of other similar places.

Wang et al. [18] proposed an MCDM model for solar panel site selection in Vietnam. In this work, the authors applied a data envelopment analysis (DEA) model for the selection of potential locations, FAHP was used for determining the weight of sub-factors, and TOPSIS was applied for ranking potential options. Wang et al. [19] proposed an MCDM model for wind power plant location selection. This study resided in the evolution of a new model that is flexible and practical for a decision-maker for site selection in the energy sector. Abdolvahhab Fetanat et al. [20] proposed a hybrid MCDM model for offshore wind power plants location selection in Iran. The evaluation factors and this model could be used in other coastal cities for promoting the progress of integrated coastal management (ICM), towards the goal of sustainability. Mostafa Rezaei-Shouroki [21] proposed an MCDM model by using a hybrid data envelopment analysis (DEA), AHP, and fuzzy TOPSIS model for wind farm location selection. The purpose of this study is to prioritize and rank 13 cities of the Fars province in Iran in terms of their suitability for the construction of a wind farm.

Fahime Heidarzade et al. [22] applied a MCDM model for wind farm site selection in Iran. In this research, a step-wise weight assessment ratio analysis (SWARA) is employed to rank factor affects to wind power plant site, and the weighted aggregates sum product assessment (WASPAS) is utilized to evaluate the decision-making unit (DMU). Halil Ibrahim Cobuloglu et al. [23] proposed a MCDM model for biomass location selection. In this work, the authors used an FANP for identifying the weight of all of the criteria. Yasir Ahmed Solangi et al. [24] integrated Delphi and AHP and fuzzy TOPSIS for the ranking and selection of renewable energy resources. The study provides important insights related to the prioritizing of RE resources for electricity generation, and can be used to undertake policy decisions toward sustainable energy planning in Pakistan. Mohammad Alhuyi Nazari et al. [25] used a TOPSIS model for analyzing solar farm location selection. The primary aim of this paper is to select suitable sites for photovoltaic installation in Iran. S. Saelee et al. [26] applied a TOPSIS multi-criteria approach for biomass type selection for boilers. Babak Daneshvar Rouyendegh et al. [27] applied intuitionistic fuzzy TOPSIS in the location selection of wind power plants in Turkey. The main purpose of the TOPSIS method is to rank the alternatives in the worst way. The intuitionistic fuzzy set (IFS) is used to reflect the approval, rejection, and hesitation of decision makers by dealing with real life uncertainty, imprecision, vagueness, and linguistic human decisions. Choudhary et al. [28] used a hybrid MCDM model including a social, technical, economic, environmental, and political (STEEP), fuzzy AHP and TOPSIS model for thermal power plant location selection in India. The paper presents a more accurate, effective, and systematic decision support tool for decision makers to conduct the evaluation process and to select optimal locations for TPPs.

Based on the literature review and experts' opinion, there are some factors that must be considered in the biomass power plant location selection process, such as economic, environmental, technical, and social-political factors, and there are many researchers who have applied the MCDM model to various fields of science and engineering—a trend that has been increasing for many years—but very few works have focused on this problem in a fuzzy environment.

3. Methodology

3.1. Research Development

Many researchers have applied the MCDM model to various fields of science and engineering, more so over the last few years. One field in the MCDM approach has identified location selection problems; thus, especially in the renewable energy sector, decision-makers have to evaluate both qualitative and quantitative criteria. Although some studies have reviewed applications of MCDM approaches in biomass power plant location selection, to the best of our knowledge, few works have focused on this problem in a fuzzy environment. This is why the authors proposed an MCDM model for optimizing location selection for biomass energy power plants in this study. There are three main steps in this research.

Step 1: All of the criteria and subcriteria affecting the site evaluation and selection processes are determined based on experts and literature reviews.

The processes of the selection criteria are shown in Figure 2.

Step 2: The FAHP model was utilized to identify the weight of all of the subcriteria in the second stage.

The FAHP method has many advantages compared with other multi-objective decision-making methods. First, many multicriteria decision-making methods face difficulties in determining the importance of each criterion, while FAHP is a well-known method of determining these weights. Therefore, FAHP can be combined with other easy methods to take advantage of each method in problem-solving. In addition, FAHP can check consistency in decision-makers' judgment. Moreover, the hierarchical analysis process is easy to understand, considering many small criteria and analyzing all of the qualitative and quantitative factors.

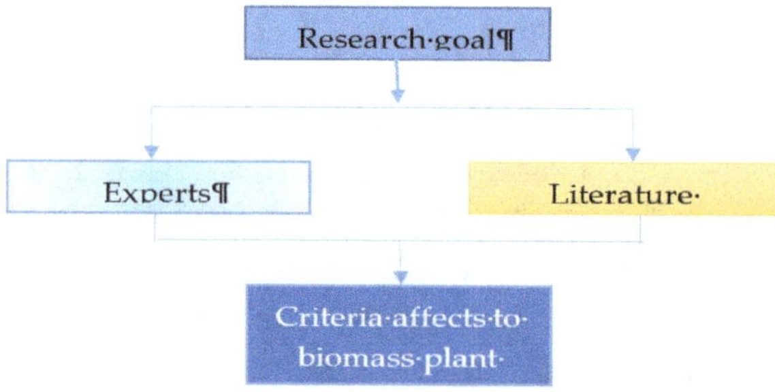

Figure 2. Processes of selection criteria.

Step 3: The TOPSIS model is used for ranking potential locations in the final stage.

The advantages of the TOPSIS methods are simplicity, rationality, comprehensibility, good computational efficiency, and the ability to measure the relative performance for each alternative in a simple mathematical form. TOPSIS is based on the concept that the chosen alternative should have the shortest geometric distance from the positive ideal solution (PIS) and the longest geometric distance from the negative ideal solution (NIS). Diagram of research as shown in Figure 3.

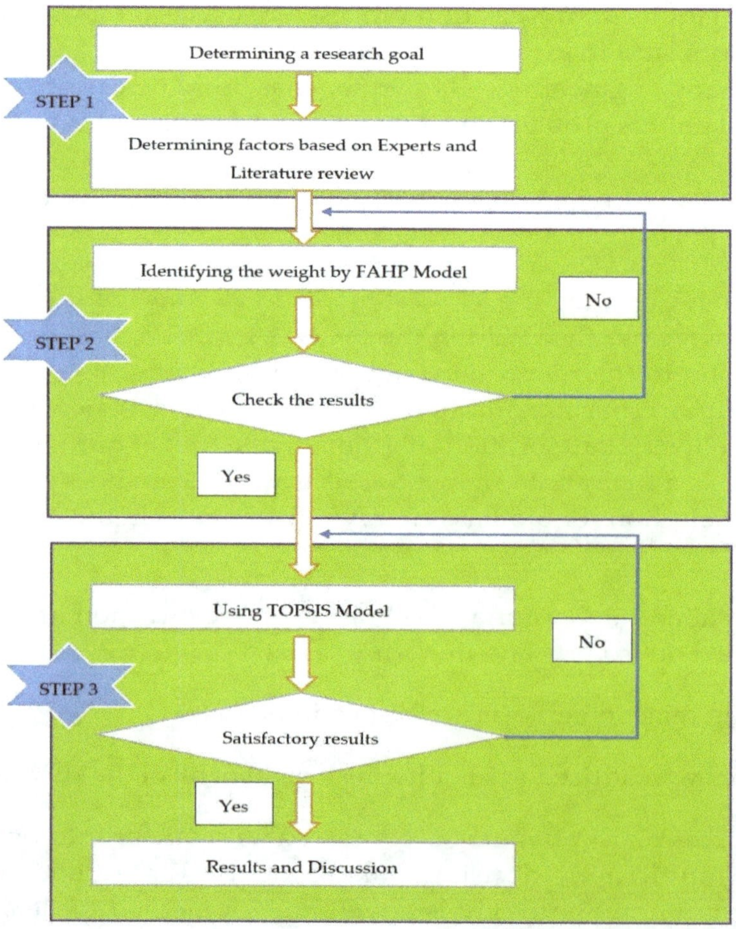

Figure 3. Diagram of research. FAHP—fuzzy analytic hierarchy process.

3.2. *Fuzzy Sets, AHP, and TOPSIS Model*

3.2.1. Fuzzy Sets and Fuzzy Number

In 1965, Lotfi A. Zadeh published the article with the title "Fuzzy Set", which describes the mathematics of "fuzzy set" and "fuzzy logic" theory. The triangular fuzzy number (TFN) can be defined as (g, p, s). TFNs are shown in Figure 4.

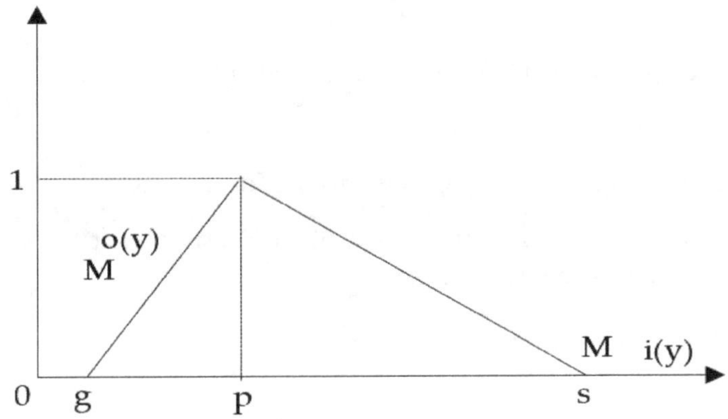

Figure 4. Traingular fuzzy number.

TFN also can be defined as follows:

$$\left(\frac{x}{\widetilde{M}}\right) = \begin{cases} 0, & x < o, \\ \frac{x-g}{p-g} & g \le x \le p, \\ \frac{s-x}{s-p} & p \le x \le s, \\ g, & x > s, \end{cases} \quad (1)$$
$$\mu$$

The representatives of each level of membership give a fuzzy number, as follows:

$$\widetilde{M} = (M^{o(y)}, M^{i(y)}) = [g + (p-k)y,\ s + (p-s)y], y\ \in [0,1] \quad (2)$$

$o(y)$, $i(y)$ indicates both the left side and the right side of a fuzzy number, respectively, as follows:

$$\begin{aligned} (g_1,\ p_1,\ s_1) + (g_2,\ p_2,\ s_2) &= (g_1 + g_2,\ p_1 + p_2, s_1 + s_2) \\ (g_1,\ p_1,\ s_1) - (g_2,\ p_2,\ s_2) &= (g_1 - g_2,\ p_1 - p_2, s_1 - s_2) \\ (g_1,\ p_1,\ s_1) \times (g_2,\ p_2,\ s_2) &= (g_1 \times g_2,\ p_1 \times p_2, s_1 \times s_2) \\ \frac{(g_1,\ p_1,\ s_1)}{(g_2,\ p_2,\ s_2)} &= (k_1/k_2,\ p_1/p_2, s_1/s_2) \end{aligned} \quad (3)$$

3.2.2. Analytic Hierarchy Process (AHP) Model

AHP is presented by Saaty. AHP is an MCDM that simplifies complex problems by sorting criteria and options in a hierarchical structure. Let $F = \{F_a | a\ = 1,2,\ \ldots,\ m\}$ is a factor set. The pairwise comparison metrics on m criteria will be presented in an m × m evaluation matric (D). Every element, h_{ab}, is the quotient of the weights of the factors, as follows:

$$\mathrm{D} = (h_{ab}), a, b = 1, \ldots, m \quad (4)$$

The relative priorities are given by the Eigenvector (l) corresponding to the largest eigenvector (λ_{max}), as follows:

$$D_l = \lambda_{max} l \quad (5)$$

The consistency is determined by the relation between the entries of D and CI, as follows:

$$\text{CI} = \frac{(\lambda_{max} - m)}{(m-1)} \tag{6}$$

CR is calculated as the ratio of the CI and the RI, as shown in Equation (7):

$$CR = \frac{CI}{RI} \tag{7}$$

CR value must less than (or equal) 0.1. If the CR >0.1, the evaluation needs to be repeated again for improving consistency.

3.2.3. Technique for Order Preference by Similarity to an Ideal Solution (TOPSIS)

TOPSIS assumes that we have m alternatives (options) and n attributes/criteria, and we have the score of each option with respect to each criterion.

- Construct the normalized decision matrix

$$e_{ij} = \frac{X_{ij}}{\sqrt{\sum_{i=1}^{m} X_{ij}^2}}. \tag{8}$$

with $i = 1, 2, \ldots, m$; and $j = 1, 2, \ldots, n$.

- Construct the weighted normalized decision matrix

$$S_{ij} = W_i e_{ij}. \tag{9}$$

with $i = 1, 2, \ldots, m$ and $j = 1, 2, \ldots, n$.

- Determine the ideal and negative ideal solutions

$$\begin{aligned} A^+ &= s_1^+,\ s_2^+, \ldots, s_n^+; \\ A^- &= s_1^-,\ s_2^-, \ldots, s_n^-; \end{aligned} \tag{10}$$

- Calculate the separation measures for each alternative

$$D_i^+ = \sqrt{\sum_{j=1}^{m} \left(s_i^+ - s_{ij}\right)^2};\ i = 1, 2, \ldots, m. \tag{11}$$

options to NIS.

$$D_i^- = \sqrt{\sum_{j=1}^{m} \left(s_{ij} - s_i^-\right)^2};\ i = 1, 2, \ldots, m. \tag{12}$$

- Calculate the relative closeness to the ideal solution (G_i)

$$G_i = \frac{D_i^-}{D_i^- + D_i^+}\ i = 1,\ 2,\ \ldots,\ m. \tag{13}$$

4. Case Study

The total potential of Vietnam's biomass energy source is more than 99 million tons/year, corresponding to the electric energy source of more than 340,000 GWh, of which the largest is the

Mekong River Delta (Mekong Delta), accounting for 33.4%, followed by the North Central Coast and South Central Coast with 21.8% [29].

According to the Vietnam Institute of Energy, agricultural byproducts from the Mekong Delta amount for about 23 million tons/year, which includes about 3.8 million tons of rice husk, nearly 17 million tons of rice straw, over 372,000 tons of corn, and nearly 1.4 million tons of bagasse [30]. Thus, the authors proposed an MCDM model for optimizing location selection for biomass energy power plants in the Mekong Delta. In a renewable energy project, selecting a location to build a power plant includes complex decision-making, which includes economic, natural, and social factors. The material source of this project is rice husk. After preliminary studies, eight potential locations in the Mekong Delta (DMUs) were selected (see Table 1 and Figure 5).

Table 1. A potential locations list. DMU—decision making unit.

No.	Location's Name	Symbol
1	Kien Giang	DMU/001
2	An Giang	DMU/002
3	Dong Thap	DMU/003
4	Soc Trang	DMU/004
5	Long An	DMU/005
6	Tra Vinh	DMU/006
7	Tien Giang	DMU/007
8	Can Tho	DMU/008

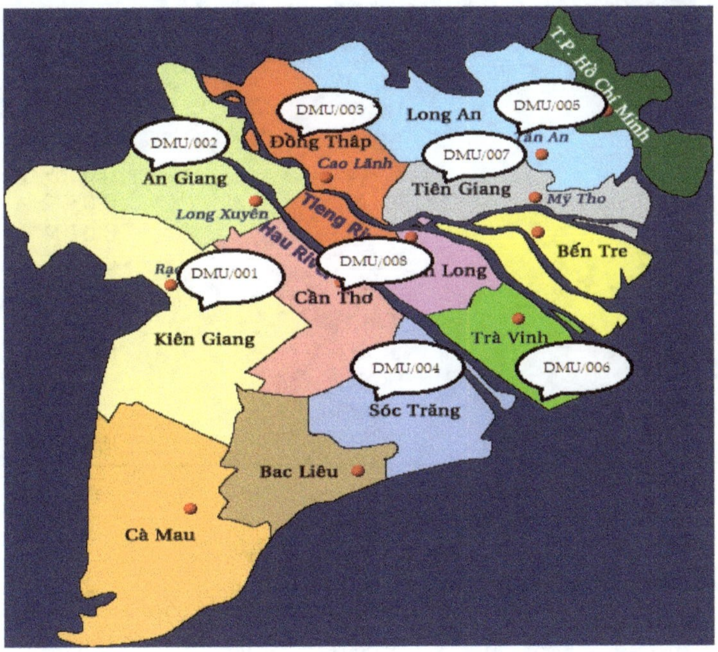

Figure 5. Mekong Delta map.

In the first stage of this work, the main factors and subcriteria affects on location selection are identified by experts and literature reviews. The general hierarchy structure of the proposed model is shown in Figure 6.

The fuzzy analytic hierarchy process (FAHP) method was utilized to identify the weight of all of the criteria in the second stage. The results are shown in Table 2.

The TOPSIS is applied for ranking potential locations in the final stage of this research. The normalized matrix and normalized weight matrix are shown in Tables 3 and 4, respectively.

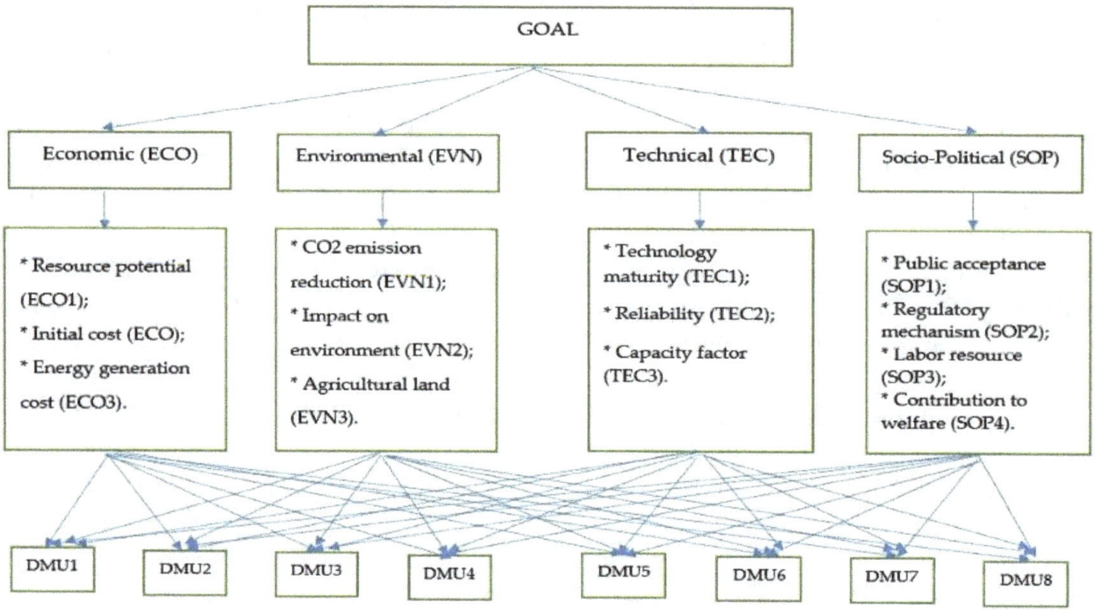

Figure 6. General hierarchy structure of proposed model.

Table 2. Weight of all of the subcriteria. ECO—economic; EVN—environmental; TEC—technical; SOP—socio-political.

No.	Criteria	Weight
1	ECO1	0.1650
2	ECO2	0.1042
3	ECO3	0.1968
4	EVN1	0.0589
5	EVN2	0.0572
6	EVN3	0.0404
7	TEC1	0.0547
8	TEC2	0.1283
9	TEC3	0.1163
10	SOP1	0.0236
11	SOP1	0.0210
12	SOP3	0.0144
13	SOP4	0.0193

Table 3. Normalized matrix.

	DMU/001	DMU/002	DMU/003	DMU/004	DMU/005	DMU/006	DMU/007	DMU/008
ECO1	0.2851	0.3326	0.3801	0.3801	0.4276	0.3326	0.2851	0.3801
ECO2	0.3567	0.3567	0.3121	0.4013	0.3567	0.2675	0.3567	0.4013
ECO3	0.3472	0.3038	0.3472	0.3906	0.3472	0.3472	0.3906	0.3472
EVN1	0.3845	0.3417	0.3417	0.2990	0.3845	0.3417	0.3845	0.3417
EVN2	0.3636	0.3636	0.3182	0.4091	0.3636	0.3182	0.3182	0.3636
EVN3	0.3219	0.3219	0.3678	0.4138	0.3219	0.2759	0.3678	0.4138
TEC1	0.3845	0.3417	0.3417	0.3417	0.2990	0.3417	0.3845	0.3845
TEC1	0.3786	0.3366	0.3786	0.2945	0.3786	0.3366	0.3786	0.3366
TEC3	0.3522	0.3522	0.3082	0.3962	0.3962	0.3522	0.3522	0.3082
SOP1	0.3581	0.3134	0.4029	0.3581	0.3581	0.3134	0.3581	0.3581
SOP2	0.3038	0.3472	0.3472	0.3906	0.3472	0.3906	0.3472	0.3472
SOP3	0.3581	0.3581	0.4029	0.3581	0.3581	0.3581	0.3134	0.3134
SOP4	0.3465	0.3898	0.3465	0.3898	0.3032	0.3465	0.3898	0.3032

Table 4. Normalized weight matrix.

	DMU/001	DMU/002	DMU/003	DMU/004	DMU/005	DMU/006	DMU/007	DMU/008
ECO1	0.0470	0.0549	0.0627	0.0627	0.0706	0.0549	0.0470	0.0627
ECO2	0.0372	0.0372	0.0325	0.0418	0.0372	0.0279	0.0372	0.0418
ECO3	0.0683	0.0598	0.0683	0.0769	0.0683	0.0683	0.0769	0.0683
EVN1	0.0226	0.0201	0.0201	0.0176	0.0226	0.0201	0.0226	0.0201
EVN2	0.0208	0.0208	0.0182	0.0234	0.0208	0.0182	0.0182	0.0208
EVN3	0.0130	0.0130	0.0149	0.0167	0.0130	0.0111	0.0149	0.0167
TEC1	0.0210	0.0187	0.0187	0.0187	0.0164	0.0187	0.0210	0.0210
TEC1	0.0486	0.0432	0.0486	0.0378	0.0486	0.0432	0.0486	0.0432
TEC3	0.0410	0.0410	0.0358	0.0461	0.0461	0.0410	0.0410	0.0358
SOP1	0.0085	0.0074	0.0095	0.0085	0.0085	0.0074	0.0085	0.0085
SOP2	0.0064	0.0073	0.0073	0.0082	0.0073	0.0082	0.0073	0.0073
SOP3	0.0052	0.0052	0.0058	0.0052	0.0052	0.0052	0.0045	0.0045
SOP4	0.0067	0.0075	0.0067	0.0075	0.0059	0.0067	0.0075	0.0059

Vietnam is a developing agricultural country and has great potential to create biomass sources for energy production. If effectively exploited, biomass energy will help reduce dependence on traditional power sources, reduce carbon emissions, reduce pollution, and bring direct profits to establishments; furthermore, farmers can participate in the biomass fuels supply chain via selling waste and agricultural byproducts. According to the calculation theory, the total potential of Vietnam's biomass energy source is more than 99 million tons/year, corresponding to the electric energy source of more than 340,000 GWh, of which the Mekong River Delta (Mekong Delta) accounts for 33.4%, followed by the North Central Coast and Central Coast with 21.8% [31]. The primary goal of this research is to propose a useful fuzzy MCDM model for biomass power plant location selection. In the first stage of this work, all of the criteria affecting the location selection are identified by experts and literature reviews; in addition, the FAHP method was utilized to identify the weight of all of the criteria in the second stage. The TOPSIS is applied for ranking the potential locations in the final stage of this research; furthermore, TOPSIS is based on the concept that the chosen alternative should have the shortest geometric distance from the PIS and the longest geometric distance from the NIS. As per the results shown in Figure 7 and Table 5, Long An (DMU/005) is found to be the best location for building biomass energy in Vietnam.

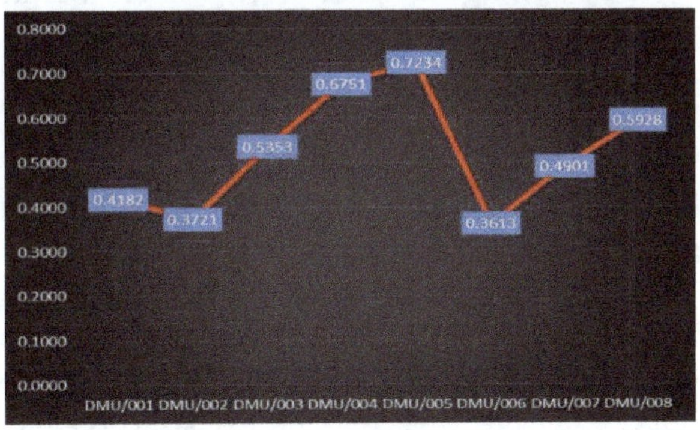

Figure 7. Final ranking list.

Table 5. Negative ideal solution (NIS) and positive ideal solution (PIS) values in the Technique for Order Preference by Similarity to an Ideal Solution (TOPSIS) model.

DMU	Di+	Di−
DMU/001	0.0265	0.0190
DMU/002	0.0255	0.0151
DMU/003	0.0192	0.0221
DMU/004	0.0145	0.0301
DMU/005	**0.0119**	**0.0312**
DMU/006	0.0254	0.0144
DMU/007	0.0252	0.0242
DMU/008	0.0170	0.0247

5. Conclusions

As an agricultural country, Vietnam is an ideal place for producing biomass energy. Agricultural waste is the most abundant in the Mekong Delta region, which makes up about 50% of the country. Major biomass resources include straw, rice husks from rice mills, bagasse from sugar mills, coffee husks from coffee-processing plants, and wood chips from wood-processing industries. Vietnam has set a target of a combined capacity of 500 MW of biomass energy by 2020, rising to 2000 MW by 2030 [32].

Therefore, Vietnam has many favorable conditions for developing biomass energy. Taking advantage of biomass energy will simultaneously provide energy for economic development, and ensure environmental protection. Although there are many studies in regard to developing methods and using different criteria for evaluating renewable energy location, research on biomass location evaluation and selection remains limited, especially in regard to the use of integrating economic and environmental criteria. In this research, the authors presented a fuzzy multicriteria decision-making model (FMCDM) for optimizing the site selection process for biomass power plants under fuzzy environment conditions. All of the criteria affecting location selection are identified by experts and literature reviews; in addition, the fuzzy analytic hierarchy process (FAHP) method was utilized to identify the weight of all of the criteria in the second stage. As a result, Long An (DMU/005) was found to be the best location for building biomass energy plants in Vietnam.

The main contribution of this work includes modeling the site selection decision process under fuzzy environment conditions. This paper also considers the evolution of a new model that is flexible and practicable to the decision-maker. This research also provides a useful guideline for biomass power plant location selection in many countries, as well as providing a guideline for location selection in related industries. Furthermore, for future research, the application of other MCDM approaches can be utilized so as to compare the results in the search of any changes.

Author Contributions: Conceptualization, C.-N.W., T.-T.T., and Y.-F.H.; formal analysis, C.-N.W. and Y.-F.H.; funding acquisition, C.-N.W.; methodology, T.-T.T.; project administration, C.-N.W. and Y.-F.H.; resources, T.-T.T.; software, T.-T.T.; supervision, C.-N.W.; writing (original draft), T.-T.T.; writing (review and editing), C.-N.W. and Y.-F.H.

References

1. Điện sinh khối-nguồn năng lượng tái tạo hữu ích. Available online: http://chatdotxanh.com/chi-tiet-tin/dien-sinh-khoi-nguon-nang-luong-tai-tao-huu-ich.html (accessed on 12 February 2019).
2. Trung, N.Q. Tổng quan về lưới điện trong hệ thống điện Quốc gia. Trung tâm điều độ hệ thống điện QG. Available online: https://www.nldc.evn.vn/newsg/1/86/Tong-quan-ve-luoi-dien-trong-he-thong-dien-Quoc-gia/default.aspx (accessed on 20 February 2019).
3. NĂNG LƯỢNG SINH KHỐI Ở VIỆT NAM: VẪN CHỈ LÀ TIỀM NĂNG. Available online: https://www.pvpower.vn/nang-luong-sinh-khoi-o-viet-nam-van-chi-la-tiem-nang/ (accessed on 18 February 2019).
4. Ward, A.; Le, H.; Tran, T.; Pham, N. *Tạo sự hấp dẫn cho năng lượng sinh khối*; Viện Tăng trưởng xanh toàn cầu: Seoul, Korea, 2018.
5. Thanh, T. Nguồn năng lượng sinh khối ở Việt Nam khoảng 150 triệu tấn/năm. Available online: https://petrotimes.vn/nguon-nang-luong-sinh-khoi-o-viet-nam-khoang-150-trieu-tannam-516007.html (accessed on 12 February 2019).
6. Phát triển năng lượng sinh khối của Việt Nam: Tiềm năng và thách thức. Available online: http://chatdotxanh.com/chi-tiet-tin/phat-trien-nang-luong-sinh-khoi-cua-viet-nam-tiem-nang-va-thach-thuc.html (accessed on 15 February 2019).
7. Pamučar, D.; Gigović, L.; Bajić, Z.; Janošević, M. Location Selection for Wind Farms Using GIS Multi-Criteria Hybrid Model: An Approach Based on Fuzzy and Rough Numbers. *Sustainability* **2017**, *9*, 1315. [CrossRef]

8. Villacreses, G.; Gaona, G.; Martínez-Gómez, J.; Jijón, D.J. Wind farms suitability location using geographical information system (GIS), based on multi-criteria decision making (MCDM) methods: The case of continental Ecuador. *Renew. Energy* **2017**, *109*, 275–286. [CrossRef]
9. Ghasempour, R.; Nazari, M.A.; Ebrahimi, M.; Ahmadi, M.H.; Hadiyanto, H. Multi-Criteria Decision Making (MCDM) Approach for Selecting Solar Plants Site and Technology: A Review. *Int. J. Renew. Energy Dev.* **2019**, *8*, 15–25. [CrossRef]
10. Chen, C.; Li, W.; Li, Y.; Zhu, Y. Biomass power plant site selection modeling and decision optimization. *Trans. Chin. Soc. Agric. Eng.* **2011**, *27*, 255–260.
11. Patomtummakan, J.; Nananukul, N. Biomass Power Plant Location and Distribution Planning System. *GMSARN Int. J.* **2018**, *12*, 11–18.
12. Cattafi, M.; Gavanelli, M.; Cagnoli, P.; Massimiliano, C. Sustainable Biomass Power Plant Location in the Italian Emilia-Romagna Region. *ACM Trans. Intell. Syst. Technol.* **2011**, *2*, 23. [CrossRef]
13. Woo, H.; Acuna, M.; Moroni, M.; Taskhiri, M.S.; Turner, P. Optimizing the Location of Biomass Energy Facilities by Integrating Multi-Criteria Analysis (MCA) and Geographical Information Systems (GIS). *Forests* **2018**, *9*, 585. [CrossRef]
14. Lv, H.; Ding, H.; Zhou, D.; Zhou, P. A Site Selection Model for a Straw-Based Power Generation Plant with CO_2 Emissions. *Sustainability* **2014**, *6*, 7466–7481. [CrossRef]
15. Wang, C.-N.; Nguyen, V.T.; Duong, D.H.; Thai, H.T.N. A Hybrid Fuzzy Analysis Network Process (FANP) and the Technique for Order of Preference by Similarity to Ideal Solution (TOPSIS) Approaches for Solid Waste to Energy Plant Location Selection in Vietnam. *Appl. Sci.* **2018**, *8*, 1100. [CrossRef]
16. Jeong, J.S.; Ramírez-Gómez, Á. A Multicriteria GIS-Based Assessment to Optimize Biomass Facility Sites with Parallel Environment—A Case Study in Spain. *Energies* **2017**, *10*, 2095. [CrossRef]
17. Isalou, A.A.; Zamani, V.; Shahmoradil, B. Alizadeh Landfill site selection using integrated fuzzy logic and analytic network process (F-ANP). *Environ. Earth Sci.* **2012**, *68*, 1745–1755. [CrossRef]
18. Wang, C.-N.; Nguyen, V.T.; Thai, H.T.N.; Duong, D.H. Multi-Criteria Decision Making (MCDM) Approaches for Solar Power Plant Location Selection in Viet Nam. *Energies* **2018**, *11*, 1504. [CrossRef]
19. Wang, C.-N.; Huang, Y.-F.; Chai, Y.-C.; Nguyen, V.T. A Multi-Criteria Decision Making (MCDM) for Renewable Energy Plants Location Selection in Vietnam under a Fuzzy Environment. *Appl. Sci.* **2018**, *8*, 2069. [CrossRef]
20. Fetanat, A.; Khorasaninejad, E. A novel hybrid MCDM approach for offshore wind farm site selection: A case study of Iran. *Ocean Coast. Manag.* **2015**, *109*, 27–28. [CrossRef]
21. Rezaei-Shouroki, M. The location optimization of wind turbine sites with using the MCDM approach: A case study. *Energy Quipsys* **2017**, *5*, 165–187.
22. Heidarzade, F.; Varzandeh, M.H.M.; Rahbari, O.; Zavadskas, E.K.; Vafaeapour, M. Placement of Wind Farms Based on a Hybrid Multi Criteria. In Proceedings of the World Sustainability Forum, Oshawa, ON, Canada, 1–30 November 2014.
23. Cobuloglu, H.I.; Büyüktahtakin, I.E. A Multi-Criteria Approach for Biomass Crop Selection under Fuzzy Environment. In Proceedings of the 2014 Industrial and Systems Engineering Research Conference, Montreal, QC, Canada, 31 May–3 June 2014.
24. Solangi, Y.A.; Tan, Q.; Mirjat, N.H.; Valasai, G.D.; Khan, M.W.A.; Ikram, M. An Integrated Delphi-AHP and Fuzzy TOPSIS Approach toward Ranking and Selection of Renewable Energy Resources in Pakistan. *Processes* **2019**, *7*, 118. [CrossRef]
25. Nazari, M.A.; Aslani, A.; Ghasempour, R. Analysis of Solar Farm Site Selection Based on TOPSIS Approach. *Int. J. Soc. Ecol. Sustain. Dev.* **2018**, *9*, 14. [CrossRef]
26. Nazari, M.A.; Aslani, A.; Ghasempour, R. Biomass Type Selection for Boilers Using TOPSIS Multi-Criteria Model. *Int. J. Environ. Sci. Dev.* **2014**, *2*, 5.
27. Daneshvar Rouyendegh, B.; Yildizbasi, A.; Arikan, Ü.Z.B. Babak Daneshvar Rouyendegh. Using Intuitionistic Fuzzy TOPSIS in Site Selection of Wind Power Plants in Turkey. *Adv. Fuzzy Syst.* **2018**, *2018*. [CrossRef]
28. Choudhary, D.; Shankar, R. An STEEP-fuzzy AHP-TOPSIS framework for evaluation and selection of thermal power plant location: A case study from India. *Energy* **2012**, *42*, 510–521. [CrossRef]
29. Nhung, H. ĐBSCL có tiềm năng năng lượng sinh khối lớn nhất nước. Tuổi Trẻ. Available online: https://tuoitre.vn/dbscl-co-tiem-nang-nang-luong-sinh-khoi-lon-nhat-nuoc-1076932.htm (accessed on 27 May 2019).

30. Nguồn năng lượng nào cho Đồng bằng sông Cửu Long? Kinh Tế Sài Gòn. Available online: https://www.thesaigontimes.vn/121298/Nguon-nang-luong-nao-cho-Dong-bang-song-Cuu-Long (accessed on 27 May 2019).
31. ĐBSCL có tiềm năng năng lượng sinh khối lớn nhất nước. Available online: http://m.icon.com.vn/vi-VN/c620/127733/DBSCL-co-tiem-nang-nang-luong-sinh-khoi-lon-nhat-nuoc.aspx (accessed on 24 February 2019).
32. Biomass Energy in Vietnam. Available online: https://www.bioenergyconsult.com/biomass-energy-vietnam/ (accessed on 10 April 2019).

11

A Time-Sequence Simulation Method for Power Unit's Monthly Energy-Trade Scheduling with Multiple Energy Sources

Liang Sun [1,2], Qi Zhang [1], Na Zhang [3], Zhuoran Song [4], Xinglong Liu [5] and Weidong Li [1,*]

1. School of Electrical Engineering, Dalian University of Technology, Dalian 116024, China; sunliang_3333@sina.com (L.S.); zhangqizq@mail.dlut.edu.cn (Q.Z.)
2. State Grid Shenyang Electric Power Supply Company, Shenyang 110081, China
3. State Grid Liaoning Economic Research Institute, Shenyang 110015, China; zhangnamoumou@126.com
4. State Grid Liaoning Electric Power Company Limited, Shenyang 110006, China; zrsong_sg@126.com
5. China Energy Engineering Group Liaoning Electric Power Survey & Design Institute Co. Ltd., Shenyang 110179, China; liuxinglong_01@126.com

* Correspondence: wdli@dlut.edu.cn

Abstract: The uncertainty of new energy output from wind power is rarely considered in the monthly energy-trade scheduling. This causes many problems since the new energy penetration level increases. The fairness of the scheduled energy for the power suppliers is difficult to guarantee. Because the actual power system operation is far away from scheduling when the monthly energy-trade schedule is carried out, unnecessary wind curtailment might occur, and even the feasibility of monthly energy-trade schedule might not be guaranteed. This affects the security and reliability of the power system operation. In this paper, a new time-sequence simulation method for the monthly energy-trade scheduling is proposed, which considers the new energy power forecasting characteristic and the computational load problem of hourly energy-trade simulation in the remaining months. The proposed method is based on a segment modelling strategy. The power generation in the scheduling month is optimized hourly, and the energy generation is optimized in the subsequent months on a monthly basis. For the scheduling month, accurate cost function is applied in the objective function, and detailed short-term operation constraints and the new energy forecasting results are considered, which can guarantee the feasibility of the new monthly energy-trade scheduling and lay a solid foundation for daily dispatching. For the subsequent months, since the load forecast accuracy is lower and no wind power forecasting results could be used, the rough cost function is applied, and only monthly constraints are considered. To ensure a balance in the execution progress of each power generating entity, the simulation time-scale is set as the remainder of the months in the study year. The new approach ensures the fairness of power execution progress and improves the new energy consumption level. A case study was used to verify the feasibility and effectiveness of the proposed method, which provides a theoretical reference for the monthly electrical energy-trade scheduling.

Keywords: monthly energy-trade scheduling; time-sequence simulation method; feasibility; fairness; consumption of renewable energy

---

1. Introduction

Among the generation scheduling processes in China, the annual contract energy planning is carried out according to the contract energy of the generating companies and the national scheduled energy. Then, the monthly energy-trade scheduling and daily generation dispatching are performed by shortening the time scale [1]. Monthly energy-trade scheduling is the intermediate link between the annual contract planning and daily dispatching. If the monthly energy-trade scheduling is improper, the annual contract energy may become difficult to complete. Therefore, the daily dispatching feasibility, operation economy, and security is impacted.

At present, most of the provincial power trading centers in China apply the average decomposition method for monthly energy-trade scheduling. In past decades, the average decomposition method was sufficient to solve the principal contradiction of the power balance. However, this method may be very simple and easy to be implement, but its predominant factors such as the operation condition are not comprehensive enough [2]. Moreover, since the end of the 20th century, with the increasing energy gap and deteriorating ecology, the other drawbacks of the average decomposition method, i.e., lack of energy conservation and emission reduction techniques, have become increasingly prominent. Realizing these drawbacks, some studies [3,4] optimized the formulation of monthly energy-trade scheduling with respect to energy-saving power generation. For instance, a study [3] proposed the combination scheme and calculation method for the non-heating thermal power units of monthly energy-trade scheduling with regard to coal consumption and pollutant emission. Another study [4] expanded the optimization space of energy savings and emission reduction benefits from the time dimension, by optimizing the monthly unit commitment and the current electrical energy distribution. Some improved methods of the conventional units based on the average decomposition method can be found in Reference [5–7]. In the monthly trade scheduling method proposed in Reference [5], the deviation of the load ratio was adjusted moderately, according to the comprehensive cost ranking results of the generation unit cost. A comprehensive consumption cost optimization method is proposed in Reference [6], which considers the impact of the monthly electric energy nonlinear fluctuation on the relational consumption cost. On the basis of the unit electric energy integrative cost diversity of different generation units, a new monthly energy-trade scheduling method is proposed in Reference [7], in which the unit electric energy integrative cost of the generation unit was weight-modified. The balance between the economic benefits, energy conservation, and emission reduction can be achieved using the method in Reference [7]. These research works considered the interests of generation units and power grid companies, as well as the social environment. As a result, the benefits of energy saving and emission reduction could be obtained, to some extent, in monthly energy-trade scheduling. However, these research works did not consider the environmental benefits of integrating renewable energy resources into the monthly power trade scheduling process.

In the existing research, the renewable energy units and the conventional units were mostly considered respectively and serially in the monthly energy-trade scheduling. In other words, the remaining load power was reserved for the traditional units in the monthly energy-trade scheduling once the renewable energy power generation is deducted, according to the total load power. This method is feasible when the penetration level of renewable energy is low. However, with the increasing penetration of renewable energy, generation scheduling of thermal power units is influenced by the high volatility, randomness, and intermittence of the renewable energy. A study [8] proposed a market clearing model for energy and reserve products, in coordinated power and gas networks with the integration of compressed air energy storage and wind energy sources (WES), based on two-stage stochastic network-constrained unit commitment. In Reference [9], a sustainable day-ahead scheduling of the grid-connected home-type micro-grids with the integration of distributed energy resources and responsive load demand was co-investigated, and an efficient energy management system optimization algorithm was studied. A novel control algorithm for joint demand response management and thermal comfort optimization in micro-grids equipped with renewable energy sources and energy storage units was presented in Reference [10]. In Reference [11], a simulation-based optimization approach for the design of an EMS in grid-connected photovoltaic-equipped micro-grids with a heterogeneous occupancy schedule was presented. The research studies above could effectively cope with the energy-optimizing problems with large-scale integration of renewable energy into the grid. However, these research studies were all for the micro-grids, which were very different from the regional power grid studied in this paper, and the optimizing time scale is also different from this paper. For the monthly energy-trade scheduling problem with the large-scale integration of renewable energy studied in this paper, the risk of the thermal power units not completing the annual contract energy planning is much higher. The fairness requirement of the deviation from the annual base power completion rate is also difficult to meet. At the same time, because the previous methods of monthly energy-trade scheduling focus on the economics of electricity decomposition, and do not consider the difficulty in implementing the subsequent daily dispatching, it might result in unnecessary water and wind curtailment when the clean energy output fluctuation is large. What is more, the operation feasibility might be difficult to guarantee, which may affect the security and reliability of the operation [12–15]. Therefore, with the large-scale integration of renewable energy into the grid, the following new formulating principles of monthly energy-trade scheduling should be followed.

- The feasibility of power generation scheduling should be ensured [16].
- Impartial and open dispatching requirements should be met [17].
- The benefits of energy conservation and emission reduction should be considered, which means priority should be given to renewable energy units [18–20].

As mentioned above, existing methods such as the average decomposition method, the deviation of load ratio method, etc. cannot meet all the above principles. Therefore, a new method tailored around three principles is needed. The time-sequence simulation method is an effective method to solve optimizing problems with multiple variables. With this method, the various complex constraints, including fairness and operation, can be comprehensively considered and higher resolutions can be achieved. In recent years, the time-sequence simulation method has been widely used in the power system area, such as in the modelling of the annual wind power trade scheduling and the low-carbon benefit evaluation [21,22]. A previous study [23] applied the time-sequence simulation method to the daily dispatching and ultra-short-term scheduling including wind power. Another study [24] optimized the monthly unit commitment model, including wind turbines with the time-sequence simulation method. The literature [25] attempted to apply the time-sequence simulation method to model monthly energy-trade scheduling. However, Reference [25] focused on optimizing the distribution of electricity during the scheduling month, whereas the economics and fairness of electricity distribution in other months of the year were not considered.

Based on the above research, this paper proposes a new time-sequence simulation method for monthly energy-trade scheduling, based on the segment-formulating strategy. The simulation time-scale is set as the remaining months of the year, so as to ensure a balance in the execution progress of each power generation entity. Optimizing convergence and efficiency can be ensured, by applying the strategy of optimizing the generation power hourly in the scheduling month and optimizing the generation energy monthly in subsequent months. The proposed method considers the power generation characteristics of various energy units and improve the consumption capability of renewable energy. Thus, it enhances the efficiency of energy conservation and emission reduction of the power grid by considering several operational constraints in the scheduling month (the following month). It can maintain the balance of the annual base electrical energy completion rate of each thermal power unit more efficiently. Thus, the fairness of execution of the generation schedule is realized. The comprehensive consideration of the system operation constraints significantly improves the feasibility of the generation schedule and provides a good foundation for subsequent daily dispatching.

2. Modelling Concept and Method

As stated previously, methods such as average decomposition and the deviation of the load ratio cannot effectively solve the problem of formulating the monthly energy-trade scheduling, when the renewable energy is absorbed by the power grid on a large scale. However, the time-sequence simulation method is a dynamic optimization method simulated by setting up an objective function and the relevant constraints. It is characterized by both long-time and short-time scales. In the long-time scale, the optimal values at each time point can be accumulated and the sum can be macroscopically restricted by the fairness constraints. In the short time scale, the optimal solutions can be obtained considering the detailed operation constraints at each time interval. The time-sequence simulation method matches with the modelling demand of monthly energy-trade scheduling. Therefore, the time-sequence simulation method is very suitable for the modelling of the monthly energy-trade scheduling.

To improve the operation economy and utilization rate of renewable energy, the objective function of the model is to minimize the combined costs including the coal consumption cost, peaking cost, start-up and shutdown cost, hydropower abandoning cost, wind power curtailment cost, and nuclear power peak shaving cost in the remaining months. To better consider multiple factors during the actual power operation, the annual constraints, monthly constraints, and short-term constraints are all considered. In the annual constraints, the energy completion rate deviation constraints are introduced to meet the power generation fairness requirement. The monthly power energy balance constraints for each generation unit are considered in the monthly constraints, to ensure reasonable allocation of the monthly scheduling energy. Operation constraints including the generator characteristic constraints, power balance constraints, and spinning reserve constraints are considered in the hourly short-term constraints, to reduce the deviation between monthly scheduling and daily dispatching, and improve the schedule feasibility. According to the typical time-sequence simulation method, the monthly energy-trade scheduling problem is modelled and simulated on an hourly basis, from the beginning of the decision point to the end of the whole year. The detailed process is shown in Figure 1.

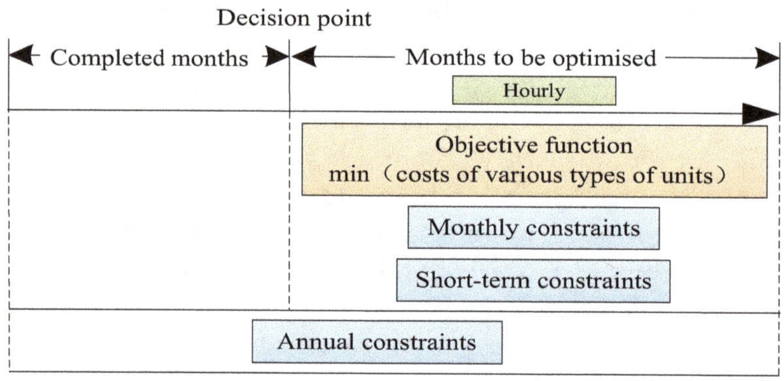

Figure 1. The monthly energy-trade scheduling model according to the typical time-sequence simulation.

However, some problems are faced when using the typical time-sequence simulation method mentioned above in the modelling of monthly energy-trade scheduling.

- During the optimizing months, the method is optimized on an hourly basis. A system operation with up to 8760 time intervals must be simulated. For large-scale power systems with hundreds of generation units, such a demand presents a massive scale optimization problem, which is difficult to solve.
- Considering the low accuracy of long-term forecasting of wind power, water inflow, and the load, the simulation results in the months farther from the decision point might be vastly different from the actual operation, which may increase the wind and water curtailment level.

Therefore, it is neither necessary nor feasible to simulate the system operation in the whole period on an hourly basis. Some works in the literature proposed preliminary solutions to this problem. For example, Reference [26] proposed a segmented idea to simplify the calculation, but adopted the method of average segmentation throughout the process, which is too simplistic and does not conform to the physical characteristics of the accuracy of the predicted value decay with time. In another example, Reference [27] divided each day into three sections—peak, flat, and valley—which did not meet the law of change in prediction. Based on these ideas, a segment modelling strategy for the monthly energy-trade scheduling is presented. The system simulation is former-accurate and after-rough. The remaining months are decomposed into the scheduling month and the subsequent months. The scheduling month is the next month after the decision point, and the subsequent months include the months after the scheduling month to the end of the year. For the scheduling month, since the load forecast accuracy is higher and wind power forecasting data could be used, the system operation is still simulated hourly, and, thus, the cost function (defined as the accurate cost function), the monthly constraints, and the short-term constraints are the same as in the typical method introduced above. In addition, for the subsequent months, since the load forecast accuracy is lower and no wind power forecasting results could be used, the system operation is simulated monthly, and the cost function (defined as the rough cost function) only contains the coal consumption cost of thermal power units, to ensure the economy of the model, and only the monthly constraints are considered. The short-term constraints, which require the prediction data with high accuracy and large-scale calculation, are abandoned in the subsequent months. The electric energy completion situation in the completed month involves the formulation for the next month as an input. Additionally, it is constrained by the annual constraints with the electrical energy of the scheduling month and the subsequent months. The process is shown in Figure 2.

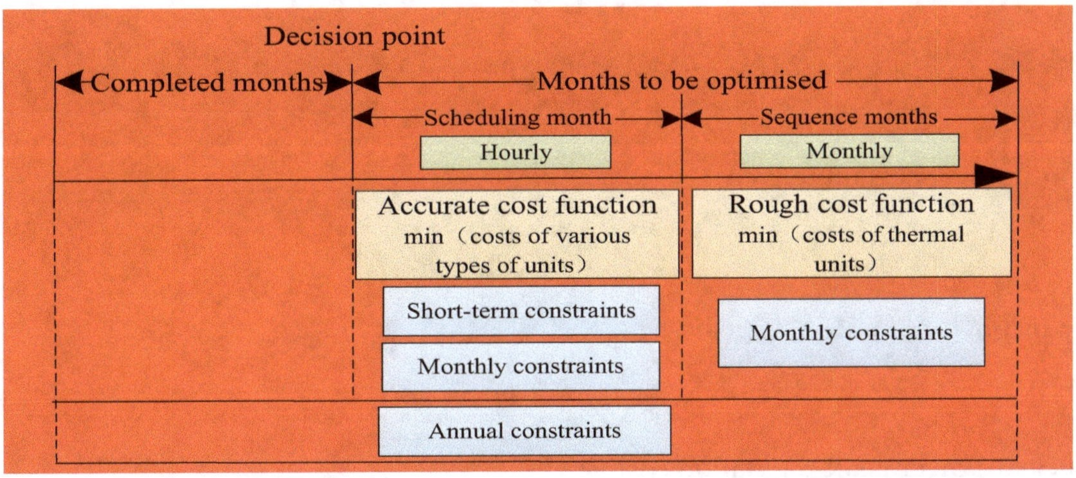

Figure 2. The monthly energy-trade scheduling model according to the improved time series simulation.

In the actual application, the monthly electrical energy-trade scheduling is rolling when scheduled from the start of the year to November of that year. During each decision period, the completed energy of each generation unit is applied to correct the energy target of the remaining months.

3. Mathematical Model

3.1. Objective Function

The objective function of the monthly energy-trade scheduling model is to minimize the sum of total costs for the remaining months. As mentioned above, to improve the feasibility of the model and the solving efficiency, the accurate cost function F_1 is used in the scheduling month, and the rough cost function F_2 is used in subsequent months.

$$\min F = \min(F_1 + F_2) \tag{1}$$

3.1.1. Accurate Cost Function for the Scheduling Month

The accurate cost function for the scheduling month is the total costs of all the different types of units in the scheduling month, which consist of the operational costs of the condensing thermal power units and operational cost, the steam extraction thermal power units, the curtailment cost of the wind power units, the abandoning cost of the hydropower units, and the peak shaving cost of the nuclear power units.

$$F_1 = \sum_{t=1}^{T}\left\{\sum_{i=1}^{N_{CN}} f_{CN}(i,t) + \sum_{k=1}^{N_{CQ}} f_{CQ}(k,t) + \sum_{l=1}^{N_w} f_w(l,t) + \sum_{s=1}^{N_m} f_m(s,t) + \sum_{v=1}^{N_n} f_n(v,t)\right\} \tag{2}$$

The nomenclature of variables can be found in Appendix A.

The operation costs of the condensing thermal power units are composed of the coal consumption cost, deep-peak-regulation cost, and start-up and shut-down costs.

$$f_{CN}(i,t) = a_i(P^t_{CN_i})^2 + b_i(P^t_{CN_i}) + c_i + M \times \Delta P^t_{CN_i} + S_{CN_i}u^t_i(1-u^t_i) \tag{3}$$

The operation costs of the steam extraction thermal power units include the coal consumption cost, deep peak-regulation cost, and the start-up and shut-down costs.

$$\begin{aligned} f_{CQ}(k,t) &= a_k(P^t_{CQ_k} + C_{v1} \times H^t_{CQ_k})^2 + b_k(P^t_{CQ_k} + C_{v1} \times H^t_{CQ_k}) + c_k \\ &\quad + M \times \Delta P^t_{CQ_k} + S_{CQ_k}u^t_k(1-u^t_k) \end{aligned} \tag{4}$$

The wind power curtailment cost is:

$$f_w(l,t) = C_w(P^t_{w_l} - \tilde{P}^t_{w_l}) \tag{5}$$

The hydropower abandoning cost is:

$$f_m(s,t) = C_m P^t_{m,g} \tag{6}$$

The nuclear power peak-regulation cost is:

$$f_n(v,t) = C_n(\bar{P}_{n_v} - P^t_{n_v}) \tag{7}$$

3.1.2. Rough Cost Function for Subsequent Months

The rough cost function is reduced to the sum costs of thermal units in the sequence months, including the condensing thermal power units and steam extraction thermal power units.

$$F_2 = \sum_{e=m+1}^{12}\left(\sum_{i=1}^{N_{CN}} W^e_{CN_i,RE} \times C_{CN_i} + \sum_{k=1}^{N_{CQ}} W^e_{CQ_k,RE} \times C_{CQ_k}\right) \tag{8}$$

3.2. Constraint Conditions

3.2.1. Short-Term Constraints

To ensure the feasibility of the monthly energy-trade scheduling, the short-term constraints limit the operation status of the units in each time interval of the scheduling month. The operation constraints of various units, and the power balance constraint and spinning reserve constraints are considered.

The short-term constraints of the condensing thermal power units include the maximum and minimum output constraints, ramp rate constraints, and minimum start-off time constraints. The constraints considered are as follows.

$$\begin{cases} u_i^t P_{CN_i,\min} \le P_{CN_i}^t \le u_i^t P_{CN_i,\max} \\ -\Delta P_{CN_i,down} \le P_{CN_i}^t - P_{CN_i}^{t-1} \le \Delta P_{CN_i,up} \\ (u_i^t - u_i^{t-1})(X_{CN_i,on}^{t-1} - T_{CN_i,on}) \le 0 \\ (u_i^{t-1} - u_i^t)(X_{CN_i,off}^{t-1} - T_{CN_i,off}) \le 0 \end{cases} \tag{9}$$

The short-term constraints of the steam extraction thermal power units include the maximum and minimum output constraints, ramp rate constraints, minimum start-off time constraints, thermoelectric relationship constraint, reserve capacity constraint, and thermal balance constraint. The constraints considered are below.

$$\begin{cases} u_k^t P_{CQ_k,\min} \le P_{CQ_k}^t \le u_k^t P_{CQ_k,\max} \\ -\Delta P_{CQ_k,down} \le P_{CQ_k}^t - P_{CQ_k}^{t-1} \le \Delta P_{CQ_k,up} \\ (u_k^{t-1} - u_k^t)(X_{CQ_k,off}^{t-1} - T_{CQ_k,off}) \le 0 \\ (u_k^t - u_k^{t-1})(X_{CQ_k,on}^{t-1} - T_{CQ_k,on}) \le 0 \\ \max(C_{m_k} \times H_{CQ_k}^t + K_k, P_{CQ_k,\min} - C_{v2_k} \times H_{CQ_k}^t) \le P_{CQ_k}^t \le P_{CQ_k,\max} - C_{v1_k} \times H_{CQ_k}^t \\ 0 \le H_{CQ_k}^t \le H_{CQ_k,\max} \\ \sum_{k=1}^{N_{CQ}} H_{CQ_k}^t = H_L^t \end{cases} \tag{10}$$

In the short-term constraints of wind turbines, the wind power output should be equal to or less than the forecasted wind power. The constraints considered are below.

$$0 \le P_{w_l}^t \le P_{w_l}^{\bar{t}} \tag{11}$$

The short-term constraints of the hydropower units are composed of the upper and lower limits of the reserve capacity, the upper limit constraint of the water flow for generating power, and the maximum output constraint. The constraints considered are below.

$$\begin{cases} V_m^{t+1} = V_m^t + f_m^t - \sum_{s=1}^{N_m} q_{m_s}^t - g_m^t \\ V_m^{\min} \le V_{m_s}^t \le V_m^{\max} \\ 0 \le q_{m_s}^t \le q_{m_s,\max} \\ P_{m_s}^t = a q_{m_s}^t h \\ 0 \le P_{m_s}^t \le \min\{P_{m_s,\max}, a q_{m_s}^t h_{m_s}^t\} \end{cases} \tag{12}$$

The nuclear power units perform the 15-1-7-1 power generation mode, which means that the nuclear power units maintain the state of full power generation for 15 h, and then decrease to the state of low power generation within 1 h. Then the nuclear power units maintain the low power generation state for 7 h, and return to the full power generation state in 1 h. The constraints considered are below.

$$\begin{cases} u_{n_v,y_1}^t + u_{n_v,y_2}^t = 1 \\ P_{n_v,y_1}^t = P_{n_v,\max} \times u_{n_v,y1}^t \\ P_{n_v,y_2}^t = \alpha P_{n_v\max} \times u_{n_v,y_2}^t \\ P_{n_v}^t = \max\{P_{n_v,y_1}^t, P_{n_v,y_2}^t\} \end{cases} \tag{13}$$

The power balance constraint of the system is:

$$\sum_{i=1}^{N_{CN}} P_{CN_i}^t + \sum_{k=1}^{N_{CQ}} P_{CQ_k}^t + \sum_{l=1}^{N_w} P_{w_l}^t + \sum_{s=1}^{N_m} P_{m_s}^t + \sum_{v=1}^{N_n} P_{n_v}^t = P_L^t \tag{14}$$

The up and down spinning reserve constraints of the system are:

$$\sum_{i=1}^{N_{CN}} P^t_{CN_i,\max} + \sum_{k=1}^{N_{CQ}} P^t_{CQ_k,\max} + \sum_{s=1}^{N_m} P^t_{m_s,\max} + \sum_{v=1}^{N_n} P^t_{n_v,\max} \geq 1.05P^t_L - 1.2P^t_w + 0.5\beta P^t_w \tag{15}$$

$$\sum_{s=1}^{N_m} P^t_{m_s,\min} + \sum_{v=1}^{N_n} P^t_{n_v,\min} \leq 0.95P^t_L - 0.8P^t_w - 0.5\beta P^t_w \tag{16}$$

3.2.2. Monthly Constraints

The monthly constraints limit the monthly scheduled energy in the scheduling month and the subsequent months, to ensure that the energy allocated in each month can be used to reasonably formulate the daily dispatching. The constraints include the upper and lower limits of the monthly power generation energy of various units and the monthly electricity balance constraint.

The monthly operation rate of the units in the monthly constraints of the condensing thermal power units is defined as:

$$\frac{W^{year}_{CN_i} - W^{ywc}_{CN_i}}{T_{wwc} \times P_{CN_i,\max} \times \eta} \leq \delta^e_{CN_i} \leq \min(1, \frac{W^{year}_{CN_i} - W^{ywc}_{CN_i}}{T_{wwc} \times P_{CN_i,\min}}) \tag{17}$$

Then, the upper and lower limit constraints of electrical energy of the condensing power units are as follows:

$$P_{CN_i,\min} \times t^e \times \delta^e_{CN_i} \leq W^e_{CN_i,RE} \leq P_{CN_i,\max} \times \eta \times t^e \times \delta^e_{CN_i} \tag{18}$$

The electrical energy balance constraints of the steam extraction thermal power units are:

$$\begin{cases} \frac{W^{year}_{CQ_k} - W^{ywc}_{CQ_k}}{T_{wwc} \times P_{CQ_k,\max} \times \eta} \leq \delta^e_{CQ_k} \leq \min(1, \frac{W^{year}_{CQ_k} - W^{ywc}_{CQ_k}}{T_{wwc} \times P_{CQ_k,\min}}) \\ P_{CQ_k,\min} \times t^e \times \delta^e_{CQ_k} \leq W^e_{CQ_k,RE} \leq P_{CQ_k,\max} \times \eta \times t^e \times \delta^e_{CQ_k} \end{cases} \tag{19}$$

The monthly electrical energy constraints of the wind turbine, hydropower unit, and nuclear power unit are:

$$0 \leq W^e_{w_l,RE} \leq W^e_{w_l,F} \tag{20}$$

$$0 \leq W^e_{m_s,RE} \leq W^e_{m_s,F} \tag{21}$$

$$0 \leq W^e_{n_v,RE} \leq W^e_{n_v,F} \tag{22}$$

The balance constraint of the monthly electrical energy is:

$$\sum_{i=1}^{N_{CN}} W^e_{CN_i,RE} + \sum_{k=1}^{N_{CQ}} W^e_{CQ_k,RE} + \sum_{l=1}^{N_w} W^e_{w_l,RE} + \sum_{m=1}^{N_m} W^e_{m_l,RE} + \sum_{v=1}^{N_n} W^e_{n_v,RE} = W^e_{L,RE} \tag{23}$$

3.2.3. Annual Constraints

The energy generated in the completed months, the scheduled energy in the scheduling month, and the scheduled energy in the subsequent months constitute the total annual energy of each thermal power unit.

$$W_{T_i,A} = \sum_{e=1}^{m-1} W^e_{T_i,H} + \sum_{t=1}^{T} P^t_{T_i} + \sum_{e=m+1}^{12} W^e_{T_i,RE}, \tag{24}$$

where the electrical energy of the completed month is the input, and the amount of electricity in the scheduling month and the subsequent months is constrained by the annual fairness constraints as the optimization value. The deviation of electrical energy in the completed month can be corrected when formulating the monthly energy-trade scheduling.

The total annual energy of the thermal power units is composed of the annual base electrical energy and the annual transaction energy.

$$W_{T_i,A} = W_{T_i,B} + W_{T_i,D} \tag{25}$$

The annual transaction energy is composed of the tie-line energy, the generation right transfer trading energy, and the trading energy of large consumers.

$$W_{T_i,D} = W_{T_i,N} + W_{T_i,O} + W_{T_i,Y} \tag{26}$$

The completion rate of the annual base electrical energy of the total thermal power units, ρ_T is defined as the ratio of the optimized annual base electrical energy of all thermal power units and the expected annual base electrical energy of all thermal power units.

$$\rho_T = \frac{\sum_{i=1}^{N_T} W_{T_i,B}}{\sum_{i=1}^{N_T} W_{T_i,BF}} \tag{27}$$

According to the National Development and Reform Committee of China's guide on strengthening and improving the regulation of power generation operation, the deviation threshold of the completion rate of the annual base electrical energy of each thermal unit is defined as $\lambda\%$. The allocated fairness constraint for the annual base electrical energy of thermal power units is as follows:

$$(1 - \lambda\%)\rho_T \le \frac{W_{T_i,B}}{W_{T_i,BF}} \le (1 + \lambda\%)\rho_T \tag{28}$$

4. Simulation and Analysis

4.1. Simulation Condition

For the solving methods in this paper, since it is actually a large-scale mixed integer quadratic programming model, the Branch Bound (BB) methods, for instance, could be applied. In the case studies, we used the IBM CPLEX Business optimization software to solve the monthly energy-trade scheduling problem in the test system.

The test system consists of four condensing thermal power units, two steam-extraction thermal power units, one 600 MW hydropower station, one 250 MW wind farm, and one 60 MW nuclear power station. The simulation process of monthly energy-trade scheduling is shown in Figure 3.

April is assumed to be the scheduling month. The electrical energy and operation parameters and the operation cost coefficient of the thermal power units are shown in Tables 1–3, respectively. The thermoelectric relationship coefficient data of the thermoelectric units are shown in Table 4. The characteristic parameters of the hydropower station and the reservoir are shown in Tables 5 and 6, respectively. The monthly predicted hydropower electrical energy (PHEQ) is shown in Table 7. The characteristic parameters of the reservoir are shown in Table 6. The cost for deep-peak regulation is 500 ¥/MWh.

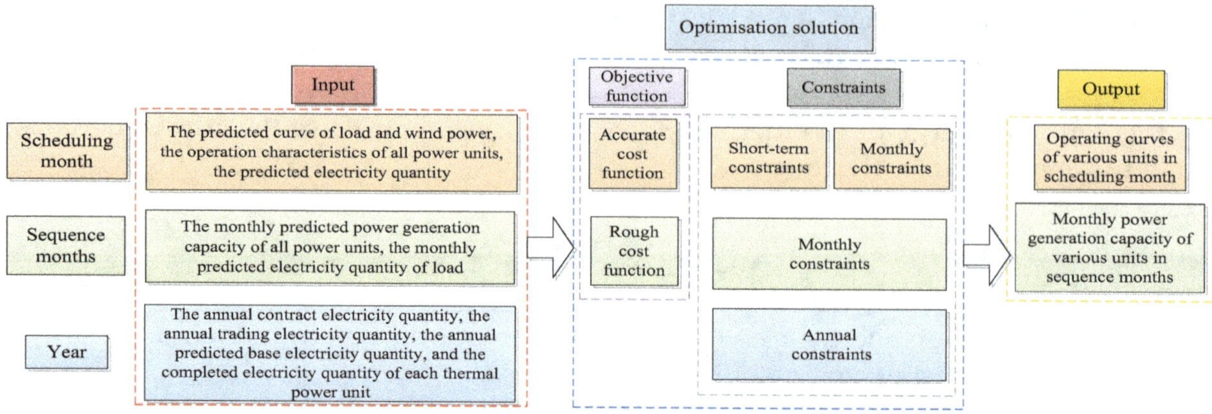

Figure 3. Simulation process of the monthly energy-trade scheduling.

Table 1. Electrical energy data of thermal power units.

Units	Annual Contract Electricity Quantity (MWh)	Annual Trading Electricity Quantity (MWh)	Annual Predicted Base Electricity Quantity (MWh)	The Completer Electricity Quantity in 1~3 Months (MWh)
1	3,139,000	2,511,200	627,800	876,740.1
2	1,569,500	1,255,600	313,900	452,014.1
3	1,307,900	1,046,320	261,580	322,918.3
4	784,800	627,840	156,960	197,038.9
5 (CHP)	1,689,800	1,351,840	337,960	441,868.7
6 (CHP)	1,109,100	887,280	221,820	325,208.2

Table 2. The operation parameters of thermal power units.

Units	Pmax (MW)	Pmin (MW)	Ton, min/Toff, min (h)	Pup (MW/h)	Pdown (MW/h)
1	600	280	8	168	168
2	350	140	5	80	80
3	250	100	5	80	80
4	150	70	6	42	42
5 (CHP)	323	150	6	90	90
6 (CHP)	212	100	6	60	60

Table 3. The operation cost coefficient of the thermal power units.

Units	S (¥/MWh)	A (¥/MW^{2}h)	B (¥/MWh)	C (¥/h)	Average Coal Consumption Cost (¥/h)
1	1,200,000	0.06	157.8	6300	203.0
2	650,000	0.048	112.8	13,440	174.6
3	500,000	0.045	130.8	8640	182.8
4	260,000	0.04	164.4	3240	195.8
5 (CHP)	600,000	0.046	163.0	11,293	218.5
6 (CHP)	500,000	0.103	162.3	6922	221.4

Table 4. The thermoelectric relationship coefficient of thermoelectric units.

Units	C_{v1}	C_{v2}	C_m	K
5 (CHP)	0.23	0.23	0.45	80.7
6 (CHP)	0.21	0.21	0.45	45.4

Table 5. The characteristic parameters of the hydropower station.

Unit	Pmax (MW)	qmax (m^3/s)	a	Annual Contract Electricity Quantity (MWh)
1	600	705.9	8.5	2,607,169

Table 6. The characteristic parameters of the reservoir.

Reservoir	Vmax (m^3)	Vmin (m^3)	V_0 (m^3)	h (m)
1	90.18×10^8	40.09×10^8	60×10^8	100

Table 7. The monthly predicted hydropower electrical energy.

Month	1	2	3	4	5	6
PHEQ (MWh)	44,847	81,109	102,817	138,739	174,379	300,159
Month	**7**	**8**	**9**	**10**	**11**	**12**
PHEQ (MWh)	359,829	429,932	350,262	307,388	161,367	156,341

Load and wind power prediction methods have always been research interests in recent decades. A lot of research results have been obtained and many methods have been proposed. The widely used prediction methods include the Depth Ridgelet Neural Network method, support vector machine, etc. This paper focuses on the formulation of monthly energy-trade scheduling. Before it is formulated, the load and wind power forecasting results have been obtained by using existing prediction methods. Therefore, the load and wind power prediction methods are not the research contents of this paper. In the case studies, the predicted wind power electrical energy (PWEE) is randomly generated based on the historical wind power data. As shown in Table 8, the error range of monthly wind electricity energy prediction in Reference [28,29] was adopted for the calculation. The actual wind generation energy in April was 61,654 MWh, which was 6.04% deviation from the predicted wind power electric energy. The actual wind power generation curve is shown in Figure 4.

Table 8. The monthly predicted wind-power electrical energy.

Month	1	2	3	4	5	6
PWEQ (MWh)	20,772	20,772	34,620	58,162	50,546	46,391
Month	**7**	**8**	**9**	**10**	**11**	**12**
PWEQ (MWh)	29,842	26,721	39,744	48,884	51,792	49,720

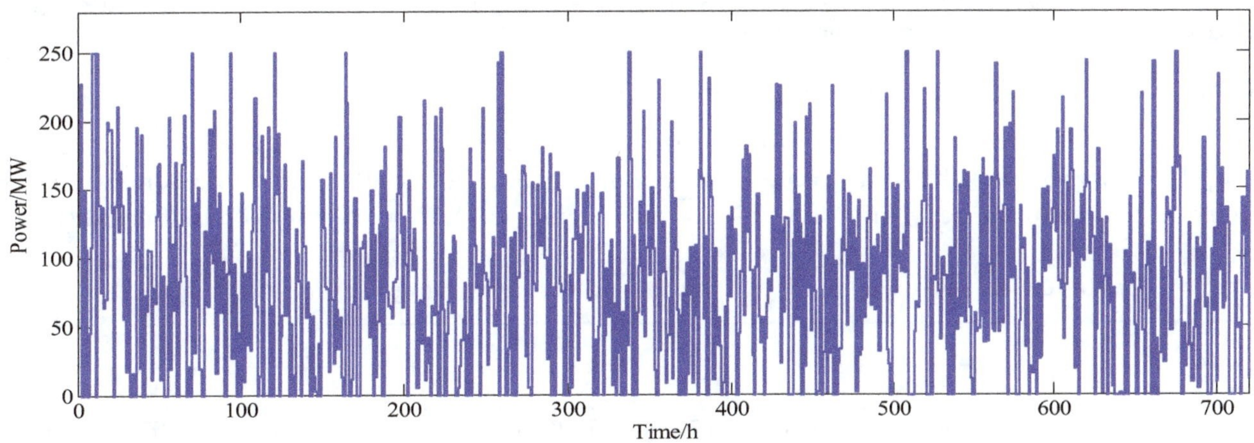

Figure 4. The actual wind power generation curve.

The annual load electrical energy is assumed to be 13.123323×10^6 MWh and the monthly load coefficients are shown in Table 9. The load curve in April (the scheduling month) is shown in Figure 5.

Table 9. The monthly load coefficients.

Month	1	2	3	4	5	6
Load coefficients	0.0949	0.0682	0.0702	0.0732	0.0752	0.0772
Month	**7**	**8**	**9**	**10**	**11**	**12**
Load coefficients	0.0992	0.0972	0.0912	0.0832	0.0722	0.0982

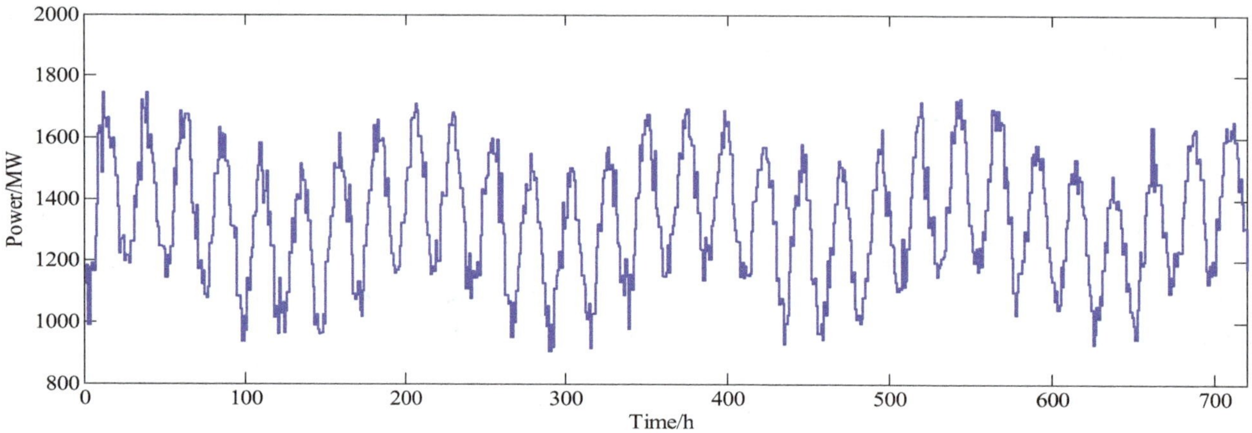

Figure 5. The load curve in April.

4.2. Simulation Results

4.2.1. Simulation Results of April (the Scheduling Month)

1. Computational Performance Analysis

To verify the computational performance of the proposed approach, a simplified test system only including the thermal power units was simulated with simple constraints. The running time of the simplified traditional model on an hourly simulation was 1024.92 s. The optimization of the traditional model did not converge because the calculation amount was excessive. However, the computational time using the proposed approach and model was 28.15 s. The computational speed was increased by 97.25%, and the computational volume was reduced effectively. The convergence of optimizing could be ensured when applying the proposed method and model.

2. Economy, Energy Conservation, and Emission Reduction Effect Analysis

The monthly scheduled energy of the thermal power units from April to December, according to the proposed method, are shown in Table 10. For comparison, the monthly scheduled energy values of the thermal power units, according to the load rate deviation method, are shown in Table 11.

Table 10. The scheduled energy of thermal power units using the proposed method.

Units	Month								
	4	5	6	7	8	9	10	11	12
1	206,840.2	204,913.7	143,514.6	148,298.4	357,120.0	345,600.0	344,914.7	187,073.0	316,955.0
2	102,170.2	63,372.9	182,371.2	208,320.0	63,372.9	67,450.3	63,372.9	201,600.0	172,327.0
3	87,763.6	55,116.8	139,164.2	148,800.0	55,116.8	144,000.0	55,116.8	144,000.0	148,800.0
4	63,404.6	89,280.0	37,705.2	60,493.8	89,280.0	86,400.0	38,962.1	37,705.2	89,280.0
5 (CHP)	177,520.1	185,531.3	77,667.0	192,249.6	165,640.7	77,667.0	80,255.9	77,667.0	192,249.6
6 (CHP)	88,819.6	126,182.4	49,856.7	117,176.2	51,518.6	49,856.7	115,656.6	49,856.7	126,182.4

Table 11. The scheduled energy of thermal power units using the load rate deviation method.

Units	Month								
	4	**5**	**6**	**7**	**8**	**9**	**10**	**11**	**12**
1	235,580.1	243,661.8	205,164.9	294,433.0	263,053.9	250,963.2	234,876.6	227,177.2	351,768.5
2	120,967.0	120,210.3	101,217.9	145,258.2	129,777.4	123,812.4	115,876.1	112,077.6	173,544.6
3	105,224.8	104,566.5	88,045.8	126,354.8	112,888.6	107,699.9	100,796.4	97,492.2	150,960.1
4	67,181.1	63,367.4	53,355.8	76,571.2	68,410.6	65,266.3	61,082.7	59,080.4	91,482.0
5 (CHP)	128,855.4	133,275.8	112,219.2	161,046.2	143,882.7	137,269.5	128,470.6	124,259.3	192,407.0
6 (CHP)	81,473.6	84,268.5	70,954.7	101,827.4	90,975.1	86,793.7	81,230.2	78,567.5	121,656.4

The thermal power generation cost was calculated according to the average coal consumption cost coefficient in Table 3. During the period from April to December, the thermal power generation costs, according to the proposed method and the load rate deviation method, were ¥ 1,388,520,214.3 (¥ 1.388 billion) ¥ and ¥ 1,397,646,068.9 (¥ 1.397 billion), respectively. Compared with the load rate deviation method, the proposed method could save ¥ 9,125,854.6 (¥9.126 million).

The total scheduled energy values of thermal power units in the scheduling month were 61,654.1 MWh when the presented time-sequence simulation method was applied in the simulation and 58,162.0 MWh when the load rate deviation method was applied. The energy of renewable energy units generated by the time-sequence simulation method and the benefits of energy conservation and emission reduction were verified.

If the operating cost of various types of units, the deep peak-regulation cost, and the start-up cost of thermal power units were considered in the scheduling month, then the total comprehensive generation cost from April to December with the time-sequence simulation method was ¥ 1,389,168,214.3 (or ¥ 1.389 billion).

3. Feasibility Analysis

The hourly power output curve of each thermal power unit in April (the scheduling month), according to the presented method is shown in Figure 6. The hourly output of each power unit in the scheduling month with the time-sequence simulation method can be obtained under the premise that the reliability of the generator and system operation is guaranteed. With the time-sequence simulation method, all the constraints together could ensure that the power output of each power unit is close to the actual operating conditions, which means a higher feasibility could be ensured.

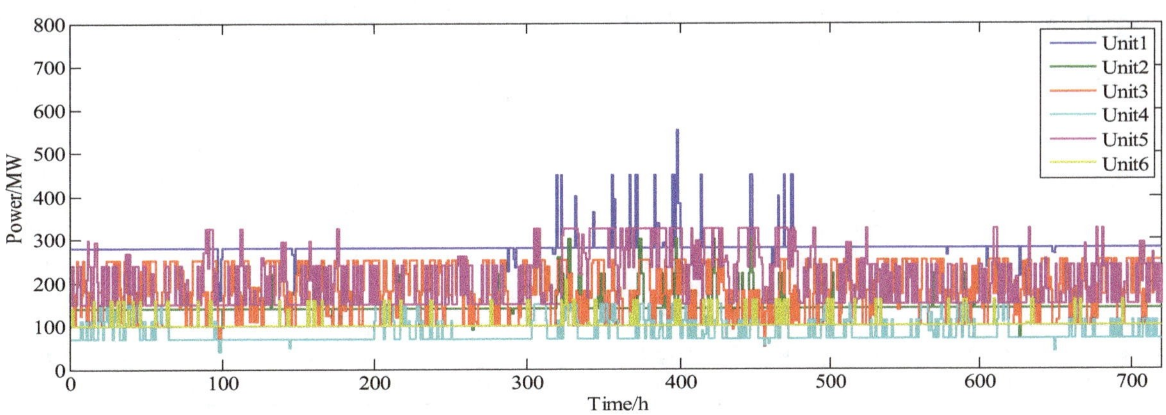

Figure 6. The hourly power output curve of each thermal power unit in April.

4. Fairness Analysis

The monthly energy-trade scheduling is simulated, according to the presented method, when considering the completed rate thresholds (CRTs) for different annual base electrical energies. The total costs based on different CRTs are shown in Table 12.

Table 12. Total cost based on different CRTs.

CRT	The Total Cost (¥)
2%	1,389,737,549.8
3%	1,389,168,214.3
5%	1,389,087,840.1
7%	1,388,881,156.7
10%	1,387,398,273.3

From Table 12, it can be seen that the relationship between the total cost and CRT are negatively correlated. With the increase in CRT, the relative optimization space of the monthly energy-trade scheduling became larger, so that there could be a more economical way to revise the monthly energy-trade scheduling. However, due to the increase in CRT, the fairness of generation scheduling was weaker. Therefore, the contradiction between the power generation units and the dispatching department was sharper. According to the current national regulations in China, the CRT was set to 3%. With the gradual improvement of market mechanism for electricity, the CRT could be adjusted more reasonably, according to the proportion of fairness and economy of the monthly energy-trade scheduling.

4.2.2. Rolling Correction Results during the Whole Year

The scheduling month was moved from January to December. The monthly energy-trade scheduling results for the whole year after the rolling correction are shown in Table 13.

Table 13. The monthly energy-trade scheduling results for the whole year.

Units		Month					
		1	2	3	4	5	6
Thermal Power plants	1	387,134.0	242,411.3	247,194.7	206,840.2	232,986.9	202,367.6
	2	198,550.9	131,918.9	121,544.3	102,170.2	140,135.5	101,553.7
	3	148,287.9	90,969.3	83,661.1	87,763.6	92,291.6	74,537.9
	4	80,500.3	58,160.3	58,378.4	63,404.6	56,976.4	57,423.3
	5 (CHP)	198,032.3	122,431.6	121,404.8	177,520.1	122,722.2	112,861.3
	6 (CHP)	118,291.4	100,518.6	106,398.2	88,819.6	78,201.0	75,727.2
Wind unit		36,659.6	32,809.7	46,058.9	61,654.1	56,428.5	54,982.0
Hydropower station		44,847.0	81,109.0	102,817.0	138,739.0	174,379.0	300,159.0
Nuclear plant		37,200.0	33,600.0	37,200.0	36,000.0	37,200.0	36,000.0
Units		Month					
		7	8	9	10	11	12
Thermal Power plants	1	280,140.4	248,503.1	232,828.2	217,412.6	248,327.6	357,120.0
	2	151,354.0	117,253.8	108,857.5	114,130.4	73,712.6	205,244.2
	3	109,850.7	85,535.5	95,751.3	113,900.4	122,991.0	148,800.0
	4	71,886.7	64,416.1	57,709.4	53,655.8	64,151.3	89,280.0
	5 (CHP)	150,314.1	154,141.0	163,445.3	122,121.1	108,187.5	163,484.7
	6 (CHP)	102,852.1	104,726.4	107,678.0	76,745.4	79,663.5	81,865.1
Wind unit		42,685.4	40,271.0	48,806.3	52,327.0	57,692.1	49,720.0
Hydropower station		359,829.0	429,932.0	350,262.0	307,388.0	161,367.0	156,341.0
Nuclear plant		37,200.0	37,200.0	36,000.0	37,200.0	36,000.0	37,200.0

5. Conclusions

A new time-sequence simulation method for monthly energy-trade scheduling is presented in this paper. The feasibility of the presented method is validated at the theoretical level using a case study. The segment modeling strategy is applied to simulate on an hourly basis for the scheduling month and for the sequence months on a monthly basis. In the power systems integrated with large-scale new energy power, multiple factors such as the generation coordinate among various types of power units, equitable distribution of electrical energy, and system operation reliability could be comprehensively considered in the monthly energy-trade scheduling process.

The results of the case study verified the feasibility and effectiveness of the proposed approach.

- The characteristics of wind power, nuclear power, hydropower, thermal power, and combined heat and power (CHP) generators were comprehensively considered. Therefore, the consumption capability of renewable energy power can be improved, according to the presented monthly energy-trade scheduling method. Thus, the energy saving and emission reduction benefits can be improved.
- By efficiently managing the balance of the annual base electrical energy completion rate of each thermal power unit, the monthly energy trade scheduling fairness can be ensured in a better way.
- Because the necessary operating constraints in the short-term time-scale could be easily introduced into the mathematical model for the scheduling month, the feasibility of the monthly energy-trade scheduling could be improved significantly. This improvement can lay a good foundation for daily dispatching.

The limitation of this study is that it does not involve an experimental study. Therefore, in future studies, this method might need to be modified according to the actual operating conditions. These conditions may be the number of power plants in a regional power grid, calculation time limit, etc., which may improve the practicality of the proposed method.

Author Contributions: Conceptualization, L.S. and W.L. Methodology, N.Z. and Z.S. Software, Q.Z. and X.L. Validation, L.S., W.L., and N.Z. Formal Analysis, Q.Z. Data Curation, Q.Z. Writing—Original Draft Preparation, Q.Z. and N.Z. Writing—Review and Editing, L.S. Visualization, X.L. Supervision, W.L. Project Administration, L.S.

Appendix A NOMENCLATURE

Table A1. Variable comparison table.

F_1	Precise objective function for the scheduling month
F_2	Rough objective function for the subsequent months
T	Time intervals in the scheduling month
i, k, l, s, v	Sequence numbers of pure condensing thermal power units, extraction steam thermal power units, wind units, hydropower stations, and nuclear plants
$N_{CN}, N_{CQ}, N_w, N_m, N_n$	The numbers of pure condensing thermal power units, extraction steam thermal power units, wind units, hydropower stations, and nuclear plants
$f_{CN}(i,t)$	Operating cost of pure condensing thermal power unit i at time t
$f_{CQ}(k,t)$	Operating cost of extraction steam thermal power unit k at time t
$f_w(l,t)$	Operating cost of wind unit l at time t
$f_m(s,t)$	Operating cost of hydropower station s at time t
$f_n(v,t)$	Operating cost of nuclear plant v at time t
a_i, b_i, c_i	Fuel cost coefficients of pure condensing thermal power unit i
$P^t_{CN_i}$	Power by the pure condensing thermal power unit i at time t
M	Average cost of deep peak regulation of thermal power units
$\Delta P^t_{CN_i}$	Power for deep peak regulation by pure condensing thermal power unit I at time t
S_{CN_i}	Start-up cost of pure condensing thermal power unit i
u^t_i	States of the pure condensing thermal power unit I at time t
a_k, b_k, c_k	Fuel cost coefficients of extraction steam thermal power unit k
$P^t_{CQ_k}$	Power by extraction steam thermal power unit k at time t
$H^t_{CQ_k}$	Thermal power by extraction-steam thermal power unit k at time t

Table A1. *Cont.*

$\Delta P^{t}_{CQ_k}$	Power for deep peak regulation of extraction-steam thermal power unit k at time t
S_{CQ_k}	Start-up cost of extraction steam thermal power unit k
u^{t}_{k}	States of extraction steam thermal power unit k at time t
C_w	Average cost of wind power curtailment
$P^{t}_{w_l}$	Consumptive power of wind unit l at time t
$\bar{P}^{t}_{w_l}$	Prediction power of wind unit l at time t
C_m	Average cost of hydropower curtailment
$P^{t}_{m,g}$	Theoretical power generated by curtail water at time t
C_n	Average cost of peak regulation by nuclear plants
$\bar{P}_{n_v}$	Rated power of nuclear plant v
$P^{t}_{n_v}$	Power by nuclear plant v at time t
m	Serial number of the scheduling month
e	Serial numbers of the subsequent months
C_{CN_i}	Average fuel cost coefficient of pure condensing thermal power unit i
$W^{e}_{CN_i,RE}$	Planned generation energy of pure condensing thermal power unit i in month e
C_{CQ_k}	Average fuel cost coefficient of extraction steam thermal power unit k
$W^{e}_{CQ_k,RE}$	Planned generation energy of the extraction steam thermal power unit k in month e
$P_{CN_i,\min}$	The maximum output power of the pure condensing thermal power unit i
$P_{CN_i,\max}$	The minimum output power of the pure condensing thermal power unit i
$\Delta P_{CN_i,down}, \Delta P_{CN_i,up}$	The maximum rate of downward ramping / upward ramping of the pure condensing thermal power unit i
$X^{t-1}_{CN_i,on}, X^{t-1}_{CN_i,off}$	The continuous starting time / downtime of the pure condensing thermal power unit I until time t-1
$T_{CN_i,on}, T_{CN_i,off}$	The minimum starting time/downtime of the pure condensing thermal power unit i
$C_{m_k}, K_k. C_{v2_k}, C_{v1_k}$	The heat-electric coefficients of the extraction-steam thermal power unit k
$H_{CQ_k,\max}$	The upper output thermal power limit of the extraction steam thermal power unit k
H^{t}_{L}	The thermal load at time t
$P_{w_l,\max}$	The rated power of wind unit l
V^{t}_{m}	Volume of water in reservoir at time t
f^{t}_{m}	Volume of water entering the reservoir at time t
$q^{t}_{m_s}$	Volume of water for power generation of hydropower station s at time t
g^{t}_{m}	Volume of abandoned water at time t
$V^{\min}_{m}$	The minimum volume for saving reservoir water
$V^{\max}_{m}$	The maximum volume for saving reservoir water
$q_{m_s,\max}$	The acceptable maximum water flow of hydropower unit s
a	The power coefficient of the hydropower unit
h	The head of the reservoir
$P_{m_s,\max}$	The rated power of the hydropower unit s
$u^{t}_{n_v,y_1}, u^{t}_{n_v,y_2}$	The states of nuclear plant v at time t
$P_{n_v,\max}$	The rated power of nuclear plant v
$P^{t}_{n_v,y_1}$	The power of nuclear plant v at time t corresponding to '$u^{t}_{n_v,y_1}$'
$P^{t}_{n_v,y_2}$	Power of nuclear plant v at time t corresponding to '$u^{t}_{n_v,y_2}$'
α	The ratio of '$P^{t}_{n_v,y_2}$' to '$P_{n_v,\max}$'
P^{t}_{L}	Load at time t
β	The confidence coefficient
$\delta^{e}_{CN_i}$	The operating rate of the pure condensing thermal power unit i in month e
$W^{year}_{CN_i}$	The annual contract electricity energy of the pure condensing thermal power unit i

Table A1. *Cont.*

$W_{CN_i}^{ywc}$	The generation energy that the pure condensing thermal power unit i has generated until decision time
T_{wwc}	The sum of scheduling month's number of hours and the subsequent months' number of hours
η	The empirical value from the annual operating rate of the thermal power unit
t^e	The number of hours in month e
$W_{w_l,RE}^e$	The generation energy of wind unit l in month e
$W_{w_l,F}^e$	The maximum generation energy of wind unit l in month e
$W_{m_s,RE}^e$	The generation energy of hydropower station s in month e
$W_{m_s,F}^e$	The maximum generation energy of the hydropower station s in month e
$W_{n_v,RE}^e$	The generation energy of the nuclear plant v in month e
$W_{n_v,F}^e$	The maximum generation energy of the nuclear plant v in month e
$W_{L,RE}^e$	The power load energy in month e
$W_{T_i,A}$	The annual planned generation energy of the thermal power unit i
$W_{T_i,H}^e$	Before the scheduling month, the generation energy that the thermal power unit i generated in month e
$P_{T_i}^t$	In the scheduling month, the power of thermal power unit i at time t
$W_{T_i,RE}^e$	In the subsequent month, the generation energy that the thermal power unit i generates in month e
$W_{T_i,B}$	The annual basic generation energy of thermal power unit i
$W_{T_i,D}$	The annual transactional generation energy of thermal power unit i
ρ_T	The completion rate of all thermal power units' annual basic generation energy
$W_{T_i,BF}$	The specified annual basic generation energy of the thermal power unit i
λ	The percentage of the annual base generation energy completion rate deviation threshold

References

1. Han, J.H. The Formulation Method of Annual Contract Energy Planning for Power Generation Unit under the Constraint of Energy Saving and Emission Reduction. Master's Thesis, Harbin Institute of Technology, Harbin, China, 2011.
2. Tang, W. Research on the Method of Compiling Monthly Energy Trade Scheduling. Ph.D. Thesis, Harbin Institute of Technology, Harbin, China, 2009.
3. Zhang, H.Q.; Chang, Y.J.; Tang, D.Y.; Wang, S.Y.; Chen, Q.; Li, Z.G.; Wang, Y.Q.; Yu, J.L. A monthly electric energy plan making method of thermal power generation unit in Energy-Saving generation dispatching mode. *Power Syst. Prot. Control* **2011**, *39*, 84–89. [CrossRef]
4. Xia, Q.; Chen, Y.G.; Chen, L. Energy-Saving power generation scheduling mode and method considering monthly unit combination. *Power Syst. Technol.* **2011**, *6*, 27–33. [CrossRef]
5. Wang, Y.; Tang, W.; Luo, H.H.; Yu, F.; Liu, Z.Y.; Jin, Z.H.; Liu, J.; Guo, Y.F.; Yu, J.L.; Liu, Z. Load Rate Deviation Method for Preparing Monthly Energy Trade Scheduling for Thermal Power Generation Units. *Power Syst. Prot. Control* **2009**, *22*, 134–140. [CrossRef]
6. Tang, W.; Wang, Y.; Yu, F.; Liu, Z.Y.; Luo, H.H.; Jin, Z.H.; Liu, J.; Guo, Y.F.; Yu, J.L.; Liu, Z. Comprehensive Cost-weighting Method for Preparing Monthly Energy Trade Scheduling for Thermal Power Units. *Power Syst. Technol.* **2009**, *33*, 167–173. [CrossRef]
7. Tang, W.; Wang, Y.; Yu, J.L.; Min, D.J.; Luo, H.H.; Guo, Y.F.; Jin, Z.H.; Liu, J.; Liu, Z. Comprehensive Consumption Optimization Method for Preparing Monthly energy trade scheduling for Thermal Power Units. *Proc. CSEE* **2009**, *29*, 64–70. [CrossRef]
8. Mohammad, A.M.; Ahmad, S.Y.; Behnam, M.; Mousa, M.; Miadreza, S.-K.; Catalão, J.P.S. Stochastic network-constrained Co-Optimization of energy and reserve products in renewable energy integrated power and gas networks with energy storage system. *J. Clean. Prod.* **2019**, *223*, 747–758. [CrossRef]

9. Mousa, M.; Hamed, A.; Seyedeh, S.G.; Hasan, U.; Terrence, F. Optimal energy management system based on stochastic approach fora home Microgrid with integrated responsive load demand and energy storage. *Sustain. Cities Soc.* **2017**, *28*, 256–264. [CrossRef]
10. Korkas, C.D.; Baldi, S.; Michailidis, L.; Kosmatopoulos, E.B. Occupancy-based demand response and thermal comfort optimization in microgrids with renewable energy sources and energy storage. *Appl. Energy* **2016**, *163*, 93–104. [CrossRef]
11. Korkas, C.D.; Baldi, S.; Michailidis, L.; Kosmatopoulos, E.B. Intelligent energy and thermal comfort management in grid-connected microgrids with heterogeneous occupancy schedule. *Appl. Energy* **2015**, *149*, 194–203. [CrossRef]
12. Yu, J.; Ji, F.; Zhang, L.; Chen, Y.S. An over painted oriental arts: Evaluation of the development of the Chinese renewable energy market using the wind power market as a model. *Energy Policy* **2009**, *37*, 5221–5225. [CrossRef]
13. Chen, Q.; Fan-Chao, G.U.; Jin, Y.Q.; Qin, C.; Chen, X.Y.; Li, Z.R. Energy-Saving Generation Dispatch of Power System Including Large-scale Wind Farms. *Proc. CSU-EPSA* **2014**. [CrossRef]
14. Chi, Y.N.; Liu, Y.H.; Wang, W.S.; Chen, M.Z.; Dai, H.Z. Study on Impact of Wind Power Integration on Power System. *Power Syst. Technol.* **2007**, *31*, 77–81. [CrossRef]
15. Rao, P.; Peng, C.H. A Research on Power Dispatch of Energy-saving and Emission-reduction Generation Based on the Improved Differential Evolution Algorithm. *J. East China Jiaotong Univ.* **2010**, *2010*. [CrossRef]
16. Ting, L.I.; Zhang, H.L.; Shao, P.; Guo, S.Q.; Guang, H.H. Economic evaluation method of monthly power generation plan considering the direct power purchase' influence. *Technol. Dev. Enterp.* **2016**. [CrossRef]
17. Wei, X.H.; Hu, Z.Y.; Yang, L. Thoughts and Suggestions about The Current Evaluation Indicators of the San Gong Dispatching. *Autom. Electr. Power Syst.* **2012**, *36*, 109–112. [CrossRef]
18. Tian, K.; Yao, J.; Wenhui, L.I. Power System Optimal Dispatching Model with Low-Carbon under Clean Energy Generation Integration. *Shaanxi Electr. Power* **2016**. [CrossRef]
19. Zhong, H.; Xia, Q.; Chen, Y.; Kang, C.Q. Energy-Saving generation dispatch toward a sustainable electric power industry in China. *Energy Policy* **2015**, *83*, 14–25. [CrossRef]
20. Lin, R.M.; Jin, H.G.; Cai, R.X. Researching Direction and Development for New Generation Energy Power System. *Power Eng.* **2003**. [CrossRef]
21. Liu, C.; Cao, Y.; Huang, Y.H.; Li, P.; Sun, Y.; Yuan, Y. The Method for Preparing Annual Wind Power Wind Power Plan Based on Time Series Simulation. *Autom. Electr. Power Syst.* **2014**, *38*, 13–19. [CrossRef]
22. Cao, Y.; Li, P.; Yuan, Y.; Zhang, X.S.; Guo, S.Q.; Zhang, C.F. New Energy Consumption Capacity Analysis and Low Carbon Benefit Evaluation Based on Time Series Simulation. *Autom. Electr. Power Syst.* **2014**, *38*, 60–66. [CrossRef]
23. Bai, L.P. The Economic Constraints and Economic Dispatch of Power Systems Considering New Energy Sources. Master's Thesis, Harbin Engineering University, Harbin, China, 2016.
24. Ji, F.; Cai, X.G. Monthly unit combination model with wind power system. *J. Harbin Inst. Technol.* **2017**, *3*. [CrossRef]
25. Sun, L.; Xu, J.; Zhang, Q.; Shen, J.K.; Liu, X.L.; Li, W.D. A time sequential simulation method for monthly energy trade planning in the power system with multiple energy resources. *J. Phys. Conf. Ser. IOP Publ.* **2018**, *1053*, 012008. [CrossRef]
26. Li, X. Research on Optimization Method of Power System Monthly Power Purchase Scheduling. Master's Thesis, Chongqing University, Chongqing, China, 2013.
27. Liang, Z.F.; Xia, Q.; Xu, H.Q.; Zhu, M.X.; Zhang, J.; Yang, M.H. Provincial grid monthly power generation plan based on multi-objective optimization model. *Power Syst. Technol.* **2009**, *13*, 90–95. [CrossRef]
28. Zhang, X.Q.; Bai, X.; Wan, Y.Z.; Peng, M.Q.; Wang, P.; Xiang, Y. Application of Grey Model in Wind Power Generation Forecast of Northwest Power Grid. *Power Syst. Clean Energy* **2011**, *4*, 66–70. [CrossRef]
29. Guo, D. Research on High Proportion of Wind Power Grid-Connected to Improve Consumption Capacity. Ph.D. Thesis, Shenyang Agricultural University, Shenyang, China, 2017.

12

Extended State Observer-Based Predictive Speed Control for Permanent Magnet Linear Synchronous Motor

Yao Wang, Haitao Yu *, Zhiyuan Che, Yuchen Wang and Cheng Zeng

School of Electrical Engineering, Southeast University, Nanjing 210096, China
* Correspondence: htyu@seu.edu.cn

Abstract: Combining the feedback of predictive function control and the feedforward of extended state observer, a composite control strategy is proposed for the permanent magnet linear synchronous motor (PMLSM). The mathematical model of the PMLSM vector control system is established based on the basic structure and operation mechanism of PMLSM. Then, a speed regulator based on predictive function control (PFC) is designed to improve the speed tracking performance of the PMLSM drive system. The state and disturbance of the PMLSM system estimated by the extended state observer (ESO) transferred to the PMLSM drive system, and the robustness of the drive system will be improved. Comparative simulation and experiment results show that the proposed method has better speed tracking performance and disturbance rejection property.

Keywords: permanent magnet linear synchronous motor (PMLSM); extended state observer (ESO); predictive function control (PFC); composite control; robustness

---

1. Introduction

Permanent magnet linear synchronous motors (PMLSMs) have been widely used in industrial applications due to the high speed, high acceleration, high accuracy, and high-power density [1]. However, the PMLSM is a nonlinear system characterized by time-varying, strong coupling, and external disturbances [2], thus the control of the PMLSM servo system is still challenging. For the sake of meeting the requirements of system static and dynamic performances performances, it is necessary to study novel control strategies and improve the robustness of the vector control system [3].

A considerable number of research literatures have been reported in the motor control community [3–10]. Generally, there are two aspects that can be considered to achieve a good dynamic performance for a PMLSM. One type of efficient method is to optimize the structure design of the PMLSM, for example, an optimal design of the auxiliary-teeth model was carried out to decrease the circulating current of PMLSM, and the effect of the unbalanced phase based on the auxiliary-teeth installation was considered [4]. However, for the PMLSM herein, another method is to design an appropriate control strategy to improve the robustness of PMLSM, which is of great significance. For the wafer stage of the lithography systems, Song et al. [3] presented an iterative learning control (ILC)-based feedforward compensation scheme. In the presence of parameter uncertainties and external disturbances, a reduced-order proportional-integral (PI)-based robust cascade control was presented to enhance the robust performance of current and speed regulations [5]. The sliding mode control (SMC) has been extensively investigated for a few decades because of its particular robustness to suppress the disturbances [6], and an extended state observer (ESO)-based SMC control for the permanent magnet synchronous motor (PMSM) servo system was presented in the work by the authors of [7]. In the framework of the active disturbance rejection control (ADRC) technique [8], Sira-Ramirez et al. [9] designed an active disturbance rejection control scheme for the angular velocity trajectory tracking.

In the work by the authors of [10], a comprehensive overview on disturbance/uncertainty estimation and attenuation techniques in PMSMs were given, where the various disturbances and uncertainties were discussed, and the relevant control methods were summarized.

In recent years, the model predictive control (MPC) technology has been well developed due to its conceptual simplicity, high dynamic response, and easy implementation etc. [11]. As mentioned in the work by the authors of [12], the MPC is one of the most popular control strategy in industrial applications, where satisfactory system performances can be guaranteed. One can regard the MPC as an optimal control method [13], which predicts the output of the state variable by employing the dynamic model of the controlled plant [14] and generates the future control input according to the optimization of the performance index function or cost function during each sampling period [15]. In the work by the authors of [16], three predictive current control schemes were compared by theoretical analyses, simulation results, and experiments. By replacing the conventional cascade structure, the speed and current controllers were combined together, and the MPC algorithm was applied to electrical drives [17]. Zhang et al. [18] presented a composite observer-based robust fault-tolerant predictive current control algorithm for PMSMs, which can eliminate the influence of motor parameter perturbation and permanent magnet demagnetization. Based on the field-oriented control (FOC) and vector control approaches, Liu et al. [19] proposed a composite speed regulation comprised by ESO feedforward compensation and predictive functional control (PFC) feedback control. In addition, an ESO can timely estimate the "lumped disturbances", including both internal unmodelled dynamics and external disturbances in the system [20], and thus the ESO-based control approaches (usually called ADRC schemes) can handle large uncertainties as well as achieve the desired performance [21]. It should be emphasized that most of the existing results are concentrated on PMSM servo control. However, the PMLSM is widely used in many applications. Thus, the investigation about designing an ESO-based MPC/PFC law to compensate the adverse effects of uncertainties and disturbances on a PMLSM is of significant importance, which remains as an open area.

The rest of this paper is organized as follows. In Section 2, the mathematical description of a PMLSM is analyzed and its simplified model is established. The main results are given in Section 3. The PFC controller is designed, and an ESO feedforward compensation is incorporated in the PFC feedback item, thus a composite speed regulate scheme is developed. Simulation results and comparative experiments are given to confirm the advantages and disturbance rejection performance of the proposed method in Section 4. Section 5 concludes the paper.

2. Mathematical Model of the Pmlsm

The objective of the servo system is tracking the reference signals as closely as possible, while guaranteeing control system and anti-disturbance performances. In this section, the mathematical model of the PMLSM is established in Laplace domain.

2.1. Mathematical Description of a Pmlsm

Generally, the basic equations of the PMLSM servo control system include the voltage equation, flux linkage equation, electromagnetic force equation, and the mechanical motion equation [22]. With the assumptions that the three-phase primary windings of the PMLSM are completely symmetrical and the traveling wave magnetic field is sinusoidally distributed [23], one can describe the PMLSM on the two-phase orthogonal synchronous rotating coordinate (d-q) as follows,

$$\begin{cases} L_{\mathrm{d}}\frac{di_{\mathrm{d}}}{dt} = u_{\mathrm{d}} - Ri_{\mathrm{d}} + \frac{\pi}{\tau}v\psi_{\mathrm{q}} \\ L_{\mathrm{q}}\frac{di_{\mathrm{q}}}{dt} = u_{\mathrm{q}} - Ri_{\mathrm{q}} - \frac{\pi}{\tau}v\psi_{\mathrm{d}} \\ F_{\mathrm{e}} = \frac{3\pi}{2\tau}\left[\psi_{\mathrm{f}}i_{\mathrm{q}} + (L_{\mathrm{d}} - L_{\mathrm{q}})\, i_{\mathrm{d}}i_{\mathrm{q}}\right] \\ M\frac{dv}{dt} = F_{\mathrm{e}} - F_L - B_{\mathrm{f}}v \end{cases} \tag{1}$$

where i_d, u_d, L_d, ψ_d, i_q, u_q, L_q, and ψ_q are the d-axis and q-axis currents, voltages, inductances, and flux linkages, respectively; ψ_f is the permanent magnet flux linkage; R is the winding resistance; τ is the pole pitch; F_e is the electromagnetic force; F_L is the load torque; B_f is the viscous friction coefficient; v and M are the velocity and the mass of linear motor rotor, respectively.

Considering a surface-mounted PMLSM, and thus $L_d = L_q = L$. We can assume that there no hysteresis, no eddy currents, no cogging, no bearing components, no magnetic saturation, no temperature rise leading to a higher resistance R and lower magnet flux, perfect magnetic alignment with an ideal symmetric 3-phase motor amplifier are exhibited [24]. In addition, we define the force constant as $K_f = 3\pi/2\tau \cdot \psi_f$. As a result, the Equation (1) can be rewritten as

$$\begin{cases} u_d = L\frac{di_d}{dt} + Ri_d - \frac{\pi}{\tau}vLi_q \\ u_q = L\frac{di_q}{dt} + Ri_q + \frac{\pi}{\tau}v\left(Li_d + \psi_f\right) \\ M\frac{dv}{dt} = K_f i_q - F_L - B_f v \end{cases} \tag{2}$$

where $\psi_q = Li_q$ and $\psi_d = Li_d + \psi_f$, respectively.

2.2. A Simplified Model of the Pmlsm

By the framework of FOC and vector control technique [25], a cascade structure comprising the speed regulation and current controllers is usually adopted in the PMLSM servo system [26]. Because the excitation flux linkage is established by permanent magnets, the current i_d is controlled to be zero in the inner loop, while the electromagnetic force is proportional to the current i_q, whose command current is determined by the outer speed loop. It should be emphasized that the dynamics of current loop is usually much faster than speed loop response, that is, the current loop has a higher bandwidth that of speed loop. As a result, the current i_q nearly equals to its expected value i_q^*, and thus the transfer function of inner current loop can be regarded as 1, i.e., $G_i(s) \approx 1$ [27]. To this end, the simplified control diagram of the PMLSM servo system is shown in Figure 1. In Figure 1, e denotes the velocity tracking error satisfying $e = v^* - v$, and v^* is the given signal.

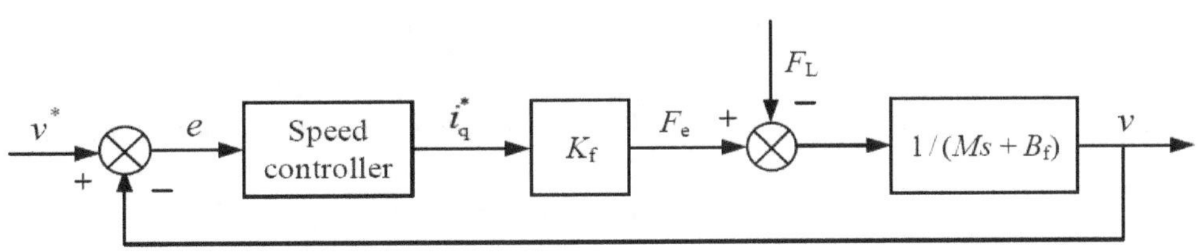

Figure 1. The simplified control diagram of the PMLSM servo system.

According to the control scheme of PMLSM given in Figure 1, the first-order system model in the Laplace domain can be expressed as

$$\Omega(s) = \frac{K_f I_q^*(s) - F_L(s)}{Ms + B_f} \tag{3}$$

3. ESO-Based PFC Design

In this section, a PFC law is firstly designed based on the mathematical model obtained in Section 2. Then, an ESO is constructed to estimate the disturbances in the PMLSM servo system, which are incorporated in the PFC feedback controller. Finally, the whole block diagram for the PMLSM control system is derived.

3.1. PFC Design

Among all the MPC approaches, PFC is characterized by the low online computation, while maintaining the advantages of the MPC method. In general, the PFC design methodology mainly comprises the following five components.

- (1) Basic functions

$$i_q^*(k+i) = \sum_{j=1}^{N} \mu_j u_{bj}(i), \qquad i = 1, 2, \ldots, P \tag{4}$$

where i_q^* is the future control variables, N is the number of base functions, μ_j is the coefficients related to the performance optimization index, $u_{bj}(i)$ is the base functions determined by the reference signals, and P is the optimization horizon.

In this paper, we select the base function as the form of step response, i.e., $u_{bj}(i) = 1$, and set $N = 1$. As a result, the following equation can be derived.

$$i_q^*(k) = \mu_1 \tag{5}$$

- (2) Prediction model

The prediction model of the plant output can be given in its discrete time form, the following formulation is easily obtained by applying the Euler discretization method to Equation (3).

$$v_m(k+1|k) = \alpha_m v_m(k) + K_m(1-\alpha_m) i_q^*(k) \tag{6}$$

where $v_m(k+1|k)$ is the predicted velocity at time $t = (k+1)T_s$ and T_s is the sample time; K_m and α_m are coefficients of the prediction model (differential equation form).

For the $(k+2)$th sampling time, we have

$$v_m(k+2|k) = \alpha_m v_m(k+1) + K_m(1-\alpha_m) i_q^*(k+1) \tag{7}$$

According to the well-known mean-level control strategy [28], the control variables are regarded as a constant during the prediction, that is,

$$i_q^*(k) = i_q^*(k+1) = \cdots = i_q^*(k+P-1) \tag{8}$$

By replacing $v_m(k+1|k)$ with $v_m(k+1)$ and combining Equations (6)–(8) we get

$$v_m(k+2|k) = \alpha_m^2 v_m(k) + K_m(1-\alpha_m^2) i_q^*(k) \tag{9}$$

Obeying the same principle, the prediction output generated at the $(k+2)$th sampling time is as follows,

$$v_m(k+i|k) = \alpha_m^i v_m(k) + K_m(1-\alpha_m^i) i_q^*(k) \tag{10}$$

Correspondingly, we can described the prediction output as the matrix form based on the Equations (5), (6), (9), and (10), which is as follows,

$$V_m(k) = V_o(k) + V_b(k)\mu_1 \tag{11}$$

where

$$\begin{aligned} V_m(k) &= \begin{bmatrix} v_m(k+1|k) & v_m(k+2|k) & \cdots & v_m(k+P|k) \end{bmatrix}^T \\ V_o(k) &= \begin{bmatrix} \alpha_m & \alpha_2^P & \cdots & \alpha_m^P \end{bmatrix}^T v_m(k) \\ V_b(k) &= K_m \begin{bmatrix} (1-\alpha_m) & (1-\alpha_2^P) & \cdots & (1-\alpha_m^P) \end{bmatrix}^T \end{aligned} \tag{12}$$

- (3) Error correction

  Because the presence of the model mismatch, unknown disturbance, parameter variation, and noise, the prediction error between prediction model output and actual output is existed. When PFC is applied to control systems with a small sampling period, it is generally believed that the error in this processing period remains constant [19]. Since the sampling period of the PMLSM control system in this paper is 10^{-4} s, it is assumed that all prediction errors are equal only within this interruption period. The error can be expressed as follows,

$$e(k) = e(k+1) = e(k+2) = \cdots = e(k+P-1) = v(k) - v_{\mathrm{m}}(k) \tag{13}$$

  where $v(k)$ is the measured velocity of the practical control system.

- (4) Reference trajectory

  A first-order reference trajectory is provided in the following form.

$$v_{\mathrm{r}}(k+i) = v^*(k+i) - \alpha_{\mathrm{r}}^i\left[v^*(k) - v(k)\right] \qquad i = 1,2,\ldots,P \tag{14}$$

  where v_{r} is the velocity reference trajectory, v^* is the set point value, $\alpha_{\mathrm{r}} = e^{-T_{\mathrm{s}}/T_{\mathrm{r}}}$ is the time constant, and T_{r} is the desired response time.

- (5) Evaluation mechanism

  Considering the tracking error between the velocity reference trajectory v_{r} and the predicted output v_{m}, and minimizing the sum of squared error e with a penalization on the control input i_{q}^*, yield the following quadratic performance index.

$$\min J = \sum_{i=1}^{P} q_i^2[v_{\mathrm{r}}(k+i) - v_{\mathrm{m}}(k+i|k) - e(k+i)]^2 + r^2 i_{\mathrm{q}}^{*2}(k) \tag{15}$$

  where q_i are the parameters imposed on each of the tracking errors and e; r is a weighting parameter of the control input.

  Define the following matrices.

$$\begin{aligned} V_{\mathrm{r}}(k) &= \begin{bmatrix} v_{\mathrm{r}}(k+1) & v_{\mathrm{r}}(k+2) & \cdots & v_{\mathrm{r}}(k+P) \end{bmatrix}^{\mathrm{T}} \\ E(k) &= \begin{bmatrix} e(k+1) & e(k+2) & \cdots & e(k+P) \end{bmatrix}^{\mathrm{T}} \\ Q &= \begin{bmatrix} q_1^2 & 0 & \cdots & 0 \\ 0 & q_i^2 & \cdots & 0 \\ \vdots & \vdots & \ddots & \vdots \\ 0 & 0 & \cdots & q_i^2 \end{bmatrix}^{\mathrm{T}}, \qquad R = r^2 \end{aligned} \tag{16}$$

  Substituting Equations (12) and (16) into (15) leads to the following cost function.

$$J = Q\|V_{\mathrm{r}}(k) - V_{\mathrm{m}}(k) - E(k)\|^2 + R i_q^{*2}(k) \tag{17}$$

  where $\|\cdot\|$ denotes the Euclidean norm.

  According to the optimization theory [29], we can obtained the following controller by calculating $\partial J / \partial i_{\mathrm{q}}^{*2} = 0$.

$$i_{\mathrm{q}}^* = (V_{\mathrm{b}}^{\mathrm{T}} Q V_{\mathrm{b}} + R)^{-1} V_{\mathrm{b}}^{\mathrm{T}} Q\left[V_{\mathrm{r}}(k) - V_{\mathrm{o}}(k) - E(k)\right] \tag{18}$$

  In a summary, the PFC law is derived as Equation (18).

3.2. ESO Design

As mentioned in the work by the authors of [30], the observer-based feedforward compensation method can improve the disturbance rejection performance. In this note, we introduce an ESO to estimate the disturbances, whose estimates will be taken the above designed PFC into account.

Extracting the mechanical motion equation from Equation (2), and rewriting it as the following differential equation form.

$$\dot{v} = \tfrac{K_f}{M} i_q - \tfrac{F_L}{M} - \tfrac{B_f}{M} v = \tfrac{K_f}{M} i_q^* + d(t) \tag{19}$$

where the lumped disturbance $d(t)$ (containing tracking error of q-axis current loop, external load torque and friction) is as follows:

$$d(t) = \frac{K_f}{M}\left(i_q - i_q^*\right) - \frac{F_L}{M} - \frac{B_f}{M} v \tag{20}$$

whose derivative is assumed to be bounded, namely, satisfying $\dot{d}(t) \leq D$.

The function $d(t)$ is generally unknown and the ESO can generate the disturbance estimate only depending on the input and output information of the control plant [20], which is free of the specific mathematical models. Here, we define the following state variables.

$$\begin{cases} x_1 = v \\ x_2 = d(t) \end{cases} \tag{21}$$

Substituting Equation (21) into (19) yields

$$\begin{cases} \dot{x}_1 = x_2 + b i_q^* \\ \dot{x}_2 = D \end{cases} \tag{22}$$

where b is a coefficient determined by the system parameters, i.e., $b = K_f/M$.

In this paper, we construct the ESO as follows,

$$\begin{cases} \varepsilon = z_1 - x_1 \\ \dot{z}_1 = z_2 - c_1\varepsilon + b_o i_q^* \\ \dot{z}_2 = -c_2\varepsilon \end{cases} \tag{23}$$

where z_1 and z_2 are the estimates of the state variables x_1 and x_2; c_1 and c_2 are the gains of designed ESO to be determined; b_o is the estimate of parameter b.

Based on the stability analysis in the work by the authors of [8], if the gains c_1 and c_2 are appropriately selected, the estimate errors between z_1, z_2 and x_1, x_2 can be guaranteed to be zero, that is,

$$\begin{cases} z_1 \to x_1 \to v \\ z_2 \to x_2 \to d(t) \end{cases} \tag{24}$$

After the total disturbance $f(t)$ is accurately estimated by the above ESO, one can be incorporated in aforementioned Equation (18) as a feedforward compensation to improve the anti-disturbance performance [31], and thus the composite speed regulator can be obtained as follows.

$$i_q^* = (V_b^T Q V_b + R)^{-1} V_b^T Q \left[V_r(k) - V_o(k) - E(k)\right] - z_2/b_0 \tag{25}$$

As a result, the whole block diagram of the proposed composite control for the PMLSM servo system is given in Figure 2.

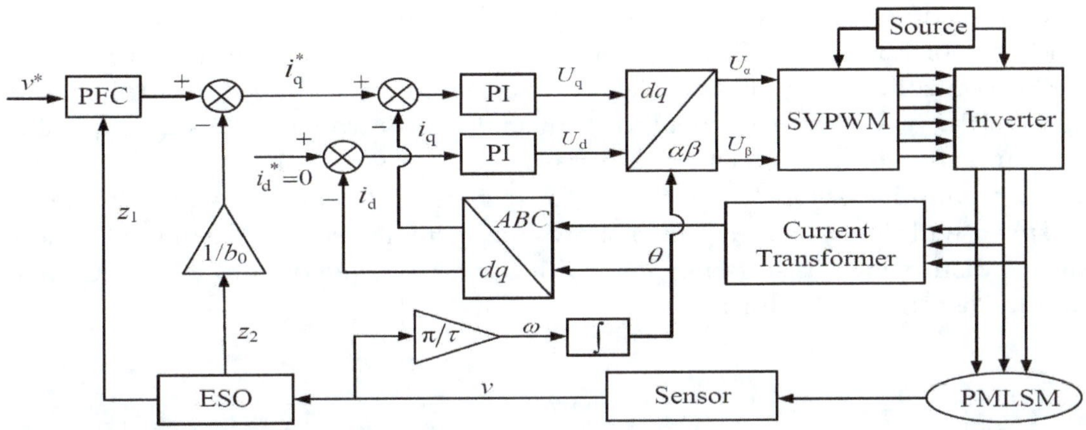

Figure 2. The ESO-based PFC control for the PMLSM.

4. Simulation and Experimental Results

In this section, some simulation results and comparative experiments will be conducted to demonstrate the effectiveness and advantages of the proposed ESO-based PFC composite control strategy. The parameters of the PMLSM are exhibited in Table 1.

Table 1. Parameters of the permanent magnet linear synchronous motor (PMLSM).

Parameter	Value	Unit
Rated voltage (U)	24	V
Pole pitch (τ)	32	mm
Winding resistance (R)	1.25	Ω
Inductance (L)	5.25	mH
Viscous friction coefficient (B_f)	2.12	N·m/s
Flux linkage (ψ_f)	0.0385	Wb
Mass (M)	14	kg

4.1. Simulation Study

In order to illustrate the advantages of the proposed composite control scheme over the conventional PID control method, the simulation comparisons are performed in this subsection. The simulation results are shown in Figures 3 and 4.

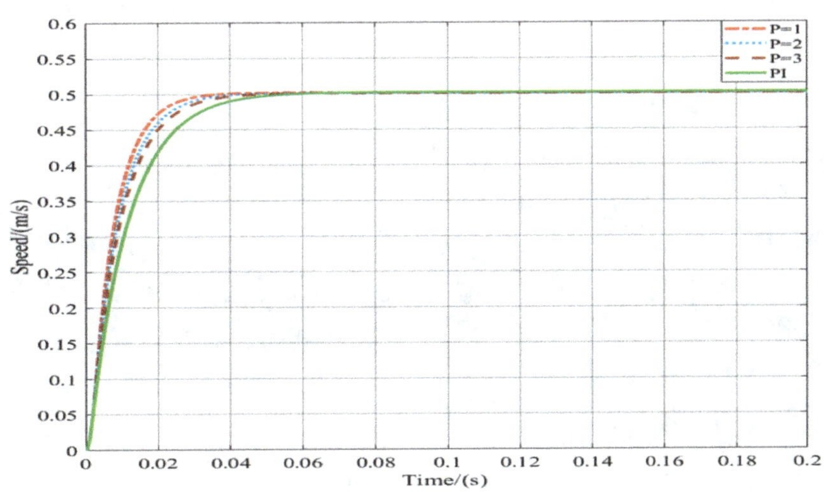

Figure 3. Speed tracking for proportional-integral (PI) and composite controllers (without load torque).

For the case of no external load torque, Figure 3 shows the velocity evolutions based on PI controller and that of ESO-based PFC controller. According to the simulation results, the speed response of PFC controller is significantly faster than that of traditional PI controller, which reflects the superiority of the predictive control algorithm. In PFC controller, the value of prediction depth (also called optimization horizon) P can be arbitrarily set. The simulation curves show that when the optimal prediction depth is set as $P = 1$, the response speed is the fastest, and with the P increasing, the response speed will slow down. Figure 4 shows the system performance of the PI and PFC + ESO when an external disturbances is suddenly loaded. It can be concluded that PFC + ESO control scheme has superior anti-disturbance performance.

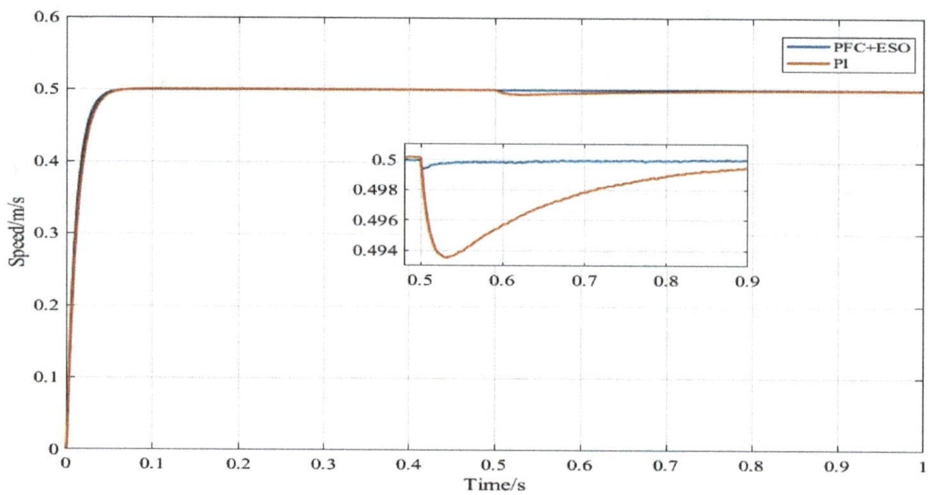

Figure 4. Performance comparison between PI and predictive function control (PFC) + extended state observer (ESO).

4.2. Experiment Comparisons

To evaluate the performance of the proposed method, the experimental setup for a PMLSM test system has been developed, which is shown in Figure 5. The overall speed regulate algorithm including the space vector pulse width modulation (SVPWM) implemented on the STM32 control board, intelligent power module (IPM) integrated driven circuit, external AD7606 conversion circuit, current and position sampling circuits (Hall-effect devices), oscilloscope, computer, Emulator, PMLSM, DC 15V and DC 24V power source, etc.

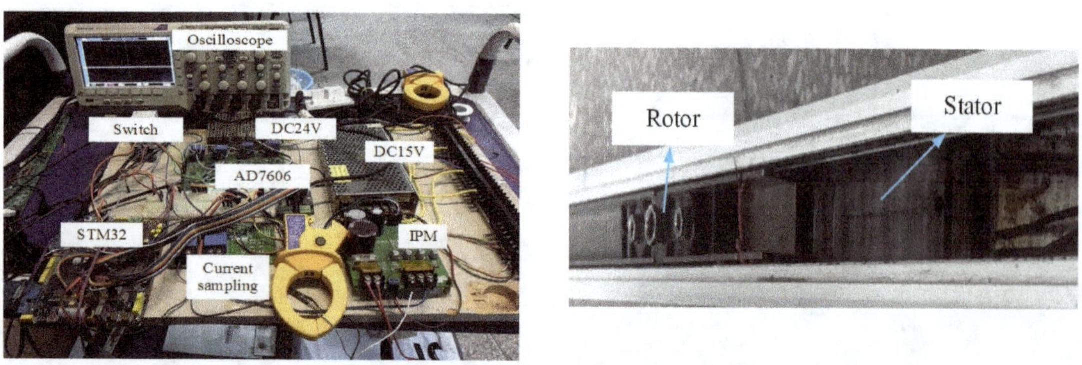

(**a**) Configuration of the test system (**b**) Permanent magnet synchronous linear motor

Figure 5. Experiment setup.

In order to verify the advantages and effectiveness of the composite control scheme, comparative simulations and experiments are established on optimized PID controller and the proposed method, respectively; with no speed overshoot as the constraint and the parameters tracking the reference speed in the shortest time are taken as the optimal parameters of the PID. The parameters of the proposed method are determined by the same method.

4.2.1. No-Load Experiments

Experiment results based on PI controller and ESO-based PFC controller (the predicted depth is set as $P = 1$) are performed without external load torque, which are shown in Figure 6. In Figure 6, the reference speed is given as $v_{ref} = 0.5$ m/s, it can be seen from the experiment results that the speed response of PFC controller is faster than that of PI controller, and the system has superior performance.

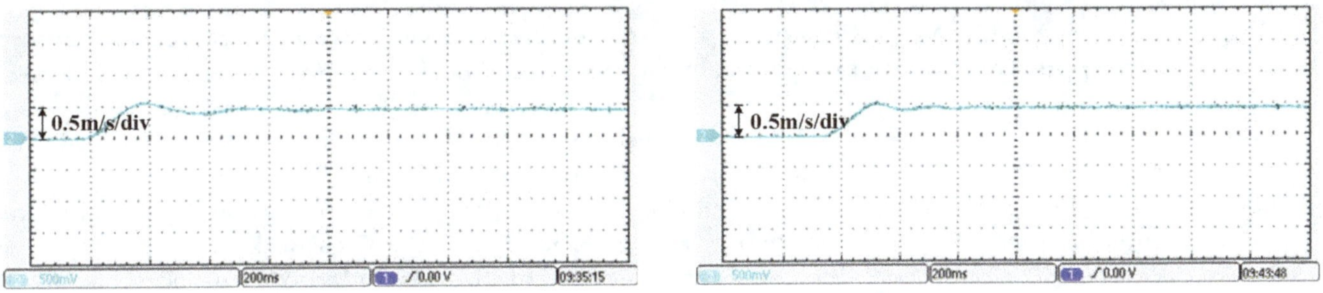

(**a**) PI control (**b**) ESO-based PFC control

Figure 6. Speed response curves (no load).

4.2.2. Load Torque Experiments

During the stable operation of the PMLSM, a load disturbance is suddenly loaded in a step form, the speed response curves based on PI and PFC + ESO controllers are shown in Figure 7. By comparing the results, it can be concluded that the proposed control scheme has faster response speed, and stronger robustness to external disturbance. In addition, the proposed ESO algorithm can well estimate the system state and disturbance.

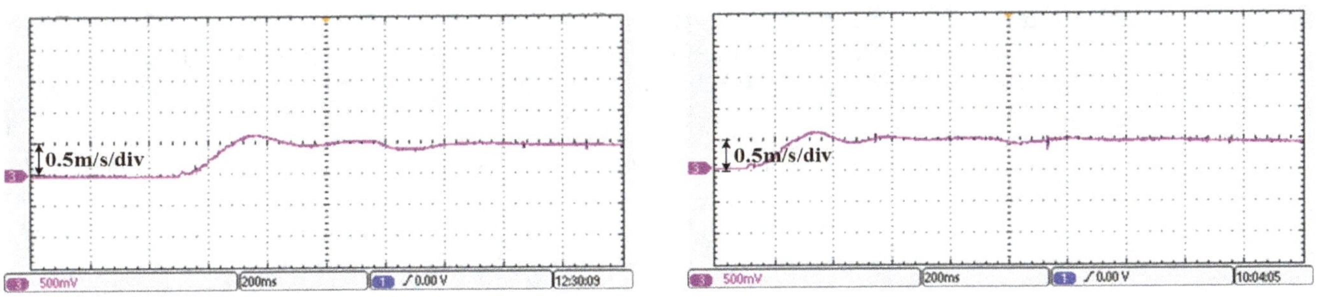

(**a**) PI control (**b**) ESO-based PFC control

Figure 7. Speed response curves.

5. Conclusions

The speed regulation problem for a PMLSM servo system has been investigated in this paper. The mathematical model of the PMLSM is established in the Laplace domain. A composite control scheme based on the PFC feedback and the ESO-based feedforward compensation has been presented. The simulation and experiment results confirm that the proposed composite control method has a superior system performance compared with the traditional PID control strategy, such as faster response speed and stronger robustness. Our future work will extend the proposed method to deal with the special force ripple and end effect problems in the PMLSM servo system.

Author Contributions: This is a joint work and the authors were in charge of their expertise and capability: Y.W. (Yao Wang) for investigation and analysis; H.Y. for validation and revision; Z.C. for methodology and data analysis; Y.W. (Yuchen Wang) for writing and revision; C.Z. for manuscript revision.

Acknowledgments: The authors would like to express their gratitude to all those who helped them during the writing of this paper. Also, the authors would like to thank the reviewers for their valuable comments and suggestions.

References

1. Yang, C.Y.; Ma, T.T., Che, Z.Y.; Zhou, L.N. An adaptive-gain sliding mode observer for sensorless control of permanent magnet linear dynchronous motors. *IEEE Access* **2018**, *6*, 3469–3478. [CrossRef]
2. Wang, M.Y.; Yang, R.; Zhang, C.M.; Cao, J.W.; Li, L.Y. Inner loop design for PMLSM drives with thrust ripple compensation and high-performance current control. *IEEE Trans. Ind. Electron.* **2018**, *65*, 9905–9915. [CrossRef]
3. Song, F.Z.; Liu, Y.; Xu, J.X.; Yang, X.F.; Zhu, Q. Data-driven iterative feedforward tuning for a wafer stage: A high-order approach based on instrumental variables. *IEEE Trans. Ind. Electron.* **2019**, *66*, 3106–3116. [CrossRef]
4. Jang, K.B.; Kim, J.H.; An, H.J.; Kim, G.T. Optimal design of auxiliary teeth to minimize unbalanced phase due to end effect of PMLSM. *IEEE Trans. Magn.* **2011**, *47*, 1010–1013. [CrossRef]
5. Son, Y.I.; Kim, I.H.; Choi, D.S.; Shim, H. Robust cascade control of electric motor drives using dual reduced-order PI observer. *IEEE Trans. Ind. Electron.* **2015**, *62*, 3672–3682. [CrossRef]
6. Yang, C.Y.; Che, Z.Y.; Zhou, L.N. Integral sliding mode control for singularly perturbed systems with mismatched disturbances. *Circuits Syst. Signal Process.* **2019**, *38*, 1561–1582. [CrossRef]
7. Li, S.H.; Zong, K.; Liu, H.X. A composite speed controller based on a second-order model of permanent magnet synchronous motor system. *Trans. Inst. Meas. Control* **2011**, *33*, 522–541.
8. Huang, Y.; Xue, W.C. Active disturbance rejection control: Methodology and theoretical analysis. *ISA Trans.* **2014**, *53*, 963–976. [CrossRef]
9. Sira-Ramirez, H.; Linares-Flores, J.; Garcia-Rodriguez, C.; Contreras-Ordaz, M.A. On the control of the permanent magnet synchronous motor: An active disturbance rejection control approach. *IEEE Trans. Control Syst. Technol.* **2014**, *22*, 2056–2063. [CrossRef]
10. Yang, J.; Chen, W.H.; Li, S.H.; Guo, L.; Yan, Y.D. Disturbance/uncertainty estimation and attenuation techniques in PMSM drives—A survey. *IEEE Trans. Ind. Electron.* **2017**, *64*, 3273–3285. [CrossRef]
11. Ganesh, H.S.; Edgar, T.F.; Baldea, M. Model predictive control of the exit part temperature for an austenitization furnace. *Processes* **2016**, *4*, 53. [CrossRef]
12. Gong, Z.; Wu, X.J.; Dai, P.; Zhu, R.W. Modulated model predictive control for MMC–based active front–end rectifiers under unbalanced grid conditions. *IEEE Trans. Ind. Electron.* **2019**, *66*, 2398–2409. [CrossRef]
13. Yan, Y.D.; Zhang, C.L.; Narayan, A.; Yang, J.; Li, S.H.; Yu, H.Y. Generalized dynamic predictive control for nonparametric uncertain systems with application to series elastic actuators. *IIEEE Trans. Ind. Inform.* **2018**, *14*, 4829–4840. [CrossRef]
14. Yang, J.; Wu, H.; Hu, L.; Li, S.H. Robust predictive speed regulation of converter-driven DC motors via a discrete-time reduced-order GPIO. *IEEE Trans. Ind. Electron.* **2019**, *66*, 7893–7903. [CrossRef]
15. Vaccari, M.; Pannocchia, G. A modifier-adaptation strategy towards offset-free economic MPC. *Processes* **2017**, *5*, 2. [CrossRef]
16. Morel, F.; Lin-Shi, X.F.; Retif, J.M.; Allard, B.; Buttay, C. A comparative study of predictive current control schemes for a permanent-magnet synchronous machine drive. *IEEE Trans. Ind. Electron.* **2009**, *56*, 2715–2728. [CrossRef]
17. Bolognani, S.; Bolognani, S.; Peretti, L.; Zigliotto, M. Design and implementation of model predictive control for electrical motor drives. *IEEE Trans. Ind. Electron.* **2009**, *56*, 1925–1936. [CrossRef]
18. Zhang, C.F.; Wu, G.P.; Rong, F.; Feng, J.H.; Jia, L.; He, J.; Huang, S.D. Robust fault-tolerant predictive current control for permanent magnet synchronous motors considering demagnetization fault. *IEEE Trans. Ind. Electron.* **2018**, *65*, 5324–5334. [CrossRef]
19. Liu, H.X.; Li, S.H. Speed control for PMSM servo system using predictive function control and extended state observer. *IEEE Trans. Ind. Electron.* **2012**, *59*, 1171–1183. [CrossRef]

20. Xue, W.C.; Bai, W.Y.; Yang, S.; Song, K.; Huang, Y.; Xie, H. ADRC with adaptive extended state observer and its application to air-fuel ratio control in gasoline engines. *IEEE Trans. Ind. Electron.* **2015**, *62*, 5847–5857. [CrossRef]
21. Wang, J.X.; Li, S.H.; Yang, J.; Wu, B.; Li, Q. Extended state observer-based sliding mode control for PWM-based DC-DC buck power converter systems with mismatched disturbances. *IET Control Theory Appl.* **2015**, *9*, 579–586. [CrossRef]
22. Chen, S.Y.; Chiang, H.H.; Liu, T.S.; Chang, C.H. Precision motion control of permanent magnet linear synchronous motors using adaptive fuzzy fractional-order sliding-mode control. *IEEE/ASME Trans. Mechatron.* **2019**, *24*, 741–752. [CrossRef]
23. Jin, H.Y.; Zhao, X.M. Complementary sliding mode control via elman neural network for permanent magnet linear servo system. *IEEE Access* **2019**, *7*, 82183–82193. [CrossRef]
24. Hu, J.G.; Liu, J.B.; Xu, L.Y. Eddy current effects on rotor position estimation and magnetic pole identification of PMSM at zero and low speeds. *IEEE Trans. Power Electron.* **2008**, *23*, 2565–2575. [CrossRef]
25. Yang, C.Y.; Liu, J.H.; Li, H.; Zhou, L.N. Energy modeling and parameter identification of dual-motor-driven belt conveyors without speed sensors. *Energies* **2018**, *11*, 3313. [CrossRef]
26. Yang, R.; Wang, M.Y.; Li, L.Y.; Zenggu, Y.M.; Jiang, J.L. Integrated uncertainty/disturbance compensation with second-order sliding-mode observer for PMLSM-driven motion stage. *IEEE Trans. Power Electron.* **2019**, *34*, 2597–2607. [CrossRef]
27. Lu, S.W.; Tang, X.Q.; Song, B.; Zheng, S.Q.; Zhou, F.X. Identification and compensation of force ripple in PMSLM using a JITL technique. *Asian J. Control* **2015**, *17*, 1559–1568. [CrossRef]
28. Mendez, J.A.; Kouvaritakis, B.; Rossiter, J.A. State-space approach to interpolation in MPC. *Int. J. Robust Nonlinear Control* **2000**, *10*, 27–38. [CrossRef]
29. Yang, C.Y.; Zhang, Q.L.; Sun, J.; Chai, T.Y. Lur'e Lyapunov function and absolute stability criterion for Lur'e singularly perturbed systems. *IEEE Trans. Autom. Control* **2011**, *56*, 2666–2671. [CrossRef]
30. Liu, M.; Zhang, L.X.; Shi, P.; Zhao, Y.X. Fault estimation sliding mode observer with digital communication constraints. *IEEE Trans. Autom. Control* **2018**, *63*, 3434–3441. [CrossRef]
31. Zhou, L.N.; Che, Z.Y.; Yang, C.Y. Disturbance observer-based integral sliding mode control for singularly perturbed systems with mismatched disturbances. *IEEE Access* **2018**, *6*, 9854–9861. [CrossRef]

Investigating the Dynamic Impact of CO_2 Emissions and Economic Growth on Renewable Energy Production: Evidence from FMOLS and DOLS Tests

Muhammad Waris Ali Khan [1,*], Shrikant Krupasindhu Panigrahi [2], Khamis Said Nasser Almuniri [2], Mujeeb Iqbal Soomro [3], Nayyar Hussain Mirjat [4] and Eisa Salim Alqaydi [5]

1 Faculty of Industrial Management, Universiti Malaysia Pahang, Gambang 26300, Malaysia
2 College of Business, University of Buraimi, Alburaimi, P.O. box 890, P.C. 512, Oman
3 Department of Mechanical Engineering, Mehran University of Engineering & Technology, SZAB Campus, Khairpur 66020, Pakistan
4 Department of Electrical Engineering, Mehran University of Engineering and Technology, Jamshoro 76062, Pakistan
5 Faculty of Health Sciences, DeMontfort University, Leicester LE8 4AA, UK
* Correspondence: waris@ump.edu.my

Abstract: Understanding the dynamic nexus between CO_2 emissions and economic growth in the sustainable environment helps the economies in developing resources and formulating apposite energy policies. In the recent past, various studies have explored the nexus between CO_2 emissions and economic growth. This study, however, investigates the nexus between renewable energy production, CO_2 emissions, and economic growth over the period from 1995 to 2016 for seven Association of Southeast Asian Nations (ASEAN) countries. Fully Modified Ordinary Least Square (FMOLS) and Dynamic Ordinary Least Square (DOLS) methodologies were used to estimate the long- and short-run relationships. The panel results revealed that renewable energy production has a significant long term effect on CO_2 emissions for Vietnam ($t = -2.990$), Thailand ($t = -2.505$), and Indonesia ($t = -2.515$), and economic growth impact for Malaysia ($t = 2.050$), Thailand ($t = -2.001$), and the Philippines ($t = -2.710$). It is, therefore, vital that the ASEAN countries implement policies and strategies that ensure energy saving and continuous economic growth without forsaking the environment. This study, as such, recommends that ASEAN countries should take measures to decrease the reliance on fossil fuels for achieving these objectives. Future research should consider the principles of circular economy and clean energy development mechanisms integrated with renewable energy technologies.

Keywords: carbon emissions; economic growth; energy; renewable energy; Fully Modified Ordinary Least Square (FMOLS); dynamic panel cointegration model

JEL Classification: C22; C33; Q20; Q43

---

1. Introduction

The promotion of sustainable development and combating climate change is altogether an integral and challenging aspect of energy planning and policy development. Over the past decades, global warming has been rising, which has emerged as one of the key challenges that humanity is facing. The increased emissions of greenhouse gases (GHGs), such as carbon dioxide (CO_2), methane (CH_4), ozone, and nitrous oxide, are causing severe damage to the global environment. Among other greenhouse gases, CO_2 emissions are considered as the principal cause of global warming, thereby toppling the climate [1–3]. The CO_2 emissions due to excessive burning of fossil fuels, such as coal, oil, and gas, along with increased deforestation, have considerably contributed to climate change. There is a great abundance of CO_2 in the atmosphere compared to other greenhouse gases and it is estimated rise in the next 100 years [4]. This situation has prompted the consideration and research on the measures to reduce CO_2 emissions. One of the key measures to reduce CO_2 emissions is to reduce fossil fuel consumption and maintain economic growth by meeting part of the energy needs by harnessing renewable energy sources [3].

In order to substantiate this measure, the literature has sufficiently maintained the establishment of relationships among renewable energy production, CO_2 emissions, and economic growth [5–7]. However, the higher production cost of renewable energy compared to fossil fuel energy hinders its full-scale commercialization and, hence, has not achieved the much-needed success so far.

The researchers are, nevertheless, attempting to establish the much required positive relationship between environmental quality and economic growth [8]. As such, it is essentially required that energy policymakers should come up with robust, sustainable policy endeavors to address the environmental and economic challenges with the effective utilization of renewable energy resources [8]. According to the Environmental Kuznets Curve (EKC) hypothesis [9], income and emissions are directly proportional to the threshold level.

In this context, various empirical studies have so far only focused on CO_2 emissions as a pollutant in the industrialized world. A review of previous studies in a similar context reveals that there is lack of studies performed on developing countries especially for the Association of Southeast Asian Nations (ASEAN) countries, including the emerging economies of the region, such as Myanmar and Vietnam, see for example [1,4,10,11]. These studies, however, have not produced clarity in their work for neglecting Myanmar and Vietnam as they have only considered five leading ASEAN countries (Malaysia, Thailand, Indonesia, the Philippines, and Singapore) for their past economic growth [10]. As such, the impact of the emerging economies of Myanmar and Vietnam for their renewable energy production has consistently gone unnoticed. As such, in this study, a summary of percentages of renewable energy production, CO_2 emissions per capita, and Gross Domestic Product (GDP) per capita for seven selected ASEAN countries for the study period of 1995–2016 is thoroughly analyzed.

The panel data plots of the CO_2 emissions per capita for ASEAN countries is shown in Figure 1a while the real GDP for the ASEAN countries as shown in Figure 1b indicate that Singapore and Malaysia have been growing consistently from 1995 until 2016. These details signify the importance of the relationships between renewable energy production, CO_2 emissions, and the economies for the selected ASEAN countries, which require essential investigations.

The dynamic relationship between energy consumption and economic growth has been thoroughly investigated in the past few decades [12]. However, the impact of harnessing renewable energy on the reduction of CO_2 emissions and thereby bringing about sustainable economic growth has been overlooked by previous studies. It is pertinent to mention that renewable energy resources are now emerging as the mainstream sources of energy in various developed economies and co-operative ownership of these resources has also increased and expanded rapidly in the past few years. The contribution of low growing economies like Myanmar and Vietnam towards renewable energy production, as shown in Figure 1c, has been consistently unnoticed [12].

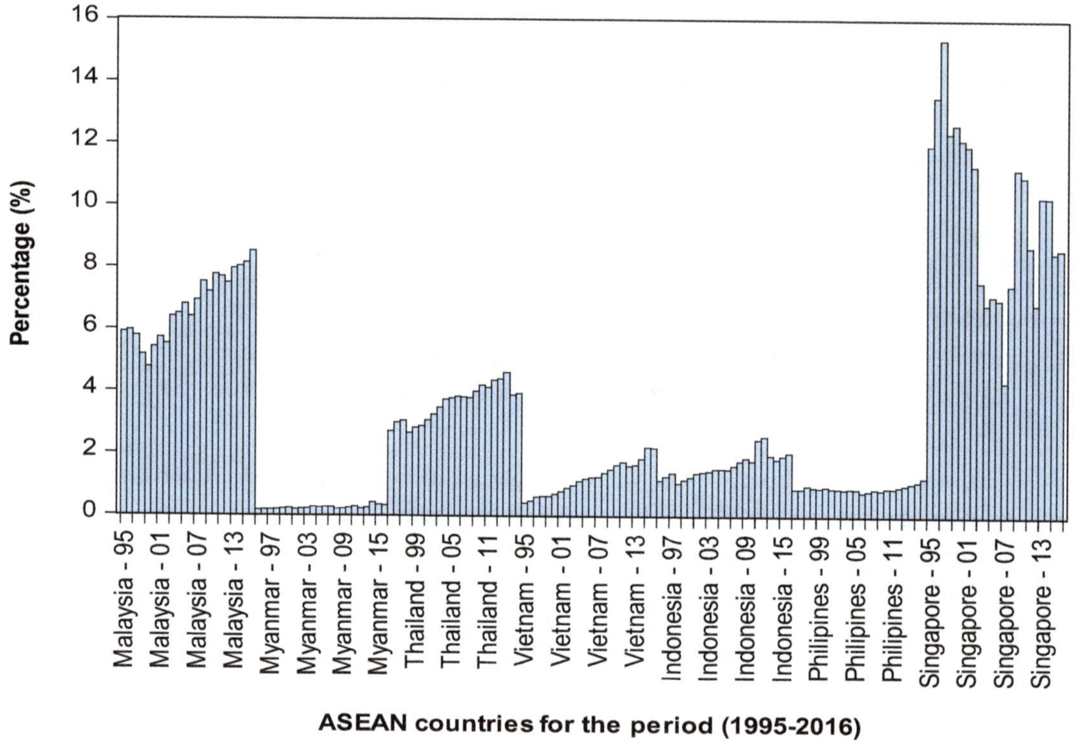

(a)

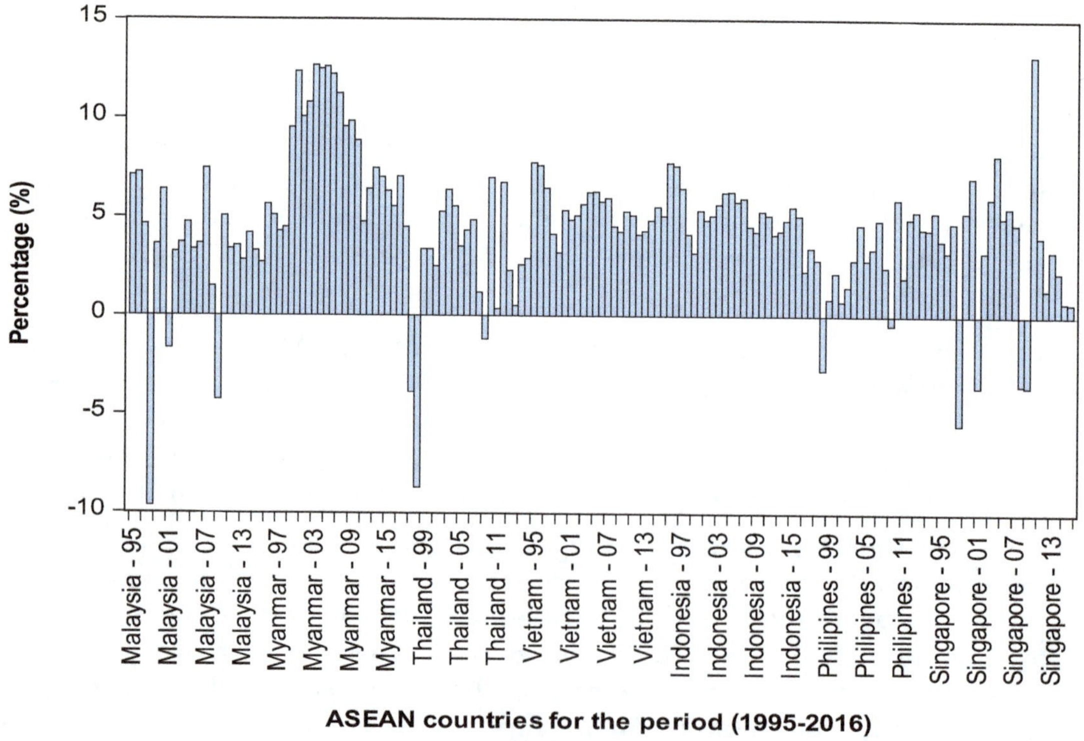

(b)

Figure 1. *Cont.*

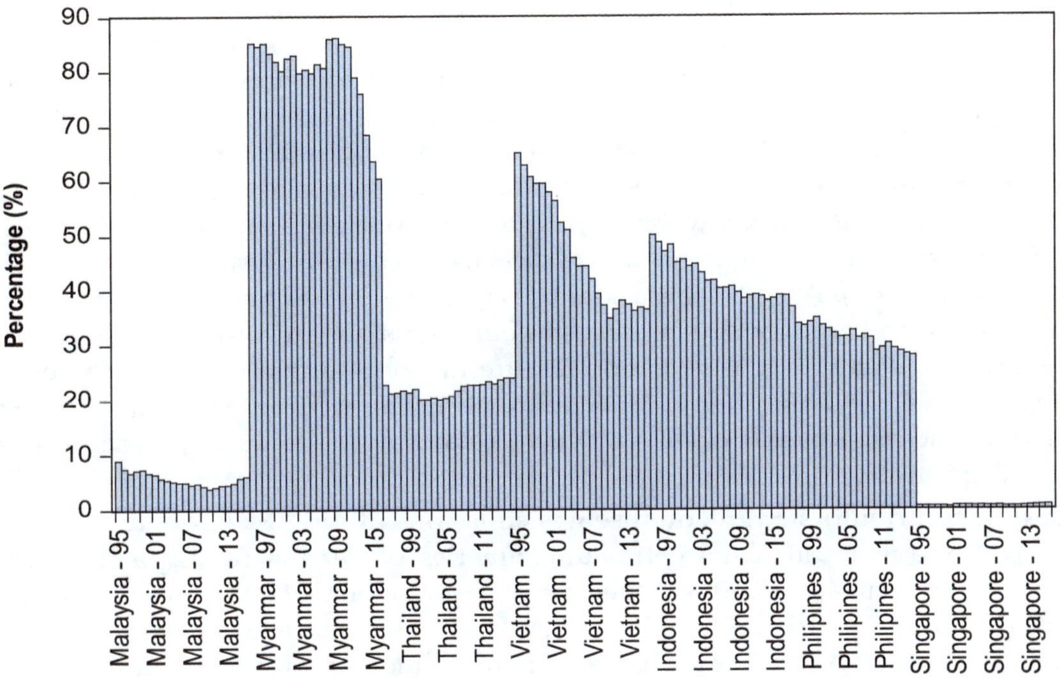

(c)

Figure 1. (**a**) Panel data plots of the CO_2 emissions per capita for Association of Southeast Asian Nations (ASEAN) countries; (**b**) Panel data plots of the real GDP (Gross Domestic Product) per capita for ASEAN countries; (**c**) Panel data plots of renewable energy production in % for ASEAN countries.

Given the above facts, the main objective of this study is to investigate the impact of renewable energy production on CO_2 emissions and economic growth for seven ASEAN countries from 1995 to 2016. This objective is achieved by employing a panel co-integration technique that followed the growth model framework. Furthermore, an attempt has been made to validate the EKC hypothesis for the selected seven ASEAN countries using panel data analysis with an approach of co-integration and Granger causality tests. The consistency of the results was assured by the inclusion of each country, thereby analyzing the environmental complexities in the respective countries. The country-specific investigation helped to merge the complexities of renewable energy production, the CO_2 emission relationship, and economic growth nexus, which has been ignored in previous studies. As such, contrary to previous studies, this study explored the causality relationships among renewable energy production, CO_2 emissions, and economic growth.

ASEAN Countries as the Context

Historically, the consumption and reliance on fossil fuels in ASEAN countries have been generally on the rise [13]. ASEAN countries are experiencing the world's most significant jump and are considered as regions vulnerable to causing climate change [14]. Every decade, since 1960, the average temperature in the Southeast Asian region has risen, which has adversely added to climate change. The economic progress of five of the ASEAN countries: Malaysia, Indonesia, the Philippines, Singapore, and Thailand, is commendable compared to the other two [4]. Nevertheless, Vietnam and Myanmar have shown the highest renewable energy production among the other regional counties. As such, these two developing economies of ASEAN, for renewable energy initiatives, were also included in this study. It is pertinent to mention that 57% of the total electricity production in Myanmar has been based on its indigenous hydropower resources, thereby tackling, effectively, the emission of greenhouse gases [15]. Moreover, Myanmar, together with Vietnam, has also focused on implementing the climate-resilient de-carbonization pathways, thus assuring its significance in the renewable energy market. It is estimated that Myanmar can further install electricity plants of 104,000 MW capacity based on the wind and hydropower plants and that it can produce another 40 TWh/year (TWh/year stands for terawatt-hour per year) from solar power plants [16]. As per a World Bank Database report (2016), Singapore (98%), the Philippines (46%), Thailand (42%), and Cambodia (33%) were the four major energy-importing ASEAN countries [17]. Meanwhile, other countries like Brunei, Indonesia, Myanmar, and Vietnam are well-known energy resource exporters. In most of the ASEAN countries, renewable energy resources are abundant and are sufficiently consumed, such as in Indonesia, Malaysia, Thailand, and Vietnam [18]. As such, these countries with the further harnessing of renewable energy resources can significantly reduce greenhouse gas emissions. On the other hand, Singapore being a highly urbanized and energy-intensive nation (Figure 1a), aims at reducing its emissions by 7%–11% by 2020 using urban renewable energy applications [19]. The environmental quality in many large ASEAN cities has also considerably declined, which is alarming, and thus requires the development of appropriate energy strategies.

The leading contributors of greenhouse gas emissions in the ASEAN counties are the cement industry, palm oil, and chemical plants, petroleum refineries, power generating plants, and wood-based industries [20]. In general, almost 90% of the energy requirements of ASEAN economies are fulfilled by fossil fuels [21]. The electricity generation using these fossil fuels is one of the key processes which results in GHG emissions. Apart from urbanization, rapid industrialization over the past decades has also caused enormous demands and consumption of electricity which, being produced from fossil fuels, has caused an environmental imbalance. Energy consumption in the selected ASEAN countries of this study grew by 261 Million Tons of Oil Equivalent (MTO) throughout 1990–2013, and the majority of this consumption was attributed to Indonesia, Malaysia, Thailand, and Vietnam [22]. The leading contributors for CO_2 emissions among ASEAN countries include Malaysia and Singapore for their rapid economic growth. It has been documented that the air quality in the capital cities of ASEAN countries, such as Bangkok, Jakarta, Kuala Lumpur, and Manila has severally deteriorated and is very poor, which is unhealthy for human beings [23]. This poor air quality in the environment can be attributed to the booming industrialization in the ASEAN countries, causing irreversible environmental damage. It is quite evident that the major industrial processes utilize non-renewable fossil fuels as their primary energy source, releasing a high amount of greenhouse gases (carbon dioxide, methane, nitrous oxide, etc.) in the surrounding environment. Moreover, some industries, such as palm oil production plants, release highly hazardous and toxic effluents in the nearby water bodies, considerably affecting the water quality and its inhabitant's survival. It is, therefore, essential to investigate and develop innovative techniques which suggest harnessing renewable and environmentally friendly energy fuels on a large scale. This study employed panel cointegration and causal relationships among renewable energy production, CO_2 emissions, and economic growth for seven key ASEAN countries for the period of 1995–2016. The recently developed Fully Modified Ordinary Least Square (FMOLS) testing approach and Granger causality tests were used to achieve the objective of this study.

The following section of the paper further provides an insight into renewable energy production, CO_2 emissions, and economic relationship analysis under this study for the selected ASEAN countries of this study.

2. Review of the Literature

The fourth industrial revolution has transformed various aspects of the industry, such as super-computing, cellular phones, robot intelligence, self-driving cars, genetic editing, and much more happening around us at an exponential speed. It is apparent that all fields of knowledge will be further revolutionized by future technologies in the next 50 years. The energy sector, therefore, has also somewhat adjusted to these changes and requires essential efforts to transform and diversify energy sources to increase energy security and reliability while reducing GHG emissions. Conservation of energy is also one of the essential aspects that the economies need to undertake to cope with climate change. In particular, with respect to mitigation efforts to contain climate change, there has been increased attention on improving energy efficiency worldwide, which can effectively reduce the energy demand [24,25]. As such, with the conservation measures realized, the consumption of energy in the future is likely to lower at 1.3% compared to 2.2% per annum during 1995–2015 [26]. However, China and India are the leading, fast-growing, and emerging economies have now almost accounted for over half of the increased global energy demand [27].

More energy will also be required for the economic growth of developing countries. It is anticipated that electricity generation capacity will grow to 90% by 2040, and around 80% of this will be from developing countries [28]. Over the next 20 years, it is also expected that energy supply growth will be sufficiently contributed by the renewable source together with nuclear and hydroelectric power.

This anticipated situation is likely to help in offsetting GHG emissions. In 2016, the top CO_2 emitters were China, India, Japan, Russia, and the United States, whereas the European nations were the top contributors to renewable power generation. However, the pace of the transition to a lower-carbon economy is insignificant and uncertain without key initiatives by leading GHG emitting countries in the future. It is anticipated that the speeding up of the said transition will also significantly impact future economic growth.

It is, therefore, expected that by 2035, non-fossil fuels and renewables will form a greater share of the overall rising energy mix. It is also anticipated that natural gas will grow twice the rate of oil and coal in the future, but the demand could be slower if coal consumption is not prioritized appropriately [29].

Various studies [30,31] have been alarmed that GHGs and, in particular, CO_2 adversely impact human activities and thus, hamper the natural ecosystem globally. Climate change and energy problems have deeply threatened mankind and their sustainable existence. Globally, the intensity of primary energy declined continuously by 30% between 1990 and 2014. Nevertheless, the global economic growth was better, resulting in a steady net growth in energy demand of 56% between 1990 and 2014 with an annual growth of 1.9% [32].

In some recent studies, the relationship between renewable energy consumption and economic growth has been investigated. In studies which examined the energy consumption of EU countries throughout 1992–2010, [33,34], using the residual cointegration test, it was established that, on average, a 1% increase in energy consumption per capita increased per capita emissions by 0.56%. A group of studies in the literature has also prioritized the economic growth and environmental pollution nexus under the EKC framework. The use of the EKC hypothesis explains the relationships between natural resources, economic growth, and pollutant emissions. Kuznets [35] hypothesized that income inequality increases to a certain level and then falls with an increased income per capita. Various studies have focused on validating the EKC hypothesis (see for example [36–38]). The main focuses of these studies have been to validate the EKC hypothesis. However, there are other groups of studies in the literature which contradict the EKC hypothesis (see for example [39,40]). There are some other studies which validate the EKC hypothesis, too [33,41,42]. Based on the contradiction of the confirmation of the EKC hypothesis, this study contributes to validating the EKC hypothesis by examining broader literature in detail.

For a review of the extensive literature on the causal relationship between energy consumption and economic growth see [20,43,44]. A summary of studies which employed the empirical linkage of CO_2 emissions, economic growth, and the energy consumption is provided in (Table 1) with the econometric method utilized by each of these studies. A review of these studies reveals that there has been a variance in the usage of the specific econometric method. Regarding the findings, some studies suggested that CO_2 emissions had a negative impact on the GDP, whereas energy consumption increased the CO_2 emissions [45].

Table 1. Summary of empirical relationships among CO_2 emissions, energy consumption, and economic growth from 2009 to 2015.

Authors	Variables	Methodology	Countries	Findings
Saidi and Hammami [46]	CO_2 emissions, energy consumption, and economic growth	Panel data using Generalized Method of moments	58 countries	Energy consumption increased economic growth; CO_2 emissions had a negative impact on economic growth
Chandran and Tang [10]	Transport energy consumption, foreign direct investment, and CO_2 emissions	Cointegration and Granger causality	Five ASEAN countries	CO_2 emissions were co-integrated only in Malaysia, Indonesia, and Thailand; economic growth contributed to CO_2 emissions
Jalil and Feridun [32]	Growth, energy and financial development, and CO_2 emissions	Autoregressive Distributed Lag (ARDL)	China	Income, trade openness, and energy consumption were able to determine CO_2 emissions
Narayan and Narayan [47]	CO_2 emissions and economic growth	Panel data	Developing countries	Individual countries showed that CO_2 emissions had fallen over the long run
Sharma [48]	Trade openness, per capita GDP, and energy consumption	Dynamic panel modeling	69 countries	Trade openness, per capita GDP, and energy consumption had positive effects on CO_2 emissions
Jaunky [49]	CO_2 emissions and income	VECM	Rich countries	Unidirectional causality from the real per capita GDP to per capita CO_2 emissions
Salahuddin, Gow [50]	Economic growth, electricity consumption, CO_2 emissions, and financial development	Dynamic OLS, fully modified OLS and dynamic fixed effect model	Gulf countries	Electricity consumption and economic growth stimulated CO_2 emissions; there was no causal link between financial development and CO_2 emissions
Ozturk and Acaravci [51]	CO_2 emissions, energy consumption, and economic growth	ARDL Cointegration test	Europe	There was a positive long-run elasticity between CO_2 emissions and economic growth
Leit [52]	Economic growth, CO_2 emissions, renewable energy, and globalization	GMM, Granger causality and ECM	Portugal	CO_2 emissions and renewable energy were positively related to economic growth
Saboori and Sulaiman [53]	CO_2 emissions, energy consumption, and economic growth	ARDL and VECM	ASEAN	There was a nonlinear relationship between CO_2 emissions and economic growth; bi-directional Granger causality between energy consumption and CO_2 emissions
Sadorsky [54]	Renewable energy consumption, CO_2 emissions, and oil prices	Cointegration	G7 countries	GDP and CO_2 emissions were major drivers of renewable energy consumption

It was apparent from the review of the literature, wherein the complex relationships among CO_2 emissions, economic growth, energy consumption, and other factors had been studied, that the evidence is still inconclusive. Furthermore, the evidence of the pollution heaven hypothesis was also limited. However, limited research efforts were the only evidence about the promotion of renewable energy sources as alternatives to fossil fuels to reduce CO_2 emissions and achieve economic growth. Finally, most of these studies utilized the panel data analysis and lack usage of extensive evidence on long-run relationships among the investigated variables.

3. Methodology

In the context of exploring the relationships among renewable energy production, CO_2 emissions, and the associated economic growth, this study employed the FMOLS model, [55], DOLS regression analysis [56], and Granger causality analysis, among others, using renewable energy production, CO_2 emissions, and economic growth as variables in the empirical model.

3.1. Empirical Model

The empirical model developed in this study examined the relationships among renewable energy production (RW_t), CO_2 emissions (CO_{2t}), and economic growth (G_t) for seven ASEAN countries: Indonesia, Malaysia, Myanmar, the Philippines, Singapore, Thailand, and Vietnam. The methodology employed for the model development was similar to earlier studies [10,11,33]. However, the parameters of this study differed from those studies.

It is an essential and well-realized fact that an effective economic policy formulation requires a better understanding of various interconnections in the economy, including the nexus among renewable energy production (dependent variable), CO_2 emissions, and economic growth. The empirical model representing the long-run relationships among these interconnections as variables is given in Equation (1) as follows:

$$RW_{it} = \pi_{0t} + \pi_1 CO_{2it} + \pi_2 GDP_{it} + \varepsilon_{it}. \quad (1)$$

Here, RW_t was the renewable energy production at time t, and for the specific country it was represented by i; CO_{2it} was the carbon dioxide emission in kilotons at time t, and for the specific country it was represented by i; GDP_{it} was the real GDP per capita at time t and for the specific country it was represented as i; and ε_{it} was the residual at time t and for the specific country it was represented as i. It was assumed that these variables were normally distributed. The long-term beta coefficient for CO_2 emissions and economic growth were π_1 and π_2 respectively. Since the study was based on panel data, as such, the long-run panel cointegration specifications applied in previous studies [57,58] were employed. The error term ε_t was assumed to be identically distributed. Finally, the signs for π_1 and π_2 were expected to be negative.

3.2. Econometric Methodology

The determination of the long-run relationships among renewable energy production, CO_2 emissions, and economic growth was undertaken in this study with the essential steps described as follows. Firstly, the stationarity properties of the panel data set variables were examined using panel unit root tests. When the data was non-stationary, the panel cointegration technique was generally used to test the co-integrating relationships in the variable series. Once the cointegration of the variables was confirmed, the long-run elasticities were estimated using the Fully Modified OLS (also known as FMOLS) test. In the third and final steps, the short- and long-run dynamics of the series were examined using panel error correction models.

3.2.1. Panel Unit Root Tests

The panel unit root tests were performed to analyze the stationarity variable with the null hypothesis of a series having a unit root. It was imperative to detect the issue of spurious correlations.

The series was assumed using the intercept, constant, and trend. The equations for the series are provided in the following relationships in Equations (2)–(9) as follows:

Without constant and trend

$$\Delta Y_{it} = \delta Y_{it-1} + \mu_{it}. \tag{2}$$

With Constant

$$\Delta Y_{it} = \alpha + \delta Y_{it-1} + \mu_{it}. \tag{3}$$

With constant and trend

$$\Delta Y_{it} = \alpha + \beta T + \delta Y_{it-1} + \mu_{it}. \tag{4}$$

H_0: $= \delta = 0$ (Unit Root)

H_1: $= \delta \neq 0$

After the first differencing (Δ) of the series, the stationarity had to be achieved [59]. The most commonly used unit root tests, which are currently in practice, are Levin and Lin (LL), Im–Pesaran–Shin (IPS) and Maddala Wu (MW) [60]. Of the three popular unit root tests, LL has not been used widely in practice due to the unrealistic nature of the hypothesis. The model developed by Im, Pesaran [61] was used in this study as below:

$$y_{i,t} = \beta_i + \gamma_i y_{i,t-1} + \varepsilon_{i,t}. \tag{5}$$

For all $i = 1, \ldots, N$ and $t = 1, 2, 3, \ldots, T$

The study performed four different unit root tests; namely, the Phillips–Perron (PP), Augmented Dicker Fuller (ADF), Levin Liu Chu (LLC) and Im–Pesaran–Shin (IPS). The IPS and LLC were used to complement the widely used ADF and PP tests to arrive at the robust results. Further, in order to choose the optimal lag, the study used the Barlett Kernel Method following [62] for the estimation.

3.2.2. FMOLS Estimator

The Fully Modified Least Square (FMOLS) was developed by Phillips and Hansen [63] in order to administer an optimal co-integrating regression estimation. However, the study used the Pedroni [64] heterogeneous FMOLS estimator for the panel cointegration regression as it has the advantage of correcting endogeneity bias and serial correlation [44]. According to Hamit-Haggar [65], FMOLS is the most suitable technique for the panel, which includes heterogeneous cointegration.

Considering that a panel FMOLS estimator for the coefficient β of model 1 was:

$$\beta^*_{NT} - \beta = \left(\sum_{i=1}^{N} L_{22i}^{-2} \sum_{i=1}^{T} (x_{it} - \overline{x}_{it})^2 \right) \sum_{i=1}^{N} L_{11i}^{-1} L_{22i}^{-1} \left(\sum_{i=1}^{T} (x_{it} - \overline{x_i}) \mu^*_{it} - T\hat{\gamma}_i \right), \tag{6}$$

where,

$$\mu^*_{it} = \mu_{it} - \frac{\hat{L}_{21i}}{\hat{L}_{22i}} \Delta x_{it}, \; \hat{\gamma}_i = \hat{\Gamma}_{21i} \hat{\Omega}^0_{21i} - \frac{\hat{L}_{21i}}{\hat{L}_{22i}} (\Gamma_{22i} + \hat{\Omega}^0_{22i}),$$

and $\hat{L}_i$ was the lower triangulation of $\hat{\Omega}_i$.

The Dynamic OLS estimator had the same asymptotic distribution as that of the panel FMOLS estimation derived by Pedroni [66]. Both the DOLS and FMOLS estimations were performed as shown to confirm the consistency of the outcome.

3.2.3. Granger Causality Test

According to the cointegration theory, if the variables are co-integrated, then the short- and long-run equilibrium can be described using an error correction model (ECM). The panel residual cointegration test confirmed that renewable energy production, CO_2 emissions, and economic growth had co-integrating relationships. However, the co-integrating relationships were unable to provide information on the direction. Therefore, a panel-based error correction model with error correction representation was used to investigate the short- and long-run causal relationships. The Granger causality test was performed within the Vector Error Correction Model (VECM) framework [67]. The Granger causality test, together with the error correction term (ECT), is stated as follows:

$$\Delta RW_{it} = \varnothing_0 + \sum_{i=1}^{p} \varnothing_{1i}\Delta CO_{2t-1} + \sum_{i=1}^{p} \varnothing_{2i}\Delta GDP_{t-1} + \sum_{i=1}^{p} \varnothing_{3i}\Delta RW_{t-1} + \rho_1\epsilon_{t-1} + \mu_{1t} \quad (7)$$

$$\begin{aligned}\Delta CO_{2it} = \; & \varnothing_0 + \textstyle\sum_{i=1}^{p} \varnothing_{1i}\Delta CO_{2t-1} + \sum_{i=1}^{p} \varnothing_{2i}\Delta GDP_{t-1} + \\ & \textstyle\sum_{i=1}^{p} \varnothing_{3i}\Delta RW_{t-1} + \rho_2\epsilon_{t-1} + \mu_{2t},\end{aligned} \quad (8)$$

$$\begin{aligned}\Delta GDP_{it} = \; & \varnothing_0 + \textstyle\sum_{i=1}^{p} \varnothing_{1i}\Delta CO_{2t-1} + \sum_{i=1}^{p} \varnothing_{2i}\Delta GDP_{t-1} + \\ & \textstyle\sum_{i=1}^{p} \varnothing_{3i}\Delta RW_{t-1} + \rho_3\epsilon_{t-1} + \mu_{3t},\end{aligned} \quad (9)$$

where, Δ and ρ denote the first difference operator and lag structure. The residuals (μ_1, μ_2, μ_3, μ_4 and μ_5) were assumed to be serially independent with a zero mean and ϵ_{t-1} was the one period lagged error correction term.

4. Statistical Results

The study used the dataset over the period from 1995 to 2016 extracted from the World Development Indicators (WDI). A total of seven ASEAN countries were selected for the study.

Table 2 provides a summary of the statistics (mean, standard deviation, minimum, and maximum) associated with the CO_2 emissions per capita for an individual country with the panel set throughout 1995–2016. The mean value for the CO_2 emissions per capita was in the range of between 0.2362 in Myanmar and 9.8729 in Singapore. As regarding the GDP per capita, Myanmar showed the least GDP mean value of 710.8, whereas Singapore achieved the highest GDP per capita of 40,232.7. However, concerning renewable energy production, Myanmar exhibited surplus production with a mean value of 79.87%, whereas Singapore was the least renewable energy producer with only 0.51%.

Table 2. Descriptive statistics for CO_2 emissions, real GDP, and renewable energy production.

Countries	Mean	S.D.	Minimum	Maximum
Panel A: CO_2 emissions per capita				
Malaysia	6.713767	1.08802	4.763997	8.530658
Myanmar	0.236277	0.063911	0.16097	0.4166
Thailand	3.610162	0.590426	2.668098	4.62186
Vietnam	1.208286	0.539194	0.404057	2.205355
Indonesia	1.624109	0.401152	1.041246	2.55975
Philippines	0.922408	0.103728	0.770906	1.235657
Singapore	9.872934	2.743237	4.342606	15.39236
Panel A: Real GDP per capita				
Malaysia	2.989936	3.881041	−9.65575	7.445581
Myanmar	8.605125	2.989286	4.274814	12.68542
Thailand	2.748366	3.755799	−8.734046	7.047772
Vietnam	5.348793	1.105453	3.213679	7.758815
Indonesia	5.348793	1.105453	3.213679	7.758815
Philippines	2.904749	2.096575	−2.74496	5.903121
Singapore	3.004872	4.330412	−5.491059	13.21649
Panel A: Renewable energy production				
Malaysia	5.638253	1.340974	3.819042	9.029613
Myanmar	79.87185	6.997041	60.4532	86.11957
Thailand	21.88286	1.305616	20.02467	24.0143
Vietnam	47.06246	10.44879	34.7959	65.12578
Indonesia	42.38311	3.78953	38.06614	50.09815
Philippines	31.93819	2.784079	28.0034	38.942
Singapore	0.517375	0.088952	0.325119	0.715843

Note: S.D., CO_2, GDP stand for standard deviation, per capita carbon dioxide emissions, and per capita real GDP. Data period was 1995–2016 for the ASEAN countries.

It was further determined that the mean value of renewable energy production was high in low-income countries, such as Myanmar, followed by middle- and high-income countries. For both the energy variables (CO_2 emissions and renewable energy production), it can be conferred that the low-income countries were more volatile with the highest coefficients of variations.

This study investigated the dynamic relationships among renewable energy production, CO_2 emissions, and economic growth over the period of 1995–2016 from a panel of seven ASEAN countries. In this context, the EKC hypothesis was employed using panel cointegration estimation methods. In the first step, to tackle homogeneity issues, the study used the unit root test as suggested by [61] for a cointegration modeling process. Once it was confirmed that there was no homogeneity issue and the variables were in the order of interest, a test of the cointegration was examined to know whether the variables were co-integrated or not [58].

The panel unit root tests undertaken in this study are provided in Table 3. The validity of the EKC hypothesis for the ASEAN countries was examined using a bivariate framework. In [68] it was argued that testing for cointegration is crucial to determine the appropriateness of the model, together with verifying the causal relationships. The study employed the Augmented Dickey Fuller (ADF) unit root test to check the integration of each series. Table 3 reports the results of the ADF test at the level and first difference. At the 1% significance level, the study found that there was no stationarity issue for any of the variables. Thus, the study proceeded to examine the presence of cointegration among renewable energy production, CO_2 emissions, and economic growth. The results also indicated that the series for each country was not spurious and had a unit root. Thus, it can be concluded that the panel data variables were characterized as I (1) process.

Table 3. Panel unit root test results for ASEAN countries.

	Unit Root Methods				
Country	**Variables**	**PP**		**ADF**	
		Level	**First Difference**	**Level**	**First Difference**
Malaysia	Renewable	−3.004	−3.885 ***	−1.404	−3.815 ***
	CO_2 emissions	0.151	−5.004 ***	0.159	−3.644 ***
	GDP	−8.489 ***	−12.199 ***	−5.327 ***	−6.260 ***
Myanmar	Renewable	0.846	−2.801 **	1.500	−2.778 ***
	CO_2 emissions	−1.599	−8.669 ***	−1.802	−6.060 ***
	GDP	−1.476	−3.675 ***	−1.433	−3.719 ***
Thailand	Renewable	−0.731	−5.439 ***	−0.733	−5.405 ***
	CO_2 emissions	−1.509	−4.928 ***	−1.509	−4.925 ***
	GDP	−3.727 ***	−9.767 ***	−3.721 ***	−5.707 ***
Vietnam	Renewable	−1.829	−3.219 **	−1.932	−3.219 **
	CO_2 emissions	1.308	−4.078 ***	0.575	−6.373 ***
	GDP	−3.039 **	−3.863 ***	−2.961 *	−3.857 ***
Indonesia	Renewable	−3.245 **	−6.827 ***	−2.704 **	−6.827 ***
	CO_2 emissions	−1.222	−7.289 ***	−1.427	−4.845 ***
	GDP	−3.039 **	−3.863 ***	−2.961 *	−3.857 ***
Philippines	Renewable	−2.577	−6.491 ***	−2.500	−5.772 ***
	CO_2 emissions	1.461	−3.720 ***	1.143	−3.857 ***
	GDP	−3.399 **	−14.868 ***	−3.399 **	−5.788 ***
Singapore	Renewable	−1.206	−5.487 ***	−1.308	−5.487 ***
	CO_2 emissions	−1.717	−5.414 ***	−1.792	−4.871 ***
	GDP	−7.688 ***	−13.506 ***	−4.692 ***	−6.483 ***

Note: *** denotes the significance level at 1%. **Δ** stands for first difference, ADF—Augmented Dicker Fuller.

The residual cointegration results are provided in Table 4 as suggested by [64,69]. The statistical results found that the majority of the tests were significant and, therefore, the null hypothesis of having no cointegration was not established. Thus, the variables were co-integrated at significant levels.

Table 4. Residual cointegration test.

Statistics	Within Dimensions		Statistics	Between Dimensions	
	Value	*p*-Value		Value	*p*-Value
Panel v-Statistic	1.264 *	0.006			
Panel rho-Statistic	−1.421	0.922	Group rho-Statistic	2.534	0.994
Panel PP-Statistic	−4.961 *	0.000	Group PP-Statistic	−1.985 **	0.023
Panel ADF-Statistic	−5.060 *	0.000	Group ADF-Statistic	−2.323 **	0.010

Note: the selection of Lag was based on the Akaike Information Criterion (AIC). The null hypothesis was that the variables were not co-integrated. * and ** denote the significance levels at 1% and 5%, respectively.

Finally, the Pedroni [64] FMOLS estimator was selected to determine the long-run relationships among the constructs. The long-run results of the FMOLS estimation for the model are provided in (Table 5) is considered in this study.

The long-run elasticities were interpreted for all the variables which were expressed in natural logarithms. The study used the FMOLS as a robustness test. The first and second differences of the variables were all stationary as provided in Table 3. The quadratic term of per capita renewable energy production was used to record a possible country-specific non-linear relationship between per capita CO_2 emissions and per-capita GDP. The main result confirmed the significant inverse relationship between CO_2 emissions and real GDP for countries like Malaysia and Singapore; whereas, there was a positive relationship between CO_2 emissions and real GDP for Myanmar. This result indicated that in the long run, the EKC hypothesis was justified. An increase of 1% in CO_2 emissions would lead to a decrease in real GDP per capita by 0.08% for Malaysia and 0.11% for Singapore. The results of this study were duly in line with the other literature findings (see for example [1,70]). The results further indicated that the coefficients of renewable energy production were statistically significant and negative, indicating a nonlinear relationship with per capita CO_2 emissions and per capita GDP in the sampled ASEAN countries considered in this study.

Table 5. Long-run elasticity results (dependent variable: renewable energy production).

FMOLS Estimation			DOLS Estimation		
Country	GDP	CO_2 Emissions	Country	GDP	CO_2 Emissions
Malaysia	1.653 [b] (2.050)	−0.593 [b] (−0.912)	Malaysia	2.006 [b] (2.684)	−1.359 [b] (−2.232)
Myanmar	0.092 [b] (2.845)	−0.126 [a] (−1.845)	Myanmar	0.064 [b] (4.678)	−0.027 [a] (−0.723)
Thailand	0.549 (−2.001)	−0.393 (−2.505)	Thailand	0.337 (−1.183)	−0.699 (−2.504)
Vietnam	0.193 [a] (−0.37)	−0.579 (−2.990)	Vietnam	1.049 [a] (−1.639)	−1.164 (−5.833)
Indonesia	−0.251 [a] (−1.254)	−0.217 (−2.515)	Indonesia	−0.485 [a] (−2.290)	−0.092 (−0.394)
Philippines	−0.392 (−2.710)	−0.042 (−0.703)	Philippines	0.592 (−2.031)	−0.288 (−3.332)
Singapore	0.18 (0.233)	−0.163 [b] (−0.904)	Singapore	−2.856 (−2.164)	−0.619 [b] (−2.638)

Notes: [a], [b], and [c] indicate the significance levels at 1%, 5%, and 10%, respectively. Intercepts and linear trends are included in the regressions. Barlett Kernel with a fixed bandwidth of 6 was used following [71]. Figures in the parentheses are the t-statistics. FMOLS—fully modified ordinary least square; DOLS—dynamic ordinary least square; GDP—gross domestic product.

The inverse U-shaped relationships between per capita renewable energy production and per capita GDP were detected from the sign of the parameters. However, the long-run estimations provided in Table 5 did not provide the direction of causality among the variables. The results also indicated that there were negative relationships between per capita renewable energy production and per capita CO_2 emissions for Indonesia, Malaysia, Myanmar, and Vietnam. These results were also supported by the previous literature [46,72]. In the globalization era, with the rapid increase in demand for energy and cleanup of the environment, there is a need for renewable energy sources as an alternative [47].

Next, by employing the FMOLS estimator, proposed by Pedroni [64], and the single equation DOLS estimator, proposed by Phillips and Loretan [73], the U-shaped relationship among renewable energy production, CO_2 emissions, and economic growth was supported and realized by the EKC hypothesis. The FMOLS and DOLS results provided in Table 5 show the parameter estimation of the model for interpreting long-run elasticities. Hence, it can be stated that renewable energy production decreased with economic growth, stabilized, and then, gradually increased. The country-specific income-energy elasticities using FMOLS and DOLS confirmed the statistical significance at 5% for the long-run relationships among the variables. Furthermore, the country-specific long-run analysis demonstrated U-shaped curves for three countries, namely, Malaysia, Thailand, and the Philippines. On the other hand, there was no significant relationship found between renewable energy production and economic growth for the four remaining ASEAN countries of this study.

This research also analyzed the results using the Granger causality panel error correction model to identify the direction of the long-run and short-run causalities in the renewable energy production, CO_2 emissions, and economic growth nexus and the interactions among them. The error correction model estimation result indicated the presence of unidirectional causality between renewable energy production and CO_2 emissions. However, in the applied research, it is obviously of interest to know the response of one variable to an impulse in another variable [74]. From the Granger causality, it was noticed that there were high but negative correlations among renewable energy production, CO_2 emissions, and economic growth. This indicated that innovation in renewable energy production would possess high effects on CO_2 emissions and economic growth. As such, increasing renewable energy production would gradually decrease the CO_2 emissions together with slowly contributing to economic growth in the long run. In the initial stages of renewable energy production, the minor pressure on the GDP may be explained by the additional production cost imposed by the renewable energy sources and their plant setups.

The study explored the causal relationship between the constructs using error correction Granger Causality test for short run Granger causality. The results for Granger causality model (see Table 6) can be summarized as:

(i) Real GDP per capita does not cause carbon emissions per capita for Myanmar, Thailand, Vietnam, Indonesia, the Philippines, and Singapore except for Malaysia.

(ii) It was found that renewable energy production Granger causes real GDP per capita for Myanmar, Indonesia, and the Philippines only; whereas it was also found that renewable energy production Granger causes carbon emissions per capita for Myanmar and Vietnam only.

(iii) There is no causal evidence between carbon emissions per capita and real GDP per capita for all the ASEAN countries; whereas only in Thailand, there was a causal relationship between carbon emissions and renewable energy production.

Table 6. Results of panel Granger causality test identifying long-run relationships between CO_2 emissions, real GDP, and renewable energy production for ASEAN countries.

Short-Run Granger Causality									
Country					**Country**				
Malaysia	Variable	ΔGDP	ΔRNW	ΔCO_2	Indonesia	Variable	ΔGDP	ΔRNW	ΔCO_2
	ΔGDP	1	0.455	4.229 **		ΔGDP	1	10.30 **	1.177
	ΔRNW	0.252	1	0.937		ΔRNW	4.367 *	1	0.902
	ΔCO_2	0.358	1.083	1		ΔCO_2	0.731	0.849	1
Myanmar	Variable	ΔGDP	ΔRNW	ΔCO_2	Philippines	Variable	ΔGDP	ΔRNW	ΔCO_2
	ΔGDP	1	0.635	0.417		ΔGDP	1	1.220	2.115
	ΔRNW	22.83 **	1	3.055 *		ΔRNW	4.590 *	1	2.627
	ΔCO_2	0.788	2.092	1		ΔCO_2	1.213	1.064	1
Thailand	Variable	ΔGDP	ΔRNW	ΔCO_2	Singapore	Variable	ΔGDP	ΔRNW	ΔCO_2
	ΔGDP	1	2.537	0.583		ΔGDP	1	1.663	1.745
	ΔRNW	1.102	1	0.691		ΔRNW	0.124	1	0.285
	ΔCO_2	1.921	21.59 **	1		ΔCO_2	2.988	0.453	1
Vietnam	Variable	ΔGDP	ΔRNW	ΔCO_2					
	ΔGDP	1	20.61 **	1.068					
	ΔRNW	1.359	1	3.319 *					
	ΔCO_2	1.190	0.121	1					

Note: The t-statistics are provided in the parentheses, respectively. *, **, and *** denote the significance levels at 1%, 5%, and 10%, respectively. Δ denotes change; GDP stands for the gross domestic product; RNW stands for renewable; CO_2 stands for carbon emissions.

Therefore, from the overall results, it can be concluded that (1) energy conservation policies have no adverse effect on climate change and GDP growth of the ASEAN countries and (2) controlling carbon emissions has no adverse effect on real GDP per capita of the ASEAN countries. Figure 2 summarizes the interactions among renewable energy production, CO_2 emissions, and economic growth developed from the findings provided in Table 6.

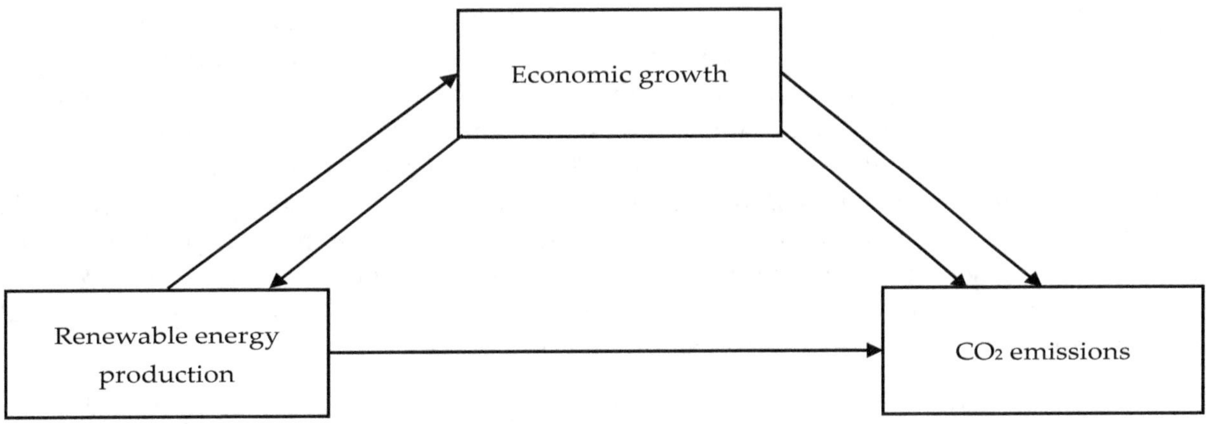

Figure 2. The long-run relationship between renewable energy production, CO_2, and economic growth for ASEAN countries.

The results corroborated the three-way link among renewable energy production, CO_2 emissions, and economic growth. Our results suggest that an increase in renewable energy usage and production will diminish CO_2 emissions, leading to clean economic growth as per the Granger causality test. Our result rejects the neoclassical assumption that energy is neutral to economic growth. Thus, this study considers renewable energy to be the most critical determinant in the long run to economic growth and vice versa. It is very crucial to take into account the adverse effect of CO_2 energy consumption on economic growth in establishing energy conservation policies.

5. Conclusions and Policy Implications

This study examined the cointegration and causal relationships among renewable energy production, CO_2 emissions, and economic growth for the ASEAN countries, namely, Malaysia, Myanmar, Thailand, Vietnam, Indonesia, the Philippines, and Singapore for the period 1995–2016. The FMOLS and DOLS analyses were undertaken to examine whether, in the long run, renewable energy production, CO_2 emissions, and economic growth were co-integrated or not. The FMOLS results suggest that renewable energy production was positive and significant in relation to real GDP at the 0.001 level for Malaysia (t value of 2.050) and for Myanmar (t value of 2.845), whereas there was a negative and significant relationship between renewable energy production and real GDP for Thailand (t value of −2.505) and the Philippines (t value of −2.710). There was no relationship between renewable energy production and real GDP for Singapore, Indonesia, and Vietnam due to limited renewable energy options. The results of the panel FMOLS and DOLS also provided significant shreds of evidence to support U-Shaped relationships among the long-run renewable energy production, CO_2 emissions, and economic growth contrary to the EKC hypothesis. The empirical results from the granger causality test (Table 6) also revealed the substantial negative nexus between renewable energy production and CO_2 emissions.

Estimation of a panel error correction model (ECM) was performed in order to define the long-run and short-run causalities. The results from the ECM indicated the presence of short-run unidirectional causality running from renewable energy production to economic growth. Whereas, in the long-run causalities, renewable energy production and CO_2 emissions were inversely related, thus confirming that an increase in the usage of renewable energy sources as an alternative to fossil fuels will offset CO_2 emissions. Besides, carbon footprints can be reduced, leading to clean energy development and economic growth. The results, therefore, implied that the efficient use of renewable energy would reduce the rate of GHG emissions in the long run.

The precedent literature, as presented in the earlier sections of the study, have generally found a positive relationship between energy consumption and CO_2 emissions; however, there is a lack of studies which performed the relationship analysis of renewable energy production and CO_2 emissions. As a novelty, this research gap was addressed and found the overall relationship between renewable energy production and CO_2 emissions, which came to be negative. The relationship between renewable energy production and CO_2 emissions for Malaysia, Thailand, Vietnam, Indonesia, and Singapore were found to be inverse in the short run. This showed that an increase in renewable energy production would lead to a reduction in CO_2 emissions. Even though, renewable energy is expensive as compared to fossil fuels or other non-renewable sources, the initiatives to produce and consume more renewable energy should be focused on in developed and developing countries, like in Singapore and Malaysia, to contain climate change by reducing environmental pollution.

The rapid growth in the renewable energy production can be driven by many factors, including cost reduction due of renewable energy technologies, dedicated policy initiatives, better access to energy resources, environmental concerns, and growing energy demand of emerging economies. Effective energy policies, therefore, can play a significant role in providing energy consumption direction and in increasing public awareness. The pedagogical innovation, therefore, suggests accelerating the renewable energy usage campaigns in the high carbon emission countries like Malaysia and Singapore for a more environmentally-friendly and sustainable future. The renewable energy sources like hydropower, solar power, wind energy, biomass, and geothermal energy would significantly condense the pollution and reduce the reliance on fossil fuels, which are the leading causes of the CO_2 emissions.

As a practical implication, energy saving and conservation policies can be adopted by the ASEAN countries quickly in order to limit environmental pollution. It is crucial that the ASEAN countries, especially Malaysia and Singapore, implement policies and strategies that ensure continuous economic growth without forsaking the environment. This study recommends that in order to decrease the reliance on fossil fuels, that have impacted the environment adversely, future research should consider and focus on the principles of the circular economy and clean energy development mechanisms integrated with renewable energy technologies.

Author Contributions: All authors contributed to this study. M.W.A.K. and S.K.P. formulated the study design. M.W.A.K., S.K.P., M.I.S. and N.H.M. conceived and designed the research methodology. M.W.A.K., S.K.P., K.S.N.A., M.I.S. and E.S.A. collected and analyzed the data. M.W.A.K., S.K.P. and N.H.M. finalized the paper.

Acknowledgments: The authors highly acknowledge the Ministry of Higher Education Malaysia for the Fundamental Research Grant Scheme (FRGS) under grant No. RDU170134 and Universiti Malaysia Pahang for their financial and technical support for completing this research work.

References

1. Heidari, H.; Katircioglu, S.T.; Saeidpour, L. Economic growth, CO_2 emissions, and energy consumption in the five ASEAN countries. *Int. J. Electr. Power Energy Syst.* **2015**, *64*, 785–791. [CrossRef]
2. Zhang, X.-P.; Cheng, X.-M. Energy consumption, carbon emissions, and economic growth in China. *Ecol. Econ.* **2009**, *68*, 2706–2712. [CrossRef]
3. Mengal, A.; Mirjat, N.H.; Das Walasai, G.; Khatri, S.A.; Harijan, K.; Uqaili, M.A. Modeling of Future Electricity Generation and Emissions Assessment for Pakistan. *Processes* **2019**, *7*, 212. [CrossRef]
4. Lean, H.H.; Smyth, R. CO_2 emissions, electricity consumption and output in ASEAN. *Appl. Energy* **2010**, *87*, 1858–1864. [CrossRef]
5. Chang, C.-C. A multivariate causality test of carbon dioxide emissions, energy consumption and economic growth in China. *Appl. Energy* **2010**, *87*, 3533–3537. [CrossRef]
6. Dhakal, S. Urban energy use and carbon emissions from cities in China and policy implications. *Energy Policy* **2009**, *37*, 4208–4219. [CrossRef]
7. Fei, L.; Dong, S.; Xue, L.; Liang, Q.; Yang, W. Energy consumption-economic growth relationship and carbon dioxide emissions in China. *Energy Policy* **2011**, *39*, 568–574. [CrossRef]
8. Hussain, N. *Development of Energy Modeling and Decision Support Framework for Sustainable Electricity System of Pakistan*; Mehran University of Eng. & Technology: Jamshoro, Pakistan, 2019.
9. Panayotou, T. Demystifying the environmental Kuznets curve: Turning a black box into a policy tool. *Environ. Dev. Econ.* **1997**, *2*, 465–484. [CrossRef]
10. Chandran, V.; Tang, C.F. The impacts of transport energy consumption, foreign direct investment and income on CO_2 emissions in ASEAN-5 economies. *Renew. Sustain. Energy Rev.* **2013**, *24*, 445–453. [CrossRef]
11. Saboori, B.; Sulaiman, J. CO_2 emissions, energy consumption and economic growth in Association of Southeast Asian Nations (ASEAN) countries: A cointegration approach. *Energy* **2013**, *55*, 813–822. [CrossRef]
12. Omri, A.; Daly, S.; Rault, C.; Chaibi, A. Financial development, environmental quality, trade and economic growth: What causes what in MENA countries. *Energy Econ.* **2015**, *48*, 242–252. [CrossRef]
13. Vilaysouk, X.; Schandl, H.; Murakami, S. Improving the knowledge base on material flow analysis for Asian developing countries: A case study of Lao PDR. *Resour. Conserv. Recycl.* **2017**, *127*, 179–189. [CrossRef]
14. IMF. Boiling Point. Finance & Development 2018. Available online: https://www.imf.org/external/pubs/ft/fandd/2018/09/southeast-asia-climate-change-and-greenhouse-gas-emissions-prakash.htm (accessed on 15 July 2019).
15. Chan, S.K.-L. Dams on a Myanmar—Thai transboundary river: Unequal hydropower exchange model in critical hydropolitical perspective. *Int. J. Dev. Issues* **2017**, *16*, 147–160. [CrossRef]
16. Amara, T. Where Will Myanmar's Energy Come From? Available online: https://thediplomat.com/2017/08/where-will-myanmars-energy-come-from(accessed on 19 January 2019).
17. Worldbank. World Development Indicators 2016. 2016. Available online: http://documents.worldbank.org/curated/en/805371467990952829/World-development-indicators-2016 (accessed on 19 January 2019).
18. Chang, T.; Fang, W.; Wen, L.-F. Energy consumption, employment, output, and temporal causality: Evidence from Taiwan based on cointegration and error-correction modelling techniques. *Appl. Econ.* **2001**, *33*, 1045–1056. [CrossRef]

19. Karthikeya, B.; Negi, P.S.; Srikanth, N. Wind resource assessment for urban renewable energy application in Singapore. *Renew. Energy* **2016**, *87*, 403–414. [CrossRef]
20. Daniel, G.R.; Wang, C.; Berthelsen, D. Early school-based parent involvement, children's self-regulated learning and academic achievement: An Australian longitudinal study. *Early Child. Res. Q.* **2016**, *36*, 168–177. [CrossRef]
21. Karki, S.K.; Mann, M.D.; Salehfar, H. Energy and environment in the ASEAN: Challenges and opportunities. *Energy Policy* **2005**, *33*, 499–509. [CrossRef]
22. Schernikau, L. *Economics of the International Coal Trade: Why Coal Continues to Power the World*; Springer: Berlin/Heidelberg, Germany, 2017.
23. Lee, H.-H.; Bar-Or, R.Z.; Wang, C. Biomass burning aerosols and the low-visibility events in Southeast Asia. *Atmos. Chem. Phys. Discuss.* **2017**, *17*, 965–980. [CrossRef]
24. Riahi, K.; Van Vuuren, D.P.; Kriegler, E.; Edmonds, J.; O'Neill, B.C.; Fujimori, S.; Bauer, N.; Calvin, K.; Dellink, R.; Fricko, O.; et al. The Shared Socioeconomic Pathways and their energy, land use, and greenhouse gas emissions implications: An overview. *Glob. Environ. Chang.* **2017**, *42*, 153–168. [CrossRef]
25. Mirjat, N.H.; Uqaili, M.A.; Harijan, K.; Das Walasai, G.; Mondal, M.A.H.; Sahin, H. Long-term electricity demand forecast and supply side scenarios for Pakistan (2015–2050): A LEAP model application for policy analysis. *Energy* **2018**, *165*, 512–526. [CrossRef]
26. Berardi, U. A cross-country comparison of the building energy consumptions and their trends. *Resour. Conserv. Recycl.* **2017**, *123*, 230–241. [CrossRef]
27. Nejat, P.; Jomehzadeh, F.; Taheri, M.M.; Gohari, M.; Majid, M.Z.A. A global review of energy consumption, CO_2 emissions and policy in the residential sector (with an overview of the top ten CO_2 emitting countries). *Renew. Sustain. Energy Rev.* **2015**, *43*, 843–862. [CrossRef]
28. Soytas, U.; Sari, R.; Ewing, B.T. Energy consumption, income, and carbon emissions in the United States. *Ecol. Econ.* **2007**, *62*, 482–489. [CrossRef]
29. Ang, J.B. Economic development, pollutant emissions and energy consumption in Malaysia. *J. Policy Model.* **2008**, *30*, 271–278. [CrossRef]
30. Ozcan, B.; Tzeremes, P.G.; Tzeremes, N.G. Energy consumption, economic growth and environmental degradation in OECD countries. *Econ. Model.* **2019**. [CrossRef]
31. Acheampong, A.O. Economic growth, CO_2 emissions and energy consumption: What causes what and where? *Energy Econ.* **2018**, *74*, 677–692. [CrossRef]
32. Jalil, A.; Feridun, M. The impact of growth, energy and financial development on the environment in China: A cointegration analysis. *Energy Econ.* **2011**, *33*, 284–291. [CrossRef]
33. Kasman, A.; Duman, Y.S. CO_2 emissions, economic growth, energy consumption, trade and urbanization in new EU member and candidate countries: A panel data analysis. *Econ. Model.* **2015**, *44*, 97–103. [CrossRef]
34. Dogan, E.; Seker, F. Determinants of CO_2 emissions in the European Union: The role of renewable and non-renewable energy. *Renew. Energy* **2016**, *94*, 429–439. [CrossRef]
35. Balsalobre-Lorente, D.; Shahbaz, M.; Roubaud, D.; Farhani, S. How economic growth, renewable electricity and natural resources contribute to CO_2 emissions? *Energy Policy* **2018**, *113*, 356–367. [CrossRef]
36. Begum, R.A.; Sohag, K.; Abdullah, S.M.S.; Jaafar, M. CO_2 emissions, energy consumption, economic and population growth in Malaysia. *Renew. Sustain. Energy Rev.* **2015**, *41*, 594–601. [CrossRef]
37. Adu, D.T.; Denkyirah, E.K. Economic growth and environmental pollution in West Africa: Testing the Environmental Kuznets Curve hypothesis. *Kasetsart J. Soc. Sci.* **2018**. [CrossRef]
38. Rasli, A.M.; Qureshi, M.I.; Isah-Chikaji, A.; Zaman, K.; Ahmad, M. New toxics, race to the bottom and revised environmental Kuznets curve: The case of local and global pollutants. *Renew. Sustain. Energy Rev.* **2018**, *81*, 3120–3130. [CrossRef]
39. Ozatac, N.; Gokmenoglu, K.K.; Taspinar, N. Testing the EKC hypothesis by considering trade openness, urbanization, and financial development: The case of Turkey. *Environ. Sci. Pollut. Res.* **2017**, *24*, 16690–16701. [CrossRef] [PubMed]
40. Twerefou, D.K.; Adusah-Poku, F.; Bekoe, W. An empirical examination of the Environmental Kuznets Curve hypothesis for carbon dioxide emissions in Ghana: An ARDL approach. *Environ. Socio-Econ. Stud.* **2016**, *4*, 1–12. [CrossRef]

41. Alam, M.M.; Murad, M.W.; Noman, A.H.M.; Ozturk, I. Relationships among carbon emissions, economic growth, energy consumption and population growth: Testing Environmental Kuznets Curve hypothesis for Brazil, China, India and Indonesia. *Ecol. Indic.* **2016**, *70*, 466–479. [CrossRef]
42. Al-Mulali, U. The impact of biofuel energy consumption on GDP growth, CO_2 emission, agricultural crop prices, and agricultural production. *Int. J. Green Energy* **2015**, *12*, 1100–1106. [CrossRef]
43. Omri, A. CO_2 emissions, energy consumption and economic growth nexus in MENA countries: Evidence from simultaneous equations models. *Energy Econ.* **2013**, *40*, 657–664. [CrossRef]
44. Özcan, B. The nexus between carbon emissions, energy consumption and economic growth in Middle East countries: A panel data analysis. *Energy Policy* **2013**, *62*, 1138–1147. [CrossRef]
45. Mi, Z.-F.; Pan, S.-Y.; Yu, H.; Wei, Y.-M. Potential impacts of industrial structure on energy consumption and CO_2 emission: A case study of Beijing. *J. Clean. Prod.* **2015**, *103*, 455–462. [CrossRef]
46. Saidi, K.; Hammami, S. The impact of CO_2 emissions and economic growth on energy consumption in 58 countries. *Energy Rep.* **2015**, *1*, 62–70. [CrossRef]
47. Narayan, P.K.; Narayan, S. Carbon dioxide emissions and economic growth: Panel data evidence from developing countries. *Energy Policy* **2010**, *38*, 661–666. [CrossRef]
48. Sharma, S.S. Determinants of carbon dioxide emissions: Empirical evidence from 69 countries. *Appl. Energy* **2011**, *88*, 376–382. [CrossRef]
49. Jaunky, V.C. The CO_2 emissions-income nexus: Evidence from rich countries. *Energy Policy* **2011**, *39*, 1228–1240. [CrossRef]
50. Salahuddin, M.; Gow, J.; Ozturk, I. Is the long-run relationship between economic growth, electricity consumption, carbon dioxide emissions and financial development in Gulf Cooperation Council Countries robust? *Renew. Sustain. Energy Rev.* **2015**, *51*, 317–326. [CrossRef]
51. Ozturk, I.; Acaravci, A. CO_2 emissions, energy consumption and economic growth in Turkey. *Renew. Sustain. Energy Rev.* **2010**, *14*, 3220–3225. [CrossRef]
52. Leit, N.C. Economic growth, carbon dioxide emissions, renewable energy and globalization. *Int. J. Energy Econ. Policy* **2014**, *4*, 391.
53. Saboori, B.; Sulaiman, J. Environmental degradation, economic growth and energy consumption: Evidence of the environmental Kuznets curve in Malaysia. *Energy Policy* **2013**, *60*, 892–905. [CrossRef]
54. Sadorsky, P. Renewable energy consumption, CO_2 emissions and oil prices in the G7 countries. *Energy Econ.* **2009**, *31*, 456–462. [CrossRef]
55. Long, X.; Naminse, E.Y.; Du, J.; Zhuang, J. Nonrenewable energy, renewable energy, carbon dioxide emissions and economic growth in China from 1952 to 2012. *Renew. Sustain. Energy Rev.* **2015**, *52*, 680–688. [CrossRef]
56. Sebri, M.; Ben-Salha, O. On the causal dynamics between economic growth, renewable energy consumption, CO_2 emissions and trade openness: Fresh evidence from BRICS countries. *Renew. Sustain. Energy Rev.* **2014**, *39*, 14–23. [CrossRef]
57. Arellano, M.; Bond, S. Some Tests of Specification for Panel Data: Monte Carlo Evidence and an Application to Employment Equations. *Rev. Econ. Stud.* **1991**, *58*, 277. [CrossRef]
58. Panigrahi, S.K. Economic Value Added and Traditional Accounting Measures for Shareholder's Wealth Creation. *Asian J. Account. Gov.* **2017**, *8*, 125–136. [CrossRef]
59. Perron, P. The Great Crash, the Oil Price Shock, and the Unit Root Hypothesis. *Econometrica* **1989**, *57*, 1361. [CrossRef]
60. Hoang, N.T.; McNown, R.F. Panel data unit roots tests using various estimation methods. *Univ. Colo. Bull.* **2006**, *6*, 33–66.
61. Im, K.S.; Pesaran, M.; Shin, Y. Testing for unit roots in heterogeneous panels. *J. Econ.* **2003**, *115*, 53–74. [CrossRef]
62. Newey, W.K.; West, K.D. *A Simple, Positive Semi-Definite, Heteroskedasticity and Autocorrelationconsistent Covariance Matrix*; National Bureau of Economic Research: Cambridge, MA, USA, 1986.
63. Phillips, P.C.B.; Hansen, B.E. Statistical Inference in Instrumental Variables Regression with I(1) Processes. *Rev. Econ. Stud.* **1990**, *57*, 99. [CrossRef]
64. Pedroni, P. Fully modified OLS for heterogeneous cointegrated panels. In *Nonstationary Panels, Panel Cointegration, and Dynamic Panels*; Emerald Group Publishing Limited: Bingley, UK, 2001; pp. 93–130.

65. Hamit-Haggar, M. Greenhouse gas emissions, energy consumption and economic growth: A panel cointegration analysis from Canadian industrial sector perspective. *Energy Econ.* **2012**, *34*, 358–364. [CrossRef]
66. Pedroni, P. *Fully Modified OLS for Heterogeneous Cointegrated Panels and the Case of Purchasing Power Parity*; Department of Economics, Indiana University: Bloomington, IN, USA, 1996.
67. Granger, C. Some recent development in a concept of causality. *J. Econ.* **1988**, *39*, 199–211. [CrossRef]
68. Perman, R. Cointegration: An Introduction to the Literature. *J. Econ. Stud.* **1991**, *18*, 3–30. [CrossRef]
69. Pedroni, P. Panel cointegration: asymptotic and finite sample properties of pooled time series tests with an application to the ppp hypothesis. *Econ. Theory* **2004**, *20*, 597–625. [CrossRef]
70. Arouri, M.E.H.; Youssef, A.B.; M'henni, H.; Rault, C. Energy consumption, economic growth and CO_2 emissions in Middle East and North African countries. *Energy Policy* **2012**, *45*, 342–349. [CrossRef]
71. Kao, C.; Chiang, M.-H.; Chen, B. International R&D Spillovers: An Application of Estimation and Inference in Panel Cointegration. *Oxf. Bull. Econ. Stat.* **1999**, *61*, 691–709.
72. Zhang, B.; Wang, Z.; Wang, B. Energy production, economic growth and CO_2 emission: Evidence from Pakistan. *Nat. Hazards* **2018**, *90*, 1–24.
73. Phillips, P.C.B.; Loretan, M. Estimating Long-Run Economic Equilibria. *Rev. Econ. Stud.* **1991**, *58*, 407. [CrossRef]
74. Rossi, E. *Impulse Response Functions*; Notes for a Lecture on Econometrics; University of Pavia: Pavia, Italy, 2004.

Permissions

All chapters in this book were first published in MDPI; hereby published with permission under the Creative Commons Attribution License or equivalent. Every chapter published in this book has been scrutinized by our experts. Their significance has been extensively debated. The topics covered herein carry significant findings which will fuel the growth of the discipline. They may even be implemented as practical applications or may be referred to as a beginning point for another development.

The contributors of this book come from diverse backgrounds, making this book a truly international effort. This book will bring forth new frontiers with its revolutionizing research information and detailed analysis of the nascent developments around the world.

We would like to thank all the contributing authors for lending their expertise to make the book truly unique. They have played a crucial role in the development of this book. Without their invaluable contributions this book wouldn't have been possible. They have made vital efforts to compile up to date information on the varied aspects of this subject to make this book a valuable addition to the collection of many professionals and students.

This book was conceptualized with the vision of imparting up-to-date information and advanced data in this field. To ensure the same, a matchless editorial board was set up. Every individual on the board went through rigorous rounds of assessment to prove their worth. After which they invested a large part of their time researching and compiling the most relevant data for our readers.

The editorial board has been involved in producing this book since its inception. They have spent rigorous hours researching and exploring the diverse topics which have resulted in the successful publishing of this book. They have passed on their knowledge of decades through this book. To expedite this challenging task, the publisher supported the team at every step. A small team of assistant editors was also appointed to further simplify the editing procedure and attain best results for the readers.

Apart from the editorial board, the designing team has also invested a significant amount of their time in understanding the subject and creating the most relevant covers. They scrutinized every image to scout for the most suitable representation of the subject and create an appropriate cover for the book.

The publishing team has been an ardent support to the editorial, designing and production team. Their endless efforts to recruit the best for this project, has resulted in the accomplishment of this book. They are a veteran in the field of academics and their pool of knowledge is as vast as their experience in printing. Their expertise and guidance has proved useful at every step. Their uncompromising quality standards have made this book an exceptional effort. Their encouragement from time to time has been an inspiration for everyone.

The publisher and the editorial board hope that this book will prove to be a valuable piece of knowledge for researchers, students, practitioners and scholars across the globe.

List of Contributors

Ana Cristina Ferreira, José Carlos Teixeira
MEtRICs Research Centre, University of Minho, 4804-533 Guimarães, Portugal

Angela Silva
Escola Superior de Ciências Empresariais, Instituto Politécnico de Viana do Castelo, 4930-600 Valença, Portugal

Senhorinha Teixeira
ALGORITMI Research Centre, University of Minho, 4804-533 Guimarães, Portugal

Rusong Han, Yinghua Jiang and Jing Wang
Department of Chemical Engineering, Xi'an Jiaotong University, Xi'an 710049, China

Lixia Kang
Department of Chemical Engineering, Xi'an Jiaotong University, Xi'an 710049, China
Shaanxi Key Laboratory of Energy Chemical Process Intensification, Xi'an 710049, China

Yongzhong Liu
Department of Chemical Engineering, Xi'an Jiaotong University, Xi'an 710049, China
Shaanxi Key Laboratory of Energy Chemical Process Intensification, Xi'an 710049, China
Key Laboratory of Thermo-Fluid Science and Engineering, Ministry of Education, Xi'an 710049, China

Hao Chang, Wei Li and Ling Zhou
Research Center of Fluid Machinery Engineering and Technology, Jiangsu University, Zhenjiang 212013, China

Ramesh K. Agarwal
Department of Mechanical Engineering and Materials Science, Washington University in St. Louis, St. Louis, MO 63130, USA

Weidong Shi
School of Mechanical Engineering, Nantong University, Nantong 226019, China

Xiang He and Pengfei Chen
School of Energy and Mining Engineering, China University of Mining and Technology, Beijing 100083, China

Ying Zhao
Institute of Engineering Thermophysics, Chinese Academy of Sciences, Beijing 100190, China
University of Chinese Academy of Sciences, Beijing 100190, China
Key Laboratory of Wind Energy Utilization, Chinese Academy of Sciences, Beijing 100190, China

Caicai Liao
Institute of Engineering Thermophysics, Chinese Academy of Sciences, Beijing 100190, China
National Laboratory of Wind Turbine Blade Research & Development Center, Beijing 100190, China
Dalian National Laboratory for Clean Energy, Chinese Academy of Sciences, Dalian 116023, China

Zhiwen Qin and Ke Yang
Institute of Engineering Thermophysics, Chinese Academy of Sciences, Beijing 100190, China
Key Laboratory of Wind Energy Utilization, Chinese Academy of Sciences, Beijing 100190, China
National Laboratory of Wind Turbine Blade Research & Development Center, Beijing 100190, China

Teng Teng
School of Energy and Mining Engineering, China University of Mining and Technology, Beijing 100083, China
State Key Laboratory of Coal Resources and Safe Mining, China University of Mining and Technology, Beijing 100083, China

Yingheng Wang
School of Electronics and Information Engineering, Tongji University, Shanghai 201804, China

Jacques Maritz
Department of Engineering Sciences, University of the Free State, Bloemfontein 9300, South Africa

Baozhu Li and Yong Li
Research Institute of Petroleum Exploration & Development, Petro China, Beijing 100083, China

Yunfu Qin and Qingyou Yan
School of Economics and Management, North China Electric Power University, Beijing 102206, China

Hongyu Lin, Zhongfu Tan, Shenbo Yang, Gejirifu De and Liwei Ju
School of Economics and Management, North China Electric Power University, Beijing 102206, China
Beijing Key Laboratory of New Energy and Low-Carbon Development, North China Electric Power University, Beijing 102206, China

Li Li
State Grid Jibei Electric Power Economic Research Institute, Beijing 100000, China

Gabriela Hernández-Luna, Rosenberg J. Romero, Antonio Rodríguez-Martínez, Jesús Cerezo Román and Guadalupe Diocelina Toledo Vázquez
Engineering and Applied Sciences Center Research—CIICAp, Autonomous University of Morelos State—UAEM, Av. Universidad 1001, Chamilpa, Cuernavaca 62210, Morelos, Mexico

José María Ponce-Ortega
Faculty of Chemical Engineering, Universidad Michoacana de San Nicolás de Hidalgo, Av. Francisco J. Múgica S/N Ciudad Universitaria, Morelia 58030, Michoacán, Mexico

Chia-Nan Wang, Tsang-Ta Tsai and Ying-Fang Huang
National Kaohsiung University of Science and Technology, Kaohsiung 80778, Taiwan

Liang Sun
School of Electrical Engineering, Dalian University of Technology, Dalian 116024, China
State Grid Shenyang Electric Power Supply Company, Shenyang 110081, China

Qi Zhang and Weidong Li
School of Electrical Engineering, Dalian University of Technology, Dalian 116024, China

Na Zhang
State Grid Liaoning Economic Research Institute, Shenyang 110015, China

Zhuoran Song
State Grid Liaoning Electric Power Company Limited, Shenyang 110006, China

Xinglong Liu
China Energy Engineering Group Liaoning Electric Power Survey & Design Institute Co. Ltd., Shenyang 110179, China

Yao Wang, Haitao Yu, Zhiyuan Che, Yuchen Wang and Cheng Zeng
School of Electrical Engineering, Southeast University, Nanjing 210096, China

Muhammad Waris Ali Khan
Faculty of Industrial Management, Universiti Malaysia Pahang, Gambang 26300, Malaysia

Shrikant Krupasindhu Panigrahi and Khamis Said Nasser Almuniri
College of Business, University of Buraimi, Alburaimi, P.C.512, Oman

Mujeeb Iqbal Soomro
Department of Mechanical Engineering, Mehran University of Engineering & Technology, SZAB Campus, Khairpur 66020, Pakistan

Nayyar Hussain Mirjat
Department of Electrical Engineering, Mehran University of Engineering and Technology, Jamshoro 76062, Pakistan

Eisa Salim Alqaydi
Faculty of Health Sciences, DeMontfort University, Leicester LE8 4AA, UK

Index

R

S

T

V

W